Peptide and Protein Delivery

Peptide and Protein Delivery

Chris Van der Walle
Strathclyde Institute of Pharmacy and Biomedical Sciences
University of Strathclyde
Glasgow, UK

ELSEVIER

AMSTERDAM • BOSTON • HEIDELBERG • LONDON
NEW YORK • OXFORD • PARIS • SAN DIEGO
SAN FRANCISCO • SINGAPORE • SYDNEY • TOKYO

Academic Press is an imprint of Elsevier

Academic Press is an imprint of Elsevier
32 Jamestown Road, London NW1 7BY, UK
30 Corporate Drive, Suite 400, Burlington, MA 01803, USA
525 B Street, Suite 1800, San Diego, CA 92101-4495, USA

First edition 2011

Front cover image: The PyMOL Molecular Graphics System, Version 1.2, Schrödinger, LLC

British Library Cataloguing-in-Publication Data
A catalogue record for this book is available from the British Library

Library of Congress Cataloging-in-Publication Data
A catalog record for this book is available from the Library of Congress

ISBN: 978-0-12-384935-9

For information on all Academic Press publications
visit our website at elsevierdirect.com

Typeset by TNQ Books and Journals Pvt Ltd.
www.tnq.co.in

Printed and bound by CPI Group (UK) Ltd, Croydon, CR0 4YY

Contents

Foreword

The discovery and first therapeutic use of insulin in Toronto in 1922 saved the lives of countless people with type 1 diabetes – and in the course of doing so provided Nobel-winning confirmation of the definitive hormonal pathophysiology of the condition. However, rather than bringing a "cure" for diabetes, those days of discovery are now seen as heralding an epoch of devastating diabetes-related complications – which we are still living through as the centenary approaches.

There are lessons in this story for the contemporary race to discover new treatments for the various other conditions that blight 21st century life. The focus within both the pharmaceutical and academic sectors is firmly on discovery of new (and ideally "blockbuster") non-peptide molecules with a prevailing assumption that non-peptide agents are preferred to avoid the parenteral route of administration. The alternative approach of using developing technologies to refine drug formulation and better target existing molecules using other routes is less glamorous, and possibly less lucrative, but may ultimately lead to greater health gains.

It is almost certainly the case that if insulin could be administered therapeutically into the appropriate anatomical location in truly physiological concentrations, which are finely-tuned in response to minute-to-minute changes in blood glucose concentration, diabetes complications would be obviated entirely. However, while such a "cure" remains elusive, intermittent subcutaneous injection of insulin into the thighs or abdomen is by far the most frequent approach worldwide. This provides the body with only a rough estimate of the amount of the hormone that may actually be required over the next few hours, delivered with a considerable degree of chronic discomfort, into the wrong physiological compartment, at supra-physiological concentrations.

Perhaps it is little wonder that, as a result, over a period of months and years adherence to insulin therapy can wax and wane in response to the challenges of life. The clinical specialties around the nursing and medical care of people with diabetes have grown up around supporting and motivating affected individuals to engage with prescribed treatment and self-care in an effort to minimize the adverse effects of therapy and the progression of complications over time. This can be an uphill and unrewarding battle.

The chapter authors of this volume are those who are currently thinking "outside the box" on these issues internationally, and grappling with the challenges of drug delivery via less "conventional" routes. As clinicians we are fortunate for the sake of those for whom we care that such work is progressing and maturing, even while a clear "breakthrough" is not yet in sight. Any new developments will of course require rigorous clinical trials for safety and efficacy. However, within the field of diabetes – and far beyond – there is little doubt that over the next decade or two alternative routes of drug delivery will change the treatments we prescribe for our patients, and their concordance with that treatment, beyond recognition. This volume is a forerunner of that new era.

John Petrie
Dr John R Petrie BSc PhD MD
Professor of Diabetic Medicine, University of Glasgow

Preface

Many pharmaceutical scientists like myself develop strong links with professionals working in the pharmaceutical industry and in clinical practice. This forms a stimulating environment which gives good oversight of the issues facing the introduction of a new drug, or drug delivery system, onto the market for clinical use. Of course, part of that challenge is learning each others "language" (or "what concerns us most") in order to make a coherent plan which addresses the particular need under discussion. An extra language has been introduced with the emergence of the biotechnology industry. The biopharmaceuticals produced within the biotechnology sector have proven to be exciting with regard to their envisaged clinical potential, but also challenging with regard to their (bio)processing and delivery.

Life science students, for example, quickly learn that "you can't administer proteins and peptides orally"; that is, there is the assumption that invasive delivery by injection is required for biopharmaceuticals. They have, of course, the appreciation that the stomach exposes drugs to very low pH, and the gastrointestinal tract as a whole secretes proteases. There is also an awareness of the barriers to the absorption of macromolecules in the intestine, such as the cell membrane, cell–cell tight junctions, glycocalyx and efflux transporters. However, it can be surprising to learn that research specifically setting out to address the problems of non-invasive delivery of biopharmaceuticals has continued apace. The large volume of work in this field has led to many collaborative approaches, spin-out companies and mature, commercialized technologies. Emergent delivery strategies have examined several routes in addition to oral administration, and each has shown promise to some degree.

The level of interest at a life-science conference can always be piqued if a presentation on "oral insulin delivery" (or other therapeutic protein) is convincing. It is primarily a curiosity concerning the delivery strategies, and how these can be applied to current therapeutic peptides and proteins, which has driven the production of this book. I hope that students studying for higher degrees in the life sciences, clinicians, scientists and industrialists in related fields will find that the chapters herein provide grounding and an up-to-date critical review of the advances made in overcoming the barriers to the non-invasive delivery of peptides and proteins. Because bioprocessing, clinical application and toxicology are so closely related to the issues of delivery, these topics are also discussed in a concise manner. As biopharmaceuticals come to play ever more important roles in modern medicine it is inevitable that the concepts detailed in these chapters will form the focus of future research and development.

I am in indebted to Ijeoma Uchegbu who first suggested this project to me and to Ravi Kumar for his insight and guidance.

Chris van der Walle
Strathclyde, Glasgow, 2010

Contributors

Ajay K. Banga College of Pharmacy and Health Sciences, Mercer University, Atlanta, Georgia

Alex Berrill Department of Biochemical Engineering, University College London, London, United Kingdom

Jamie Biddlecombe MedImmune Ltd., Granta Park, Cambridge, United Kingdom

Daniel Bracewell Department of Biochemical Engineering, University College London, London, United Kingdom

Paolo Caliceti Department of Pharmaceutical Sciences, University of Padua, Padova, Italy

Hak-Kim Chan Faculty of Pharmacy, The University of Sydney, New South Wales, Australia

Woei Ping Cheng School of Pharmacy, University of Hertfordshire, Hatfield, United Kingdom

Christine Dufès Strathclyde Institute of Pharmacy and Biomedical Sciences, University of Strathclyde, Glasgow, Scotland

Anushree Herwadkar College of Pharmacy and Health Sciences, Mercer University, Atlanta, Georgia

Philip Chi Lip Kwok Faculty of Pharmacy, The University of Sydney, New South Wales, Australia

Orest Olejnik Allergan Inc., Dupont Drive, Irvine, California

Shivani Rai Paliwal Department of Pharmaceutical Sciences, Dr. Hari Singh Gour University, Sagar, Madhya Pradesh, India

Rishi Paliwal Department of Pharmaceutical Sciences, Dr. Hari Singh Gour University, Sagar, Madhya Pradesh, India

Patrícia Pereira Faculdade de Farmácia, Universidade de Lisboa, Lisboa, Portugal

Margus Pooga Institute of Molecular and Cell Biology, University of Tartu, Tartu, Estonia

Helin Räägel Institute of Molecular and Cell Biology, University of Tartu, Tartu, Estonia

M. R. Rekha Biosurface Technology Division, Sree Chitra Tirunal Institute for Medical Sciences & Technology, Thiruvananthapuram, Kerala, India

Garnpimol C. Ritthidej Department of Manufacturing Pharmacy, Chulalongkorn University, Bangkok, Thailand

Stefano Salmaso Department of Pharmaceutical Sciences, University of Padua, Padova, Italy

Elke Schmidt Strathclyde Institute of Pharmacy and Biomedical Sciences, University of Strathclyde, Glasgow, Scotland

Chandra P. Sharma Biosurface Technology Division, Sree Chitra Tirunal Institute for Medical Sciences & Technology, Thiruvananthapuram, Kerala, India

Colin Thompson School of Pharmacy and Life Sciences, The Robert Gordon University, Schoolhill, Aberdeen, Scotland

Christopher F. van der Walle Strathclyde Institute of Pharmacy and Biomedical Sciences, University of Strathclyde, Glasgow, Scotland

Suresh P. Vyas Department of Pharmaceutical Sciences, Dr. Hari Singh Gour University, Sagar, Madhya Pradesh, India

Chapter 1

An Overview of the Field of Peptide and Protein Delivery

Christopher F. van der Walle [1] and Orest Olejnik [2]

[1] *University of Strathclyde, Strathclyde Institute of Pharmacy and Biomedical Sciences, 161 Cathedral Street, Glasgow, G4 0RE, Scotland,* [2] *Visiting Professor, University of Strathclyde, Allergan Inc., 2525 Dupont Drive, Irvine, California 92623, USA*

Chapter Outline

1.1. INTRODUCTION

The so called "biologics portfolio" of biotechnology and pharmaceutical companies represents an area of strong market growth. Biologics may include a variety of therapeutics, such as vaccines, recombinant proteins, genes, synthetic tissues and viruses, which have emerged from research in molecular and cell biology, in response to unmet clinical needs and expanding indications. While there is keen interest in areas such as gene therapy and tissue engineering, the recombinant protein sector, particularly that of the monoclonal antibodies (mAbs), provides much of the focus as regards protein delivery. The therapeutic antibodies market is complex and continually under review [1], but the sample of approved antibodies that is shown in Table 1.1 gives some indication of the rate of growth in this area.

It is important at this stage to outline some of the key events that have led to the current level of interest in peptide and protein drugs, and to define the terms "peptide" and "protein". A consensus has not been properly reached

Peptide and Protein Delivery. DOI: 10.1016/B978-0-12-384935-9.10001-X

TABLE 1.1 Examples of therapeutic monoclonal antibodies

Generic name	Brand name	Year of approval	Pharmacological target	Clinical indication	Type
Muromonab-CD3	Orthoclone OKT3	1986	T cell CD3 Receptor	Immune disorders, transplant rejection	murine
Abciximab	ReoPro	1994	Glycoprotein IIb/IIIa	Cardiovascular disease	chimera
Rituximab	Rituxan	1997	B-lymphocyte antigen CD20	Oncology (Non-Hodgkin lymphoma)	chimera
Infliximab	Remicade	1998	Tumor necrosis factor (TNF)-α signaling	Immune disorders	chimera
Cetuximab	Erbitux	2004	Epidermal growth factor (EGF) receptor	Colorectal cancer, head and neck cancer	chimera
Trastuzumab	Herceptin	1998	Human epidermal growth factor receptor 2 (HER2/neu)	Breast cancer	humanized
Palivizumab	Synagis	1998	Respiratory syncytial virus (RSV) protein F	RSV infection	humanized
Gemtuzumab	Mylotarg	2000	Transmembrane receptor CD33	Acute myelogenous leukemia	humanized
Efalizumab	Raptiva	2002	Integrin, alpha L subunit (CD11a)	Psoriasis	humanized

Omalizumab	Xolair	2004	Immunoglobulin E (IgE)	Allergy-related asthma	humanized
Natalizumab	Tysabri	2006	Integrin alpha-4 (α4) subunit	Multiple sclerosis, Crohn's disease	humanized
Ranibizumab	Lucentis	2006	Vascular endothelial growth factor A (VEGF-A)	Macular degeneration	humanized
Eculizumab	Soliris	2007	Complement protein C5	Paroxysmal nocturnal hemoglobinuria	humanized
Certolizumab pegol	Cimzia	2008	Tumor necrosis factor (TNF)-α inhibitor	Crohn's disease	humanized
Ustekinumab	Stelara	2009	Cytokines IL-12 and IL-23	Psoriasis	human
Canakinumab	Ilaris	2009	Activated cytokine IL-1b (catabolin)	Cryopyrin-associated periodic syndrome (autoimmune disorder)	human
Golimumab	Simponi	2009	Tumor necrosis factor (TNF)-α	Rheumatoid arthritis, ankylosing spondylitis	human

with respect to the use of these latter two terms, with the 51 amino acid, mature human insulin being a good example of the ambiguity, since it is generally described as a peptide but also as a protein by some. As a guide, peptides can be considered to be up to 50 amino acids in length, with proteins being larger than this. This boundary corresponds approximately to the upper limit of routine peptide synthesis in the solid phase, i.e. that which is achievable using iterative cycles of "activation" and "deprotection" for the growing peptide chain attached to functionalized resin beads (though much longer chains can be synthesized) [2]. There are competing recombinant technologies for the heterologous expression of peptides, employing a DNA cassette encoding a series of concatenated peptides linked by methionine residues, which are then cleaved by cyanogen bromide [3]. However, solid phase peptide synthesis remains the production method of choice for many small peptide hormones of around 20 amino acids in length.

The first peptide-based drugs approved for clinical use were insulin, thyroid hormone and factor VIII, introduced from the 1920s onwards. However, the trajectory of the field changed dramatically as molecular biology techniques developed in the 1960s. These techniques allowed proteins to be engineered through recombinant means at the level of the gene, doing away with the need for extraction and purification from animal or human tissue. The best example of this is human insulin, which in the late 1970s and early 1980s was synthesized through heterologous expression in *Escherichia coli* by Genentech and Eli Lilly. The rise of molecular biology as a tool by which to generate biopharmaceutical drugs – those that include proteins, DNA, conjugates, viruses, etc., initially far outpaced the development of delivery technologies. These biomacromolecules almost inevitably do not survive the stomach and intestinal environment due to pH and the presence of proteases. Nor do they readily transit the epithelial barrier due to the presence of cell–cell tight junctions, the semi-permeable cell membrane and, for intestinal epithelia, efflux proteins such as P-glycoprotein and the cell glycocalyx. Furthermore, biomacromolecules are very much larger than organic drugs, have short plasma half-lives, are involved in active transport processes (e.g. receptor mediated endocytosis), are susceptible to chemical and physical degradation (including aggregation) and are very potent. The need to overcome these challenges to their delivery has resulted in huge volumes of research, drawn from many disciplines including pharmaceutical materials, chemical engineering, biophysics, analytical methodology, cell and molecular biology and *in vivo* studies. Recently however, there is evidence for optimism in the general field of peptide and protein delivery, which hopefully is reflected in the various chapters of this book.

Following on from the difference in the number of amino acids that differentiate a peptide and a protein, it is necessary to consider differences in structure beyond the primary sequence. Secondary structure, which typically involves β-sheets, β-turns and α-helices is not necessarily found in peptides around 20 amino acids in length. When we reach chains of about 50 amino acids, secondary structure is likely, particularly if stabilized by disulphide bonds, as is the case for insulin. Tertiary structure, however, remains the preserve of proteins, since it involves the precise organization of secondary structural elements into 3-dimensional (3-D) domains. Quaternary structure, similarly, will involve non-covalent bonding between domains and amino acids which may be very distant in their primary structure but are brought together through the intricate folding of the peptide backbone. There is of course much information on protein structure, and the reader is referred to other key texts [4]. It is important to realize that, whereas the function of a small peptide will be dependent only on the functional groups of various amino acids, the function of a protein will additionally be dependent on maintenance of a precise 3-D structure. One can therefore appreciate that delivery technologies relating to proteins have the harder task of maintaining the non-covalent bonding pattern of this 3-D structure. This may be one reason that the encapsulation of peptides into polyester microspheres for controlled delivery is an established technology, but encapsulation of proteins remains largely at the research stage, albeit promising [5].

1.2. CURRENT FORMULATION DEVELOPMENT

Proteins such as monoclonal antibodies (mAbs) are most commonly administered by subcutaneous injection. It is often suggested that patient compliance would improve if a non-invasive route of administration were available. However, this need not always be the case. For example, a depot injection of a particular proteinaceous drug, administered once every month or less, may be more convenient than the requirement for repeated, daily, oral or pulmonary doses. Thus, on the consideration of patient compliance alone, it is likely that invasive delivery will remain one route of administration. Of course, the need for daily injections of the drug insulin has in itself stimulated much of the research regarding non-invasive protein delivery, particularly oral and pulmonary delivery. The story of inhalable insulin, marketed by Pfizer Inc. as Exubera, is now widely known: although Exubera was greeted with much enthusiasm in 2006, around a year later it was removed from the market due to "lack of acceptance", and considerable uncertainties over patient compliance, long-term safety and cost remain [6]. Nevertheless, research is still ongoing in these areas of protein delivery and many others – such as ocular, dermal, central nervous system and nasal. It is these topics that form the initial focus of this book, in each chapter the case is made for approaches which have shown promise in overcoming the inherent barriers to delivery mentioned above. The same approaches must of course also consider stabilization and the overall immunogenicity of the therapeutic product. For this reason, this book finishes with chapters on bioconjugation, then moves to formulation for medical devices, and finally manufacture and supply.

In addition to developing a stable formulation that ensures the desired safety and efficacy, there are clinical, commercial and regulatory requirements. The preformulation studies for purified peptides and proteins must take into account the unique attributes of these molecules. That is:

(i) high molecular weight, e.g. the cyclic peptide cyclosporine A may be only 1.2 kDa but a typical mAb will be 144 kDa;
(ii) multiple functional groups with a wide range of pK_a values, leading to chemical and physical instability, in turn effecting the shelf life and the decision to make liquid versus solid state (lyophilized) formulations;
(iii) maintenance of the 3-D structure: Figure 1.1 of an example mAb [7] shows how intricate the folding of the peptide backbone and disulphide bridging can be;
(iv) high water solubility but increase in viscosity at concentrations above 50 mg/ml, particularly with regard to subcutaneous injection;

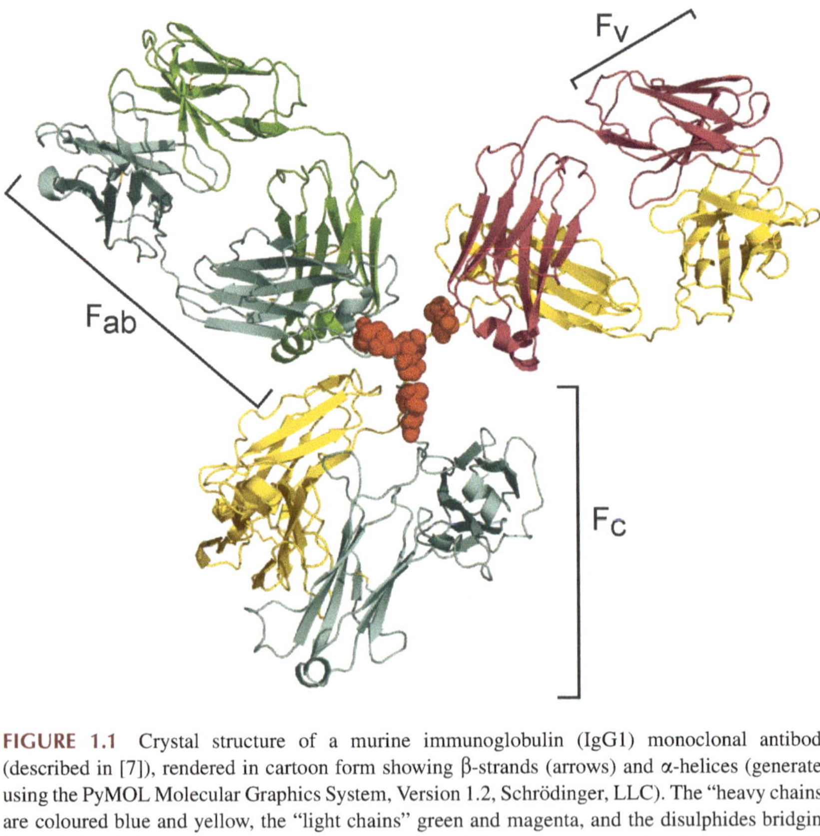

FIGURE 1.1 Crystal structure of a murine immunoglobulin (IgG1) monoclonal antibody (described in [7]), rendered in cartoon form showing β-strands (arrows) and α-helices (generated using the PyMOL Molecular Graphics System, Version 1.2, Schrödinger, LLC). The "heavy chains" are coloured blue and yellow, the "light chains" green and magenta, and the disulphides bridging heavy/light chains as red spheres. Fv, Fragment variable; Fab, Fragment antigen binding; Fc, Fragment crystallization.

(v) interactions with packaging, especially silicone-coated glass syringes; and
(vi) adsorption to surfaces.

Stabilization of the product must be compatible with aseptic processing and manufacture, which may use non-conventional technologies such as freeze/spray drying. Various analytical techniques are central in preformulation and will include methods such as:

(i) chromatography and light scattering, to determine purity and aggregation;
(ii) microscopy, to observe particulates 2 to $\geq$ 10 μm [8,9];
(iii) circular dichroism and infra-red spectroscopy, for conformational changes;
(iv) iso-electric focusing, for determination of the iso-electric point (Pi);
(v) mass spectrometry, for chemical degradation;
(vi) calorimetry and fluorescence, for determination of unfolding/stabilization; and
(vii) rheometry.

The various formulation parameters must be closely monitored since degradation may occur in response to changes in buffer (pH and ionic strength), metal impurities, temperature, irradiation, gases (particularly oxygen and moisture), shear and surface adsorption.

The excipients added to stabilize proteins during formulation, processing and to increase shelf life typically are non-reducing sugars (especially sucrose and trehalose) and polyethylene glycol (PEG); these excipients are representatives of lyoprotectants and cryoprotectants. Lyo- and cryo-protection afforded by sucrose is thought to occur through its preferential exclusion from the protein/polypeptide surface [10], thermodynamically favoring the folded state [11], or via hydrogen bonding with polar groups on the protein surface [12]. The level of stabilization afforded by sugars is usually dependent on their bulk concentration. In contrast, the stabilization mechanism of PEG on proteins is less well understood and may occur through steric hindrance of protein–protein interactions, increased solution viscosity and/or decreased protein structural mobility [13]. There are several other categories of excipient, including non-ionic surfactants (to reduce adsorption and aggregation), amino acids and various salts and buffers (e.g. phosphate salts may be avoided due to differential crystallization of the acid/base). Screening programs are then required to determine changes in protein stability, conformation, aggregation, degradation, etc., using the analytical techniques mentioned above. Depending on the desired product, excipient compatibility must be screened for in the liquid and/or solid state. For solid state formulations, careful optimization of the lyophilization cycle is required [14]. Although the primary goal is product stability, the cost of the freeze–dry cycle is also a consideration, particularly on scale-up [15,16]. The glass transition temperature of the frozen solution (T_g')

and the collapse temperature of the lyophilized cake (T_c) are important parameters. A detailed overview of lyophilization is outside the scope of this book and the reader is referred to some excellent reviews [17,18].

1.3. CASE STUDIES OF DELIVERY APPROACHES: CYCLOSPORINE A AND INSULIN

It should be pointed out that these examples have been selectively chosen. Cyclosporine A (CsA, also written as cyclosporin and ciclosporin) is a small, cyclic peptide with a molecular weight of the same order of magnitude as many organic drugs. Its solubility and bioavailability are poor and so it would be a Class IV drug according to the Biopharmaceutics Classification System (BCS) (Table 1.2) [19]. In both respects CsA is perhaps atypical of many proteins, which may be three orders of magnitude larger in size but of high aqueous solubility and low permeability (Class III).

Nevertheless, CsA is an interesting case, since its numerous formulations illustrate what can be achieved, and their development provides inspiration for other peptide delivery systems. CsA is very hydrophobic and was first formulated with the oil cremophor EL as an irregular oil/water mixture which undergoes emulsification and digestion *in situ* with bile salts and pancreatic enzymes, prior to absorption (Sandimmune®, Novartis, Switzerland). Patient variability does affect the therapeutic outcome of CsA, which increases the risk of graft rejection. A later dosage form (Neoral®, Novartis), available as a soft gelatin capsule or oral solution containing micro-emulsion pre-concentrate, was designed to rapidly emulsify in the presence of aqueous fluid, rather than depend on the action of bile salts and enzymes. This reduced interpatient variability and improved bioavailability. Thus, to the field of drug delivery, CsA represents an irresistible challenge, the overall aim being the development of non-invasive delivery strategies for doses targeting its narrow therapeutic index. Table 1.3 shows the depth of research and outlines the potential for alternative routes of administration [20], each of which are subsequently dealt

TABLE 1.2 The classification of drugs according to the Biopharmaceutics Classification System (BCS)

Class	Aqueous solubility	Permeability	Example
I	High	High	
II	Low	High	
III	High	Low	Immunoglobulins
IV	Low	Low	Cyclosporin A

TABLE 1.3 An overview of cyclosporin A (CsA) delivery systems. Reprinted from reference [20] with permission of Elsevier

Route	Formulation/Delivery system	Summary
Oral	Emulsion pre-concentrate for soft gelatin capsule and oral solution	Higher solubilizing capacity
	Microemulsion pre-concentrate for soft gelatin capsule and oral solution	Improved bioavailability, reduced variability of absorption, marketed by Novartis as Neoral®
	Nanoparticles of polycaprolactone (PCL) and polylactic acid-polyethylene glycol (PLA-PEG)	Polymer matrix controlled release, enhanced lymphocytic uptake, improved bioavailability without increase in toxicity
	pH sensitive nanoparticles of Eudragit® S100 and hydroxypropyl methylcellulose phthalate	pH dependent release and improved bioavailability
	Positively charged nanoparticles of chitosan	Improved bioavailability with lower variability
	Solid dispersions containing polyoxyethylene stearate	Improved CsA dissolution and bioavailability comparable to that of Neoral®
	Cyclodextrins	Improved bioavailability with reduced variability in absorption
Parenteral	Solution of CsA in cremophor EL (polyoxyl 35 castor oil) and ethanol	Improved solubilization, marketed by Novartis as Sandimmun®
	PLA and PLA-PCL microspheres for intramuscular and subcutaneous injection	Sustained drug release and improved immunosupression
Pulmonary	Propylene glycol aerosols by nebulization and dry powder formulations	Undergoing clinical trials for prevention of graft lung transplant rejection
	Liposomal aerosol	Higher CsA retention in the lungs
Dermal	Topical penetration enhancer monoolein	Improved dermal delivery
	Iontophoresis with lecithin vesicles, or electroporation combined with ethanol	Enhanced transdermal delivery

(*Continued*)

TABLE 1.3 An overview of cyclosporin A (CsA) delivery systems. Reprinted from reference [20] with permission of Elsevier—cont'd

Route	Formulation/Delivery system	Summary
Ocular	Topical anionic microemulsion	Marketed by Allergan Pharmaceuticals as Restasis® Ophthalmic Emulsion for treating dry eye syndrome
	Topical micelles of polyoxyl 40 stearate	Higher corneal permeation
	Topical chitosan nanoparticles	Longer retention with higher CsA levels in cornea and conjunctiva
	Poly(lactic-co-glycolic acid) (PLGA) microspheres and implants for intraocular and subconjunctival injection	Higher corneal levels and CsA and improved corneal graft suvival, controlled release

with in detail in this book. The associated formulations of CsA that remain under investigation for potential translation to clinical use/the market range from emulsion systems and liposomes to biodegradable particulates and penetration enhancers [21]. There is no single ideal formulation, which underscores the need for delivery strategies tailored to the individual molecule, i.e. it is unlikely that a generic platform technology will be applicable to all peptides and proteins. For instance, emulsion systems are clearly suitable for CsA but are not conducive to proteins because of unfolding and aggregation at the oil/water interface.

Insulin is usually administered by subcutaneous injection or a pump and cannula, the latter affording greater control when needed. The subcutaneous route is more convenient, and is inevitably used for control of type I diabetes. However, as discussed above, the drive remains to provide non-invasive routes for insulin dosing and, despite the set-back with Exubera, pulmonary delivery systems are the subject of ongoing clinical studies, albeit with mixed results [22]. An added challenge to insulin delivery is the need for a pulsatile release system which would follow the normal post-prandial response of a sudden rise in blood glucose levels. Oral insulin delivery is perhaps the primary target, which must overcome enzymatic degradation and poor permeability in the gastrointestinal tract. Although these obstacles were once considered insurmountable, there are currently several companies pursuing oral insulin delivery technologies which are in various stages of clinical trials. For example, one Phase III trial due for completion 2014 will report on the potential role of oral insulin to delay or prevent type 1 diabetes in patients who have high levels of insulin autoantibodies [23]. Beyond current commercialization activities, there lies a large volume of research employing microemulsions and nanoparticulates composed of polysaccharides/polyesters/Eudragit to achieve oral insulin delivery; though it should be noted that oral bioavailability remains stubbornly around 10% or less [24–27]. Nasal, buccal, rectal and vaginal delivery systems have also been explored for insulin delivery. Of these, nasal and buccal delivery have generated the most attention [22]. The nasal cavity is highly vascularized and bypasses the hepatic metabolism but suffers from a low surface area (compared to the lung), mucociliary clearance and the presence of proteolytic enzymes. Chitosan derivatives are commonly used materials for these routes [28], in part because chitosan acts as a permeation enhancer by transiently loosening the epithelial tight junctions. Sublingual delivery has the advantage of ease of administration, but the challenge includes the multilayered structure of the buccal mucosa and the constant flow of saliva. However, there remains scope in typical insulin treatment regimes to include these novel delivery systems as adjuvants. For example, following a regular insulin regime in type I diabetics, oral insulin spray caused similar changes in blood glucose compared to subcutaneous insulin injection [29]. Note that much of the research involves the use of animal models (notably streptozotocin-induced diabetic rats), and scaling factors and the extrapolation of results to human subjects are important topics for examination. Because of this, the promise of earlier research has not been fully realized and much work still remains before non-invasive insulin delivery becomes a reality.

1.4. CONTROLLED DELIVERY

It is important to distinguish between non-invasive delivery and controlled delivery, since the development of both may draw from polymeric materials technologies. The use of biodegradable polyester microspheres for controlled delivery is not limited to a specific route of administration, which may be either invasive (e.g. subcutaneous/depot injection) or non-invasive (e.g. oral, inhaled). Conversely, polymers used in non-invasive buccal delivery may be required for mucoadhesion but not controlled delivery. There is a well established track record in the use of poly(lactic-co-glycolic acid) (PLGA) microspheres for the controlled delivery of peptide hormones (Table 1.4) [30]. However, peptide encapsulation into PLGA microspheres using an emulsification–solvent evaporation protocol continues to represent a technological challenge in terms of manufacture and product quality. It is notable that the transfer of microencapsulation technologies for proteins from the laboratory bench to the clinic and market has proven far more difficult than anticipated. Many research papers initially showed promising results for protein encapsulation in PLGA microspheres, in part because model proteins (namely lysozyme and albumins) were used that were unusually stable under

TABLE 1.4 Commercially available biodegradable drug delivery systems

Name of product	Dosage form	Active ingredient	Biodegradable polymer
Lupron® Depot	Microspheres	Leuprolide	PLGA
Sandostatin LAR® Depot	Microspheres	Octreotide	PLGA
Neutropin® Depot	Microspheres	Somatropin	PLGA
Trelstar® Depot	Microspheres	Triptorelin	PLGA
Decapeptyl®	Microspheres	Triptorelin pamoate	PLGA
Parlodel® Depot	Microspheres	Bromocriptine mesylate	PLGA
Enantone® Depot	Microspheres	Leuprolide	PLGA
Zoladex®	Rod	Goserelin	PLGA

emulsion conditions on account of multiple disulphide bridges and strongly amphipathic helices [5]. Optimization of an encapsulation protocol is therefore dependent on the physicochemical characteristics of the peptide/protein and the polymer itself (monomer composition, molecular weight, glass transition, etc.).

The selection of the right encapsulation technique must ensure the satisfaction of three main criteria:

(i) the structure of the peptide/protein remains intact, particularly with respect to aggregation and neutralizing antibodies [31];
(ii) the size of the spheres has to support the administration route aimed for – from microns for injectable and inhalable particles, to nanoparticles for oral administration; and
(iii) the drug release profile must be appropriate to the intended medical application (vaccine, long-term release, delayed release, etc.).

Fortunately, for short peptides, little consideration needs to be made of secondary structure, and microencapsulation is now a standard route to achieving a controlled or sustained delivery profile. Achieving an appropriate release profile can be tricky since the so-called "burst release" is often observed in initial encapsulation-release experiments, wherein the majority of the drug entrapped in the microsphere is subsequently released within the first few hours upon immersion in aqueous media. Similarly, maximizing the encapsulation efficiency is essential to ensure the process remains cost-effective. Despite these challenges, the field has progressed rapidly in the last decade, and this method of release may one day become routine for proteins. Current research in this area involves scale-up of organic solvent-free encapsulation technologies [32] and encapsulation of complex, self-assembling systems such as bacteriophages, which are expected to play a role in anti-bacterial therapies [33].

1.5. FUTURE TRENDS

The case studies described above outline future opportunities, and one objective of each of the following chapters is to describe or predict future trends. It is useful to broadly categorize the strategies into recombinant approaches and materials technologies. The former includes the engineering of protein/peptide "pro-drugs" and/or "chimeras" which may involve appending a peptide or protein domain to a therapeutic protein in order to target and bind a particular receptor. Bacterial exotoxins (as distinct from endotoxins) are useful examples, since they have a general AB (or AB_5) structure; the A subunit brings about the enzymatic activity of the toxin while the B subunit binds to a specific cell surface receptor, triggering endocytosis and internalization. The key to drug delivery therefore is to isolate the B subunits to bring about internalization without toxic effects. Another example of chimeric proteins is to fuse a protein drug to the end of the neonatal Fc domain (cf. antibody structure, Figure 1.1). The intention is for these chimeras to bind the neonatal Fc receptor (FcRn), which remains expressed in adults in the lung epithelium, in order for FcRn-mediated endocytosis to internalize the bound drug, intact. Similarly, the humanization of murine mAbs (those originally raised in mice) to reduce immunogenicity and prolong blood levels similarly involves complete substitution of the Fc and partial substitution of the Fab domains; Trastuzumab (Herceptin®) is an example of this technology (Table 1.1) [34]. Site directed mutagenesis is a central, simple tool in molecular biology but very effective in protein engineering since specific amino acids can be targeted to alter protein pharmacokinetics and physical/chemical/enzymatic stability. One such example is Vitatropin™ (human growth hormone), but the many recombinant mAbs are also modified in this manner. Of particular interest to the area of protein engineering will be the incorporation of *artificial* amino acids into peptides and proteins, possibly expanding their function or improving bioprocessing. Recently, the tools required to efficiently encode and direct the incorporation of unnatural amino acids into proteins were developed, yielding a novel protein cross-link [35]. This is an exciting development which will drive protein engineering towards custom-designed protein function.

Materials technology will become an increasingly important area for the development of particulates. Their utility in the delivery of biomolecules has already been noted above with respect to invasive (e.g. subcutaneous) and non-invasive (e.g. pulmonary and oral) routes. It should be remembered that these same particulate technologies will also play a role in the bioprocessing of proteins. This represents a slight departure from the focus on delivery strategies but it is important to remember the downstream processes that must occur before a pure product can be formulated, especially steps involving purification, concentration and storage. A new technology which straddles both delivery and bioprocessing is Protein Coated MicroCrystals (PCMCs). This involves coprecipitation of an aqueous solution of protein in near saturated concentrations of a solute (e.g. an amino acid) mixed rapidly with a non-solvent. This promotes very rapid protein dehydration and, as the two components coprecipitate, the protein (or other biomolecule) is immobilized onto the surface of the crystalline excipient support. Many proteins are compatible with the PCMC technique, irrespective of size and pI, e.g. hormones, cytokines, mAbs and antigens. The microcrystals can be engineered to be of a diameter consistent with inhalation but the process is also suitable for rapid concentration of dilute protein solutions and subsequent storage as a dry powder at standard temperature [36]. These stable powders can be rapidly redissolved back into aqueous media to release the protein in a native form.

Protein stability and aggregation are major, and still largely unsolved, issues affecting the development and production of biopharmaceuticals. Beside their impact on development costs, the safety of biopharmaceuticals is also significantly affected by these issues. Protein aggregation is a major element behind the immunogenicity, and perhaps also the toxicity, of many bioactive polypeptides.

1.6. TOXICITY PROFILES

The complex nature of proteins, peptides and other compounds derived through biotechnological processes present many challenges when it comes to understanding their physicochemical and therapeutic behavior. Their specific primary, secondary, tertiary and quaternary structures, as previously mentioned, play key roles in defining the integrity and biological activity of biomacromolecules. This is the case whether they are directly incorporated into simple, formulated systems, or use more sophisticated delivery platforms. Either way their safety and toxicity must be evaluated through the use of scientifically sound methods as a matter of paramount importance and necessity. It does not mean that a routine study design is adopted. In fact, the intricacies of these compounds require a customized and question based approach [37]. Even after implementing a well designed approach questions remain; are the safety margins sufficient, are the risk-to-benefit ratios acceptable? A level of uncertainty will always exist – as exemplified by a life-threatening experience involving TGN1412, a compound which was investigated for the treatment of chronic inflammatory conditions [38]. This was a fully humanized monoclonal antibody (mAb) with activity directed against the CD28 costimulatory receptor on T-cells. No adverse effects were expected, especially since genetically engineered monoclonal antibodies were considered not only to be an exciting class of compounds for their potential to target diseases, but also because they would not harm healthy cells. After all, numerous mAb drug products are approved – such as Herceptin® for treating breast cancer (including its new indication for advanced HER2-positive stomach cancer) and Remicade® for treating inflammatory disorders, providing treatment benefit to patients. There are still many more mAbs undergoing clinical trials. What separated TGN1412 from the mode of action of other antibody treatments was that, in all other cases, biological reactions were inhibited and shut down, but TGN1412 was designed to do the opposite. The intent was to activate CD4+ effector memory T-cells which would in turn overstimulate any rogue T-cells, causing them to apoptose. Although the evidence is inconclusive, it was believed that activation of the T-cells by TGN1412 was responsible for eliciting a "cytokine storm" which produced serious adverse effects in the human volunteers [39]. Absence of CD28 expression on the CD4+ effector memory T-cells in the species used for non-clinical safety testing provided something of an explanation for the failure to predict why a cytokine storm occurred in humans. From a rational perspective, this experience underscores the need for an integrated safety assessment through appropriate species selection and relevant animal models, including appropriate use of biomarkers.

In stark contrast to the investigation of an unintentionally toxic compound, it is interesting to review the anecdotal history of the discovery and intentional development of a known toxic biomacromolecule. It was the 19th century German poet and physician Justinus Kerner (1786–1862) who first provided comprehensive details of the clinical symptoms of botulism associated with ingestion of unhygienically produced sausages [40]. Kerner's description of what he termed "sausage poison" covered symptoms, duration, and other findings, and was written decades before the causative botulinum toxin was identified. His somewhat heroic, and extremely risky, scientific approach was to self-ingest extracts of tainted sausages to gain direct knowledge and experience of the symptoms. This provided from his perspective the most accurate assessment of the effects of the "sausage poison". Although not advocated here, Kerner's unorthodox approach led him to insights concerning the aetiology and pathophysiology of the poison. He deduced that the toxic substance acted by interfering with the signal transmission within the peripheral, sympathetic and parasympathetic nervous system, while leaving the sensory signal transmission intact. He further developed ideas about using botulinum toxin as a remedy for various diseases that involve an overactive nervous system such as chorea, as well as speculating on its use in other disorders.

Unbeknownst to Kerner, he was clearly dealing with *Clostridium botulinum*, an anaerobic, gram positive bacterium which produced highly potent protein complexes called neurotoxins. These toxins were responsible for the serious food poisoning observed and the neuroparalytic effects of botulism. Seven generally immunologically distinct botulinum neurotoxins have been characterized from different strains of *C. botulinum*, these being respectively botulinum neurotoxin (BoNT) serotypes A, B, C.sub.1, D, E, F and G, each of which is distinguished by neutralization with type-specific antibodies [41]. The different serotypes of botulinum toxin vary in the animal species that they affect and in the severity and duration of the paralysis they evoke. For example, it has been determined that botulinum toxin type A is 500 times more potent, as measured by the rate of paralysis produced in the rat, than is botulinum toxin type B. Additionally, botulinum toxin type B has been determined to be non-toxic in primates at a dose of 480 U/kg, which is about 12 times the primate LD_{50} for botulinum toxin type A. Notwithstanding, it is botulinum toxin type A which is considered to be the most lethal natural biological agent known to humankind. To put this into perspective, on a molar basis, botulinum toxin type A is about 1.8 billion times more lethal than diphtheria, about 600 million times more lethal than sodium cyanide, about 30 million times more lethal than cobra toxin and about 12 million times more lethal than cholera. Rhetorically, one wonders whether, had Kerner been privy to this information, would he still have pursued a path of self-experimentation?

Therapeutic proteins inherently have the potential to elicit antibody production leading to an immunogenic response. Antibody responses can vary appreciably across a range of biopharmaceutical products in both incidence and clinical sequelae [42]. Detectable antibodies from less than 1% (e.g. Actimmune®, Rituxan®) to over 80% (e.g. OKT3, Fabrazyme®) show the disparity of incidence between products. For clinical sequelae the variability can be from no effect or reduced efficacy to allergic reactions, thrombocytopenia and anemia. The serious adverse events causing these unintended consequences are a constant area of concern, which have resulted in FDA mandated "black box warnings" for certain products that were originally deemed very safe; such as Remicade®. The infliximab product was reported to contribute to tuberculosis and other serious opportunistic infections including histoplasmosis, listeriosis and pneumocystosis, after findings in both the clinical research and post-marketing surveillance settings. Despite advances in gene sequencing, expression systems and post-expression techniques such as PEGylation, the worry surrounding the clinical immunogenicity of therapeutic proteins has not gone away [43]. Moreover, it raises the question about the usefulness and ability of non-clinical data in predicting human immunogenicity.

1.7. REGULATORY MATTERS

The regulation of biotechnology, from research through to product approval and eventual marketing of the product, is not a simple process. There are several series of complex laws and guidance documents. It can be confidently stated that the regulations will continue to evolve and no doubt increase as scientific knowledge advances, novel technologies are developed and new questions are asked. Unequivocally, the progression to product licensure must rely heavily on sound scientific approaches, comprehensive data and analysis, with conformance to the presiding governing agencies' regulations. Note the plurality of "agencies". The strategic intent of marketing a product in a single country is not a preferred option. The desire to provide global access to beneficial therapeutic products, and of course to maximize return on investment, requires that product development should meet most, if not all, global regulatory expectations. This establishes an expectation in understanding at minimum the requirements of the major agencies, namely FDA (US), EMA (Europe) and PMDA (Japan).

The challenge has been dealing with the inevitably country-centric laws. In an effort to resolve the differences between them, there has been a concerted effort to harmonize worldwide requirements through the International Conference on Harmonization (ICH). This was a multiple step process, involving consensus-building between relevant government bodies and research-based industries, release of drafts for wider consultation, regulatory consultations across regions, agreement on the ICH guidelines, adoption by the regulators and finally tripartite implementation [44]. The published guidances are available through the ICH official website (www.ich.org), which also provides information on document annexes. These documents are also used by countries other than the ICH members. However, it must be acknowledged that for the non-ICH regulatory authorities who were not involved in the process, full compliance may not always be ensured. The World Health Organization has adopted a proactive role in raising awareness and in addressing any areas of concern [45].

The ICH guidance documents are principally categorized into three areas; safety, efficacy and quality. Aside from their obvious regulatory compliance implications, they aid the development scientist in identifying which critical elements need to be studied. This in turn indicates what information needs to be collected in order to establish the specifications that become the basis for defining a product. As stated in the ICH guidance Q6B concerning test procedures and acceptance criteria for biotechnological and biological products, specifications are:

"A list of tests, references to analytical procedures, and appropriate acceptance criteria which are numerical limits, ranges, or other criteria for the tests described. It establishes the set of criteria to which a drug substance, drug product, or materials at other stages of its manufacture should conform to be considered acceptable for its intended use. Conformance to specification *means that the drug substance and drug product, when tested according to the listed analytical procedures, will meet the acceptance criteria. Specifications are critical quality standards that are proposed and justified by the manufacturer and approved by regulatory authorities as conditions of approval."*

The importance of this is discussed in section 1.8.

Whilst the area of regulations is broad, simply attaining authorization to market a product is only one of its aspects. There is another important area of regulatory law that should never be overlooked, i.e. patents. From the literature it is clear that advancements in the field of proteins and protein delivery remain unabated, and within this abundant information valuable intellectual property and inventions must have been generated. Unfortunately, all too often data appears in the public domain well before the invention is recorded. Publications become prior art, such that any opportunity to patent, and hence metaphorically protect the fruits of one's labor, is essentially lost. Awareness of the patent process and the merit in receiving an officially issued invention can indeed be rewarding. Even back in the 18th century, the US government recognized its worth as mentioned in Article 1, Section 8 of the US Constitution:

"Congress shall have the Power … to promote the Progress of Science and Useful Arts by securing for limited Times to Authors and Inventors the exclusive Right to their respective Writings and Discoveries" [46].

1.8. COMMERCIALIZATION CONSIDERATIONS

The fundamental properties of biological materials impose significant and unique constraints on bioreactions, downstream bioprocessing stages and final fill-finish product manufacture. The complexity and sensitivity of these materials dictates a dependency on defined methods and equipment for all processing stages; from raw materials through to the therapeutic protein. This also includes the development of the final formulation carrier platform, and selection of the delivery system whether it be by vial, syringe or other appropriate applicator system. Any change or deviation in the raw material, equipment geometry, or processing and storage conditions can produce unpredictable consequences in the performance of the product. A good example of this is Eprex® (Epoetin alfa). This was manufactured for many years with a good safety record using human serum albumin (HSA) as the formulation stabilizing vehicle. After HSA was substituted with polysorbate 80 and glycine, there was an increase in the incidence of pure red cell aplasia associated with erthropoeitin neutralizing antibodies [47]. It was postulated that the polysorbate surfactant caused leachable materials to be released from the uncoated rubber stopper, which was an integral component of the Eprex syringe. Switching to Teflon® coated stoppers apparently minimized the problem, although other contributory factors were mentioned, involving cold chain storage controls, as well as protein aggregation [48,49]. Another example is seen in the divergence in physical characteristics observed in botulinum neurotoxin, due to the differences in manufacturing processes used to produce, purify and stabilize it [50]. Materials that may be considered comparable are in fact uniquely different in biological activity. It is a reason why regulatory approvals for changes are not straightforward, even where no effects are predicted or observed.

Complexity means that challenges in processing and manufacturing technology transfers are always considerable. Strict emphasis must be placed on determining the essential processing parameters which can affect the quality of the product. The quality attributes of the product must be identified and understood, since knowing these attributes leads to establishing specifications which define the product quality. A review of the principles that should be considered in setting specifications is outlined in the ICH Q6B guidance document, as mentioned previously. It is clear that much weight is given to conducting appropriate tests in order to justify these specifications. Since imposing specifications on all quality attributes would be onerous and impractical, the focus must be on those determined to be critical through appropriate planning, and design of experimentation. It necessitates a comprehensive and disciplined approach by implementing strategies such as quality by design (QbD), design space, process analytical technology (PAT), failure mode effects and criticality analysis (FMECA), and other risk assessment practices [51,52].

When assured that all appropriate knowledge has been gained, the steps in the process are completely defined and under control, and successful trial (demonstration) batches have been run, the next stage is formal validation. At this point, a pre-determined number of batches are manufactured, with protocols and standard operating procedures in place to prove that the process can reproducibly meet the established critical quality attributes. If conducted correctly, a successful outcome is expected – otherwise the sequence repeats. Continued failure may indicate problems with operating procedures, operator skills, undefined nuances in the equipment, or an unconsidered critical quality attribute. Hopefully, the application of sound scientific and operational practices will lead to positive results. A successful outcome is obviously rewarding and satisfying.

REFERENCES

[1] A.K. Pavlou, M.J. Belsey, The therapeutic antibodies market to 2008, Eur. J. Pharm. Biopharm. 59 (2005) 389–396.

[2] J. Jones, Amino Acid and Peptide Synthesis, The Bath Press, Bath, 2002.

[3] K. Sun-Chang, L.J. Hyun, K.M. Hyung, K.J. Hyun, H. Seung-Suh, L. Hyun-Soo, Method for mass production of antimicrobial peptide. Patent applicant: Samyang Genex Co. Ltd., Korea Advanced Inst Sci & Tech, Korea. Publication: WO9854336 (1998)–12–03.

[4] T.E. Creighton, Proteins: Structures and Molecular Properties, W.H. Freeman, New York, 1993.

[5] C.F. van der Walle, G. Sharma, M. Ravi Kumar, Current approaches to stabilising and analysing proteins during microencapsulation in PLGA, Expert Opin. Drug Deliv. 6 (2009) 177–186.

[6] C. Black, E. Cummins, P. Royle, S. Philip, N. Waugh, The clinical effectiveness and cost-effectiveness of inhaled insulin in diabetes mellitus: A systematic review and economic evaluation, Health Technol. Assess. 11 (2007) 1–126.

[7] L.J. Harris, E. Skaletsky, A. McPherson, Crystallographic structure of an intact IgG1 monoclonal antibody, J. Mol. Biol. 275 (1998) 861–872.

[8] U.S. Pharmacopeia. Particulate Matter in Injections In The United States Pharmacopeia – National Formulary, General Chapter <788>. (2010).

[9] S.K. Singh, N. Rathore, A. McAuley, A.S. Rathore, Best practices for formulation and manufacturing of biotech drug products: how to maintain product stability and prevent particulates, BioPharm Int. 22 (6) (2009).
[10] B.S. Kendrick, B.S. Chang, T. Arakawa, B. Peterson, T.W. Randolph, M.C. Manning, et al., Preferential exclusion of sucrose from recombinant interleukin-1 receptor antagonist: Role in restricted conformational mobility and compaction of native state, Proc. Natl. Acad. Sci. U.S.A. 94 (1997) 11917–11922.
[11] Y. Jung, H.J. Kang, J.M. Lee, S.O. Jung, W.S. Yun, S.J. Chung, et al., Controlled antibody immobilization onto immunoanalytical platforms by synthetic peptide, Anal. Biochem. 374 (2008) 99–105.
[12] S.D. Allison, B. Chang, T.W. Randolph, J.F. Carpenter, Hydrogen bonding between sugar and protein is responsible for inhibition of dehydration-induced protein unfolding, Arch. Biochem. Biophys. 365 (1999) 289–298.
[13] W. Wang, Instability, stabilization, and formulation of liquid protein pharmaceuticals, Int. J. Pharm. 185 (1999) 129–188.
[14] X. Tang, M.J. Pikal, Design of freeze-drying processes for pharmaceuticals: Practical advice, Pharm. Res. 21 (2004) 191–200.
[15] S.L. Nail, J.A. Searles, Elements of quality by design in development and scale-up of freeze-dried parenterals, BioPharm Int. 21 (1) (2008).
[16] H. Gieseler, G. Lee, Effect of freeze-dryer design on drying rate of an amorphous protein-formulation determined with a gravimetric technique, Pharm. Dev. Technol. 13 (2008) 463–472.
[17] J.F. Carpenter, M.J. Pikal, B.S. Chang, T.W. Randolph, Rational design of stable lyophilized protein formulations: Some practical advice, Pharm. Res. 14 (1997) 969–975.
[18] L.Q. Chang, M.J. Pikal, Mechanisms of protein stabilization in the solid state, J. Pharm. Sci. 98 (2009) 2886–2908.
[19] M.V.S. Varma, S. Khandavilli, Y. Ashokraj, A. Jain, A. Dhanikula, A. Sood, et al., Biopharmaceutic classification system: A scientific framework for pharmacokinetic optimization in drug research, Curr. Drug Metab. 5 (2004) 375–388.
[20] J.L. Italia, V. Bhardwaj, M.N. Kumar, Disease, destination, dose and delivery aspects of ciclosporin: The state of the art, Drug Discov. Today 11 (2006) 846–854.
[21] H. Liu, Y. Wang, S. Li, Advanced delivery of ciclosporin A: Present state and perspective, Expert Opin. Drug Deliv. 4 (2007) 349–358.
[22] R.I. Henkin, Inhaled insulin-intrapulmonary, intranasal, and other routes of administration: Mechanisms of action, Nutrition 26 (2010) 33–39.
[23] National Institute of Diabetes and Digestive and Kidney Diseases (NIDDK). Oral Insulin for Prevention of Diabetes in Relatives at Risk for Type 1 Diabetes Mellitus. ClinicalTrials.gov Identifier: NCT00419562, ClinicalTrials.gov. (2007).
[24] F. Cui, K. Shi, L. Zhang, A. Tao, Y. Kawashima, Biodegradable nanoparticles loaded with insulin–phospholipid complex for oral delivery: Preparation, in vitro characterization and in vivo evaluation, J. Control. Release 114 (2006) 242–250.
[25] C. Damge, P. Maincent, N. Ubrich, Oral delivery of insulin associated to polymeric nanoparticles in diabetic rats, J. Control. Release 117 (2007) 163–170.
[26] B. Sarmento, A. Ribeiro, F. Veiga, D. Ferreira, R. Neufeld, Oral bioavailability of insulin contained in polysaccharide nanoparticles, Biomacromolecules 8 (2007) 3054–3060.
[27] G. Sharma, K. Wilson, C.F. van der Walle, N. Sattar, J.R. Petrie, M.N. Ravi Kumar, Microemulsions for oral delivery of insulin: Design, development and evaluation in streptozotocin induced diabetic rats, Eur. J. Pharm. Biopharm. 76 (2010) 159–169.

[28] L.H. du Plessis, A.F. Kotze, H.E. Junginger, Nasal and rectal delivery of insulin with chitosan and N-trimethyl chitosan chloride, Drug Deliv. 17 (2010) 399–407.
[29] J. Guevara-Aguirre, M. Guevara-Aguirre, J. Saavedra, G. Bernstein, A.L. Rosenbloom, Comparison of oral insulin spray and subcutaneous regular insulin at mealtime in type 1 diabetes, Diabetes Technol. Ther. 9 (2007) 372–376.
[30] F. Mohamed, C.F. van der Walle, Engineering biodegradable polyester particles with specific drug targeting and drug release properties, J. Pharm. Sci. 97 (2008) 71–87.
[31] S. Hermeling, D.J.A. Crommelin, H. Schellekens, W. Jiskoot, Structure-immunogenicity relationships of therapeutic proteins, Pharm. Res. 21 (2004) 897–903.
[32] F. Jordan, A. Naylor, C.A. Kelly, S.M. Howdle, A. Lewis, L. Illum, Sustained release hGH microsphere formulation produced by a novel supercritical fluid technology: In vivo studies, J. Control. Release 141 (2010) 153–160.
[33] U. Puapermpoonsiri, J. Spencer, C.F. van der Walle, A freeze-dried formulation of bacteriophage encapsulated in biodegradable microspheres, Eur. J. Pharm. Biopharm. 72 (2009) 26–33.
[34] P. Carter, L. Presta, C.M. Gorman, J.B. Ridgway, D. Henner, W.L. Wong, et al., Humanization of an anti-p185HER2 antibody for human cancer therapy, Proc. Natl. Acad. Sci. U.S.A. 89 (1992) 4285–4289.
[35] H. Neumann, K. Wang, L. Davis, M. Garcia-Alai, J.W. Chin, Encoding multiple unnatural amino acids via evolution of a quadruplet-decoding ribosome, Nature 464 (2010) 441–444.
[36] B.D. Moore, J. Deere, R. Edrada-Ebel, A. Ingram, C.F. van der Walle, Isolation of recombinant proteins from culture broth by co-precipitation with an amino acid carrier to form stable dry powders, Biotechnol. Bioeng. 106 (2010) 764–773.
[37] J.A. Cavagnaro, Preclinical safety evaluation of biotechnology-derived pharmaceuticals, Nat. Rev. Drug Discov. 1 (2002) 469–475.
[38] G. Suntharalingam, M.R. Perry, S. Ward, S.J. Brett, A. Castello-Cortes, M.D. Brunner, et al., Cytokine storm in a phase 1 trial of the anti-CD28 monoclonal antibody TGN1412, N. Engl. J. Med. 355 (2006) 1018–1028.
[39] D. Eastwood, L. Findlay, S. Poole, C. Bird, M. Wadhwa, M. Moore, et al., Monoclonal antibody TGN1412 trial failure explained by species differences in CD28 expression on CD4+ effector memory T-cells, Br. J. Pharmacol. 161 (2010) 512–526.
[40] F.J. Erbguth, M. Naumann, Historical aspects of botulinum toxin: Justinus Kerner (1786–1862) and the "sausage poison", Neurology 53 (1999) 1850–1853.
[41] K.R. Aoki, Pharmacology and immunology of botulinum neurotoxins, Int. Ophthalmol. Clin. 45 (2005) 25–37.
[42] E. Koren, L.A. Zuckerman, A.R. Mire-Sluis, Immune responses to therapeutic proteins in humans – clinical significance, assessment and prediction, Curr. Pharm. Biotechnol. 3 (2002) 349–360.
[43] P.J. Bugelski, G. Treacy, Predictive power of preclinical studies in animals for the immunogenicity of recombinant therapeutic proteins in humans, Curr. Opin. Mol. Ther. 6 (2004) 10–16.
[44] S.K. Branch, Guidelines from the International Conference on Harmonization (ICH), J. Pharm. Biomed. Anal. 38 (2005) 798–805.
[45] World Health Organization. The Impact of Implementation of ICH Guidelines in Non-ICH Countries – Regulatory Support Series No. 9, (2002), pp. 1–30.
[46] M.A. Voet, The Generic Challenge: Understanding Patents, FDA and Pharmaceutical Life-cycle Management, 2nd ed.). BrownWalker Press, Boca Raton, USA, 2008.

[47] B.-L. Chen, CMC issues and regulatory requirements for biosimilars, Trends in Bio/Pharmaceutical Industry 5 (2009) 19–26.
[48] H. Schellekens, W. Jiskoot, Eprex-associated pure red cell aplasia and leachates, Nat. Biotechnol. 24 (2006) 613–614.
[49] B. Sharma, M.H. Ryan, K. Boven, Reactions to Eprex's adverse events, Nat. Biotechnol. 24 (2006) 1199–1200.
[50] A. Brashear, Clinical comparisons of botulinum neurotoxin formulations, Neurologist 14 (2008) 289–298.
[51] A.S. Rathore, H. Winkle, Quality by design for biopharmaceuticals, Nat. Biotechnol. 27 (2009) 26–34.
[52] R.B. Shah, J.T. Park, E.K. Read, M.A. Khan, K. Brorson, Quality by design (QbD), biopharmaceutical manufacture, in: M.C. Flickinger (Ed.), Encyclopedia of Industrial Biotechnology: Bioprocess, Bioseparation, and Cell Technology, Wiley-Blackwell, Oxford, UK, 2010, pp. 1–36.

Chapter 2

Pulmonary Delivery of Peptides and Proteins

Philip Chi Lip Kwok and Hak-Kim Chan
Advanced Drug Delivery Group, Faculty of Pharmacy, Building A15, The University of Sydney, New South Wales 2006, Australia

Chapter Outline

Peptide and Protein Delivery. DOI: 10.1016/B978-0-12-384935-9.10002-1

2.1. INTRODUCTION

Proteins are polymers consisting of amino acids covalently linked by peptide bonds. Peptides are small proteins composed of up to a few dozen amino acids [1]. The molecular weight of larger proteins may range from thousands to several million atomic masses, depending on the number of amino acids in the chain [2]. Due to their large sizes, absorption through the epithelial barriers in the gastro-intestinal tract is low [3]. Furthermore, proteins are rapidly degraded by digestive enzymes. Thus, oral bioavailability without advanced delivery systems is generally poor. The most common route of administration for pharmaceutical proteins is parenteral, including intravenous, subcutaneous and intramuscular injections. However, the pulmonary route has also been used for protein delivery because the lungs possess a large surface area and an extensive vascular network that favors absorption [3]. Inhalation delivery is non-invasive. This is especially of value for pharmaceutical proteins that are to be administrated over the long-term, such as insulin for diabetes mellitus, or recombinant human deoxyribo-nuclease (rhDNase), also known as dornase alfa, for cystic fibrosis.

2.2. CLINICAL APPLICATIONS OF PULMONARY PROTEIN DELIVERY

2.2.1. Local and Systemic Delivery via Inhalation

Pharmaceutical aerosols are inhaled through the mouth and into the respiratory tract for local or systemic delivery. In sequential order the route consists of the oropharynx, larynx, trachea, bronchi, bronchioles and alveoli. The larger bronchioles and the upper respiratory tract have ciliated epithelia, forming the mucociliary escalator. Gas exchange occurs in the distal bronchioles and alveoli, which are extremely thin and are bordered by a profuse network of capillaries [4]. The alveolar sacs are composed of two types of cells: (i) Type I flat squamous cells for rapid gas exchange; and (ii) Type II cells that produce pulmonary surfactants which line the alveolar epithelium [5]. These surfactants are mainly dipalmitoylphosphatidylcholine and small amounts of surfactant proteins A, B and C [6]. In order to deposit successfully in the airways and alveoli, the aerodynamic diameter of particles and droplets should be < 5 μm [7]. The smaller the aerodynamic diameter, the deeper the site of deposition. Particles of > 5 μm diameter are deposited mainly in the oropharynx by inertial impaction and are then swallowed into the gastro-intestinal tract [7].

Pulmonary delivery offers direct drug targeting for local diseases such as asthma, chronic obstructive pulmonary disease (COPD) and cystic fibrosis. Inhalation therapy may also be applied to treating systemic diseases, as it has some advantages over other routes of administration. The lungs provide a large surface area and high blood supply for drug absorption into the systemic circulation. It also avoids the first pass hepatic metabolism that orally delivered drugs undergo. Furthermore, inhalation is less invasive than injection, which is the conventional delivery method for many proteins. Thus, interests in developing inhaled proteins and peptides for systemic treatments have been increasing in recent decades.

2.2.2. Absorption and Bioavailability from the Lungs

As discussed above, the respiratory route has favorable properties for the absorption of proteins and peptides. On the other hand, the lungs also have pulmonary clearance and metabolic pathways to guard against foreign macromolecules. These natural defense mechanisms can inadvertently oppose therapeutic protein delivery and need to be overcome if absorption is to increase.

Mucociliary clearance is the primary clearance mechanism for removing foreign particles in the airways. The epithelial cilia mechanically transport the particles along the mucous layer upwards toward the oropharynx, where the particles are subsequently swallowed or expectorated. Coughing is the usual mechanism for removing large particles that deposit in, and irritate, the throat. In the alveolar region, macrophages may engulf and destroy particulate materials. Besides the generic particle elimination mechanisms discussed above, the lungs also have specific clearance mechanisms for proteins and peptides. A wide range of pulmonary peptidases have been found, both in humans and the mammals employed for *in vivo* inhalation research (e.g. rats, rabbits and dogs). These include elastase, collagenase [8–11], chymotrypsin [8], prolyl endopeptidase, aminopeptidase P, some species of carboxypeptidase, angiotensin converting enzyme (ACE), neutral endopeptidase (enkephatinase), cathepsin B [12–19], mast cell proteases and neutrophil proteases [20]. Some of these enzymes are known to act in particular ways on physiological proteins. For instance, the ACE activates angiotensin I and deactivates bradykinin. Proteolytic activities are regulated by proteinase inhibitors such as antileukoprotease and α-1-antitrypsin to prevent potential damage to lung tissues. These proteolytic enzymes can affect the delivery of therapeutic proteins to various extents. Peptides containing less than 30 amino acids generally have poor bioavailability due to destruction by pulmonary peptidases [21]. On the other hand, those with molecular sizes between 6–50 kDa are less susceptible to peptidases and are better absorbed [21]. However, the metabolism of proteins in the lungs may be different from that in the gastro-intestinal tract. Thus the suitability of any given protein for pulmonary delivery must be assessed specifically.

There are two major pathways of protein absorption from the lungs, namely, transcytosis and paracellular transport. Transcytosis is the transport of macromolecules from the outside of the cell to the inside, which may or may not be mediated by receptors [22]. Proteins and peptides may be absorbed directly into epithelial cells through transporters on the cell membranes [23]. For example, the epithelial transferrin receptor has been targeted for the uptake of horseradish peroxidase [24]. A β-lactam-transporting peptide transporter has also been proposed to effect receptor-mediated delivery of certain peptides [25]. In paracellular transport, the molecule moves through junctions between two or more cells, as well as gaps in the membrane formed after the shedding of dead cells [22]. It has been found that the paracellular movement of cationic peptides is inversely dependent on molecular size [26,27]. The current theory concerning protein absorption from the airways proposes that paracellular transport may be the major pathway for peptides < 40 kDa in molecular size, such as insulin and calcitonin [28]. On the other hand, those > 40 kDa are mainly absorbed by transcytosis [28]. Although diffusional absorption is more common for small molecules, it has been observed for some peptides, such as thyrotropin-releasing hormone [29]. As previously mentioned, macromolecules in the alveolar region may be engulfed by alveolar macrophages. Other factors that may affect protein absorption are physicochemical, such as the formulation ionic strength, pH, concentration and solubility. These can influence the charge, conformation, stability, and consequently the absorption of the protein.

A range of excipients broadly known as absorption enhancers can be formulated with protein and peptides to increase their absorption. These include protease inhibitors, surfactants, liposomes and cyclodextrins [30]. As the name indicates, protease inhibitors act by decreasing the enzymatic activity of proteases. The extent of absorption enhancement depends on the specificity and potency of the protease inhibitor, as well as the susceptibility of the protein concerned to proteolytic attack [30]. In increasing order of effectiveness sodium glycocholate, soybean trypsin inhibitor, aprotinin and bacitracin have been observed to inhibit the hydrolysis of insulin in lung homogenate [31]. The apparent relative bioavailability of insulin was approximately doubled when co-administered intratracheally with nafamostat mesilate, which inhibits plasmin, kallikrein and trypsin [32]. The activities of a wide range of other pulmonary protease inhibitors have been studied, including chymostatin, amastatin, bestatin, Tos-Phe-chloromethylketone, Tos-Lys-chloromethyketone, diisopropyl flurophosphate and phosphoramidon, amongst others [30].

It has been proposed that surfactants may increase transcellular and/or paracellular permeability through interactions with the cell membrane and tight junctions, respectively [30]. The effects of various bile acids, bile salts, fatty acids and non-ionic surfactants on pulmonary protein absorption have been reported [30]. For instance, the bioavailability of insulin in beagle dogs was significantly increased when nebulized with 2–32 mM sodium taurocholate, which is a bile salt [33]. This was due to a reduction in the transepithelial electrical resistance across Caco-2 cell monolayers, which increased the permeability of insulin. The bile salt, at a concentration of 2–15 mM, also progressively dissociated the hexameric insulin into smaller oligomers, hence facilitating their absorption [33]. Sodium glycocholate, n-lauryl β-D-maltopyranoside and linoleic acid-HCO60 (HCO60, polyoxyl 60 hydrogenated castor oil) mixed micelles enhanced the pulmonary absorption of $(Asu^{1,7})$-eel calcitonin (a calcitonin analog) in rats [34]. Similarly, octyl-β-D-glucoside and unsaturated fatty acids such as oleic, palmitoleic and linoleic acids increased the pulmonary absorption of salmon calcitonin [35,36].

Liposomes have been used to encapsulate drug molecules for controlled release and toxicity reduction. They may also enhance the absorption of proteins and peptides from the lungs. It has been proposed that the presence of exogenous surfactant molecules of liposomes in the lungs increases the recycling of pulmonary surfactants, which leads to increased systemic uptake of the proteins [30]. Proteins and peptides residing in the hydrophilic cores of liposomes are also shielded from enzymatic degradation [37]. Liposomes are usually made from lecithins (phosphatidylcholines), phosphatidylserines, phosphatidylethanolamines, phosphatidylglycerols, phosphatidylinositols and sphingomyelins [37]. The liposomal composition, charge, size, drug loading and other conditions of the formulation can affect the drug encapsulation capacity and release rate. Several proteins, such as insulin [38], cyclosporine A [39], interleukin-2 [40], catalase [37], superoxide dismutase [37] and ricin vaccine [41], have been formulated with liposomes for pulmonary delivery. In these studies, the liposomes increased the absorption and therapeutic effects, lengthened the residence time, or reduced the toxicity of the protein concerned.

Cyclodextrins are cyclic glucose oligomers with a hydrophobic cavity that can form complexes with drug molecules. Due to the large size and overall hydrophilicity of proteins and peptides, they cannot fit entirely into the cavity. Rather, interactions may be restricted to certain hydrophobic amino acid groups on the protein [30]. Such protein–cyclodextrin interactions may change the conformation, and consequently the physicochemical properties, of the macromolecule, leading to enhanced absorption [30]. Other proposed mechanisms include direct disruption of the alveolar epithelial membrane through interactions between the cyclodextrins and pulmonary lipids and proteins, and/or reduction of protease activity [30]. The ranking of cyclodextrin derivatives in the order of increasing transmucosal protein transport and toxicity are dimethyl-β-cyclodextrin > α-cyclodextrin > β-cyclodextrin > γ-cyclodextrin > hydroxypropyl-β-cyclodextrin [42]. Dimethyl-β-cyclodextrin has been shown to increase the pulmonary absorption of insulin [43] and salmon calcitonin [36] in rats. Besides these studies, the application of cyclodextrins in pulmonary protein delivery is currently limited.

2.2.3. Safety of Inhaled Proteins and Peptides

Immunogenicity is a potential issue for proteins and peptides, because they may be antigenic and trigger local or systemic immune responses. This is especially of concern for exogenous peptides derived from synthetic or biological sources. Since pulmonary delivery is a relatively new route of administration for proteins, the long-term adverse effects of usage over decades are uncertain. However, many proteins and peptides have been shown to be generally safe for inhalation in the short-term, at least they are not more harmful than subcutaneous injections [5]. An obvious example is the successful approval and marketing of the dornase alfa product, Pulmozyme®. Inhaled insulin has also been demonstrated to be safe over a two-year period [44].

Besides the protein, the safety of the excipients is also important. Except for lecithin (phosphatidylcholine), all other absorption enhancers discussed above are not yet approved for inhalation [37]. Furthermore, the safety aspects of many of these compounds have not been studied extensively. Of the few that have been examined, some proved favorable while some did not. For instance, certain absorption enhancers (n-lauryl β-D-maltopyranoside, sodium glycocholate and linoleic acid-HCO60 mixed micelles) showed low pulmonary toxicity on rats at specific concentrations [34]. On the other hand, polycationic liposomes have been observed to induce oxygen radical-mediated pulmonary toxicity in mice [45]. In another study, sodium taurocholate decreased the viability of Caco-2 cells *in vitro* at high concentrations [33]. Polymeric materials employed in controlled release formulations may accumulate in the peripheral lung over long-term use, where mucociliary clearance is absent [37]. The safety profiles of many other novel excipients need to be assessed thoroughly for regulatory approval before their clinical application.

2.2.4. Types of Proteins and Peptides for Inhalation

Many pharmaceutical proteins and peptides have been formulated as inhalation aerosols for local and systemic diseases (Table 2.1). The formulations are at various stages of development, ranging from Phase I studies to approved products. Exubera® was a marketed insulin dry powder inhaler but was withdrawn in October 2007 due to poor sales [46].

2.3. STABILITY OF PROTEINS AND PEPTIDES IN PULMONARY DELIVERY FORMULATIONS

The biological activity of proteins is strongly dependent on their molecular structure, which involves several organization levels [1]. The primary structure is the amino acid sequence, which ultimately dictates the non-covalent interactions that the molecule undergoes when forming higher order structures. The secondary structure is the periodic spatial arrangement of the polypeptide chain backbone due to hydrogen bonding between the C=O and N−H groups. Alpha-helices and beta-sheets are typical secondary structures within a protein. The tertiary structure is the 3-dimensional conformation of the whole molecule, including the positions of all amino acid side chains. Some proteins may consist of multiple peptide chains grouped together by non-covalent inter-molecular interactions. The arrangement of these subunits relative to each other constitutes the quaternary structure. Alterations in the protein structure at any level may lead to a change or loss of biological activity. The mechanisms of degradation can be biochemical or physical in nature.

TABLE 2.1 Examples of proteins and peptides for inhalation (adapted from references [22,37,47])

Local diseases	Compound
α-1-Antitrypsin deficiency	α-1-Proteinase inhibitor
Asthma	Anti-IgE Mab Interleukin-1 receptor Interleukin-4 Lactoferrin Vasoactive intestinal peptide
Anti-tuberculosis vaccine	Muramyl dipeptide
Bronchospastic pulmonary diseases	Calcitonin gene-related peptide
Cancer/Pneumocystis carnii	Interleukin-2
Chronic bronchitis	Uridine triphosphate derivatives
Cystic fibrosis	rhDNase (approved product: Pulmozyme®) Secretin Targeted genetics adeno-associated virus for cystic fibrosis
Emphysema/Cystic fibrosis	α-1-Antitrypsin Secretory leukoprotease inhibitor
Idiopathic pulmonary fibrosis	Interferon-γ
Lung transplant	Cyclosporine A
Oxidative stress	Catalase Superoxide dismutase
Respiratory distress syndrome	Surfactant-associated proteins from natural bovine lung extract (approved product: Survanta®)
Systemic diseases	**Compound**
Anemia	Erythropoietin
Anti-coagulation	Heparin
Cancer	Interleukins Luteinizing hormone-releasing hormone
Diabetes insipidus	1-Deaminocysteine-8-D-arginine vasopressin Desmopressin

(Continued)

TABLE 2.1 Examples of proteins and peptides for inhalation (adapted from references [22,37,47])—cont'd

Local diseases	Compound
Diabetes mellitus	Insulin (approved product: Exubera®; discontinued end 2007) Exendin-4 Glucagon-like peptide
Endometriosis	Leuprolide
Growth deficiency	Human growth hormone
Hemophilia	Factor IX
Hypoglycemia	Glucagon
Infertility	Follicle-stimulating hormone
Multiple sclerosis	Interferon-β
Neutropenia	Recombinant human granulocyte-colony stimulating factor (rhG-CSF)
Obesity	Peptide YY
Osteoporosis	Calcitonin Parathyroid hormone
Viral infections	Interferon-α Ribavirin

2.3.1. Biochemical Stability

Biochemical degradation involves changes in the covalent bonds in the polypeptide chain, mainly through hydrolysis, isomerization, deamidation, oxidation and disulphide bridging [3,48]. Hydrolysis is the cleavage of peptide bonds, usually under extreme acidic or alkaline conditions. However, even at neutral pH, the asparagine–proline and asparagine–glycine bonds are relatively labile [3]. Except for glycine, all naturally-occurring amino acids are chiral and are of the L-form. They can isomerize to the D-form under certain circumstances [3,48]. Deamidation is the transformation of the amide groups in asparagine and glutamine side chains into carboxylic acid groups, consequently forming aspartic and glutamic acids, respectively [3,48]. The side chains of cysteine, methionine, tyrosine, tryptophan and histidine are prone to oxidation [3,48]. The sulfur-containing side chains of cysteine and methionine can also form intra- or intermolecular covalent disulfide bridges [3,48]. All such biochemical degradations will disrupt the primary protein structure.

2.3.2. Physical Stability

Physical degradation refers to changes in the non-covalent interactions within or between protein molecules. It can occur independently, or via biochemical degradation, and results in alterations to the higher order structures (secondary and above). Common types of physical degradation include denaturation, aggregation, precipitation and adsorption [3,48]. Denaturation is the unfolding of a molecule from its native conformation and may or may not be reversible [48]. Denatured proteins may also associate with each other to form molecular aggregates [3,48]. Precipitation is essentially aggregation on a macroscopic scale, when dissolved proteins come out of solution due to a reduction in solubility [48]. Protein molecules are surface-active and may adsorb onto various interfaces [48].

Owing to their complex structure, proteins are much more fragile than small drug molecules. The risks of biochemical and physical degradation depend on many physicochemical factors, such as temperature, pH, storage humidity, formulation constituents, delivery device and manufacturing process, amongst others. These must be well studied and controlled to maintain the stability and efficacy of protein products.

2.4. TYPES OF PULMONARY DELIVERY FORMULATIONS

2.4.1. Nebulizers and the AERx® Pulmonary Delivery System

Nebulizers produce droplets from a bulk liquid using either compressed air (jet nebulizers) or ultrasonic waves (ultrasonic nebulizers). They generate continuous aerosol streams over a period of minutes, thus their aerosol output is much higher than those from dry powder inhalers and metered dose inhalers (see below). In jet nebulizers, compressed air is forced through a small hole or series of holes on the tip of a tube protruding upwards from the liquid reservoir [49]. This creates a low pressure zone at the tip of the tube. Consequently, liquid is drawn up the tube by the Venturi effect, and is then rapidly dispersed as droplets by the air jet [49]. Large droplets impact on a baffle inside the nebulizer and flow back into the reservoir. Smaller droplets exit through the mouthpiece for inhalation. The droplet size depends on the nebulizer construction and air pressure but generally ranges from 2–5 μm. The liquid cools and concentrates during operation due to evaporation in the air jet. Some examples of commercial jet nebulizers are shown in Figure 2.1. Ultrasonic nebulizers have a piezoelectric crystal below the liquid reservoir. The crystal vibrates rapidly to generate ultrasonic waves through the liquid to produce droplets for inhalation [50]. The piezoelectric energy input increases the liquid temperature during use and may affect the stability of the protein [51].

Protein nebulizer formulations are mainly aqueous solutions and are relatively easy to develop. The factors that may affect their stability are similar to those for general protein liquid formulations, such as ionic strength, pH and the

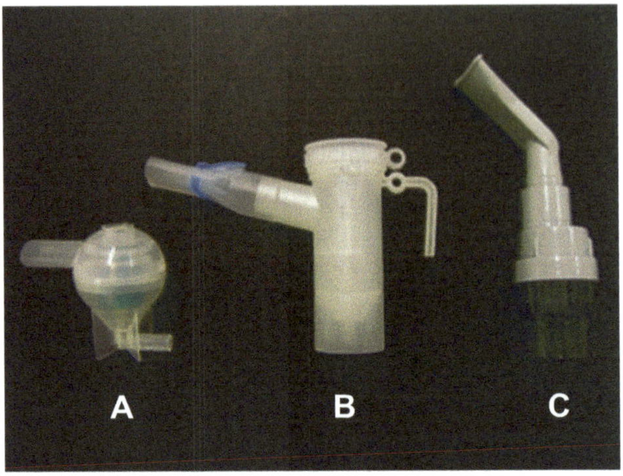

FIGURE 2.1 Commercial jet nebulizers: (A) Pari IS-2®; (B) Pari LC® Plus and (C) Sidestream®. Pari LC® Plus is recommended for the administration of Pulmozyme®.

type and concentration of other dissolved compounds [52]. However, there are some potential problems that are unique to nebulizers. Atomizing air pressure, shear, recirculation and progressive reduction of solution volume in jet nebulizers may lead to protein degradation [53]. Formulations without suitable protein stabilizers may also be subject to molecular adsorption to plastic surfaces [53]. The degradation rate of nebulized lactate dehydrogenase was found to increase with air pressure, nebulization time and low starting solution volumes [54]. The aggregation rate of rhG-CSF also increased with nebulization time. However, these adverse effects were significantly reduced by the addition of polyethylene glycol (PEG) 1000 in the formulation [54]. In another study, the stability of nebulized aviscumine, a dimeric protein, was improved by adding various surfactants, cryoprotectants and buffer salts to the reconstitution medium [55–57].

High energy vibrations and heat generated in ultrasonic nebulizers may denature proteins, as was observed on rhDNase [51]. Therefore, the only jet nebulizers so far approved are for the administration of Pulmozyme® [58]. The formulation of Pulmozyme® is very simple, with only sodium chloride, calcium chloride dihydrate and water for injection as excipients [58,59], illustrating that not all protein nebulizer formulations require complex additives for stabilization. Thus the stability of the protein of interest should be assessed on a case-by-case basis and excipients employed where appropriate.

The AERx® pulmonary delivery system devised by Aradigm (California, USA) generates aerosols by mechanically extruding the liquid from a unit-dose reservoir through micron-sized orifices [60]. Precise delivery control is achieved by internal mechanical components that trigger the extrusion mechanism upon inhalation by the patient. This device has been successfully used for the delivery of a number of proteins, including rhDNase [61], insulin [62,63] and interleukin-4 receptor [64].

2.4.2. Metered Dose Inhalers

Metered dose inhalers (MDIs) are popular respiratory devices due to their portability and simple operation. Although easy to use, the MDI is sophisticated in design (Figures 2.2 and 2.3). The drug(s) are suspended or dissolved in a liquefied propellant system, which may also contain excipients such as co-solvents or surfactants. The formulation is maintained under pressure in a small canister, which is crimped with a metering valve. Upon actuation through an actuator, the valve opens and the metered dose is dispensed as an aerosol spray by the expansion and vaporization of the propellant under ambient conditions [65].

The major issue in MDI protein formulation is the conformational stability of these macromolecules in liquefied propellants such as the hydrofluoroalkanes (HFA), which are less polar than water. Fourier transform Raman spectroscopy has been applied to investigate the secondary structure of hen's egg lysozyme suspended in HFA-134a (1,1,1,2-tetrafluoroethane) and HFA-227 (1,1,1,2,3,3,3-heptafluoropropane) [66]. Conformational data on the peptide backbone, C–C stretching and disulfide bonds were obtained. The technique is simple and non-destructive as the formulations can be analyzed in crimped glass vials [66].

Various proteins have been formulated as MDI suspensions [67,68]. Nonionic surfactants that are soluble in propellant were used to improve the ability of the protein particles to be kept in suspension [67,68]. These included Triton X-100, Triton X-405, Laureth 9, Brij 30, Brij 97, Brij 98, Tween 80, Nonidet-40 and diethylene glycol monoethyl ether for HFA-134a [67,68]. An

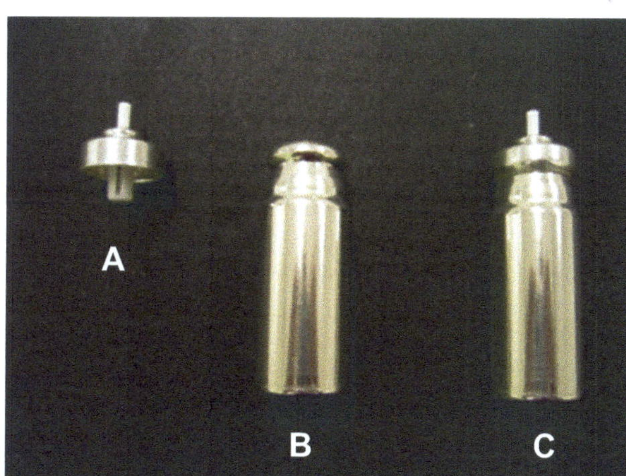

FIGURE 2.2 MDI components: (A) metering valve; (B) aluminium canister and (C) crimped MDI.

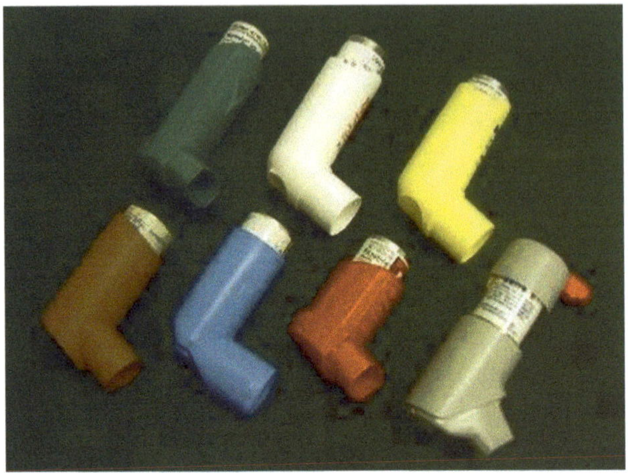

FIGURE 2.3 Commercial MDI products with their actuators.

aqueous solution of the protein and surfactant was first freeze dried to produce the required particles, which were subsequently suspended in HFA-134a alone or a mixture of this propellant and dimethylether [67] or ethanol [68]. The propellant mixtures were found to improve the aerosol performance over the pure propellant. Kos Pharmaceuticals (Florida, USA) has developed an insulin MDI suspension in HFA-134a [69,70]. This formulation has been shown to provide comparable blood glucose control to subcutaneous insulin injections in type 2 diabetic patients [69].

It is more difficult to develop MDI solutions for proteins than suspensions due to the non-polar nature of the propellants. Nevertheless, a stable HFA-134a solution containing leuprolide acetate was successfully formulated incorporating water and an undisclosed co-solvent [71].

2.4.3. Dry Powder Inhalers

There are many marketed dry powder inhalers (DPIs) with various devices and formulations (Figure 2.4). These products provide single- or multi-doses via oral inhalation, depending on the design of the powder reservoir and metering components [72]. Unlike the MDIs discussed above, DPIs do not contain propellants. The energy of powder dispersion is usually derived from the inhalation effort applied by the patient. Powders for pulmonary delivery are often cohesive, due to the high specific surface area (surface area-to-mass ratio) of micron-sized particles. Thus the powders must be formulated appropriately such that the active ingredients can be easily dispersed into fine particles upon inhalation.

Formulating protein powders for pulmonary delivery is especially challenging as it requires not only that the powders can flow and disperse, but also

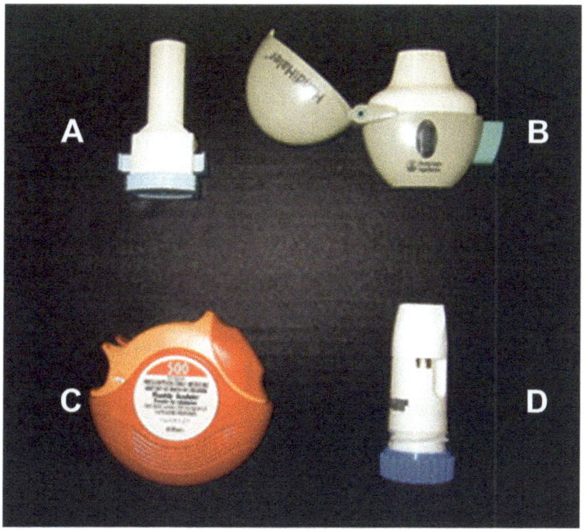

FIGURE 2.4 Commercial DPI products: (A) Aerolizer® (Novartis); (B) Handihaler® (Boehringer Ingelheim); (C) Accuhaler® (GlaxoSmithKline); (D) Turbuhaler® (AstraZeneca).

stability of the proteins. To satisfy the latter requirement, proteins are usually formulated as amorphous glasses which are, however, physically unstable and tend to crystallize with inter-particulate bond formation and loss of powder dispersibility. In addition, the stability requirements limit the manufacturing processes that can be used for protein powder production.

Removal of water from proteins during powder production can cause significant molecular conformational damage that can lead to further biochemical degradation such as aggregation, deamidation and oxidation during storage (cf. Chapter 13, pages 314–318). Amorphous glassy excipients, mainly carbohydrates, have been widely used to stabilize proteins for inhalation, such as lactose for rhDNase [73, 74]; trehalose, lactose and mannitol for recombinant humanized anti-IgE monoclonal antibody (rhuMAbE25) [75]; and mannitol and raffinose for insulin [76]. Other suitable excipients include polymers (e.g. polyvinylpyrrolidone), proteins (e.g. human serum albumin), peptides (e.g. aspartame), amino acids (e.g. glycine) and organic salts (e.g. citrates). Although lactose has been widely used for inhalation products for small molecule drugs, it may not be compatible with proteins. Being a reducing sugar, lactose reacts with lysine, and protein glycation has indeed been found in both rhDNase and rhuMAbE25 [75, 77]. The exact mechanism for protein stabilization by excipients is as yet unclear. However, contributing factors may include:

(i) formation of a glassy state of the protein-excipient system;
(ii) hydrogen bonding between the protein and excipient molecules;

(iii) crystallinity of the excipients; and
(iv) residual water content.

The diffusion rate and mobility of the protein molecules in the glassy state are much lower than those in the rubbery state. Therefore, any physicochemical reactions leading to biochemical instability will be reduced [78]. Crystalline excipients such as mannitol are known to decrease the stability of proteins [79]. However, mannitol can be used in the amorphous form, for instance, in the presence of glycine [80]. Fourier transform infrared spectroscopy offers information on protein secondary structures, and has provided evidence for protein stabilization by hydrogen bonding [81]. Water promotes the instability of proteins by enhancing molecular mobility [82], as shown by nuclear magnetic resonance spectroscopy [83]. The crystalline or amorphous state of the excipients is crucial because it controls the distribution of water between the protein and the excipient in a powder [84].

While glassy materials are desirable for protein biochemical stability, one of their drawbacks is their physical instability. Moisture uptake by fine particles of hydrophilic amorphous materials can be very fast, due to their large specific surface area and high energy state. For example, water uptake by spray dried rhDNase plus lactose induces crystallization that adversely affected powder dispersibility [73,74]. During crystallization, water acts as a plasticizer to lower the glass transition temperature (T_g) (approx. 10°C reduction per 1% water in sugar-containing formulations), which will enhance molecular mobility required for nucleation if the T_g approaches the storage temperature [85]. It is thus important to keep the powder dry in order to maintain the high T_g, or to use excipients with a high T_g, or to store the powders at a low temperature. It has been proposed to store fragile glasses 50°C below the T_g to minimize crystallization [86]. This method would be practical for room temperature storage for amorphous materials with a T_g above 70°C. However, crystallization of amorphous excipients at temperatures well below the T_g is still possible, which consequently induces protein degradation [87]. The aggregation of salmon calcitonin when co-spray dried with 30% mannitol (completely amorphous) is higher than that with 70% mannitol (mostly crystalline) at a given relative humidity (RH) [87]. This indicates that the concentration of the peptide in the amorphous phase and the water content relative to the amorphous content are crucial in predicting the stability of the formulation. This is because the peptide is mixed at a molecular level only with the amorphous excipient and most of the absorbed moisture interacts with the amorphous phase [87]. The crystallization of the excipient will also cause morphological changes, such as crystalline bridging, on the particles, which will adversely affect dispersibility. The effect of moisture on powder dispersion can be instantaneous [88]. The hygroscopic effect can be reduced by using hydrophobic excipients, such as L-isoleucine [89].

Amorphous materials are usually cohesive because they have higher surface energy and are more hygroscopic than the crystalline forms. Thus powder dispersibility is strongly dependent on the structure of the solid state. Most protein drugs are formulated with excipients for improved biochemical stability, as discussed above. However, the distribution of protein and excipient/s in a particle may not be uniform. When a protein–excipient solution droplet undergoes drying to form a particle, the outer surface tends to be enriched with proteins or macromolecules that are surface active, whereas the small molecule excipients diffuse rapidly into the particle core. Under special circumstances, small excipient molecules can crystallize on the particle surface [90]. Various formulation approaches have been employed to improve protein powder dispersibility. The particle size distribution can be controlled to decrease cohesion [90]. Proteins can be co-spray dried with a suitable excipient to modify the surface energy and morphology of the resultant particle, as is the case with rhDNase with sodium chloride [90]. Blending of protein drugs with inert carrier particles (e.g. lactose) can improve the dispersibility of the product [90]. Large (mean diameters > 5 μm) and small (3–5 μm; PulmoSphere™) porous particles containing proteins have excellent aerosol performance due to their low cohesion and small aerodynamic diameters [91–93]. Surface corrugation can also decrease cohesion by reducing the inter-particulate contact areas. Wrinkled, nonporous, solid, bovine serum albumin particles were reported to disperse significantly better than their smooth, spherical counterpart (Figure 2.5) [94–97].

Inhalable protein powders can be produced by various methods, including spray drying [93–100], spray freeze drying [92], lyophilization followed by milling [101] and solvent precipitation with supercritical fluids [102–105]. Precipitation with regular anti-solvents has been carried out using high-gravity controlled precipitation for insulin [106] and confined impinging jet mixing for cyclosporine A [107,108] to yield nano-suspensions. The suspensions were spray dried to obtain micron-sized agglomerates of nanoparticles [106,108] A laboratory-scale spray dryer is shown in Figure 2.6.

Some issues regarding dry powder production techniques for protein formulations should be noted. Spray drying exposes the protein to mechanical shear and air–liquid interfacial denaturation. The hot air for drying also subjects the protein to thermal stress and denaturation. Aggregation of spray dried recombinant human growth hormone was suppressed by adding Zn^{2+} ions and surfactant polysorbate 20 [98,99]. Compared to spray drying, spray freeze drying produces porous and light particles with superior dispersibility and the production yield is almost 100%. However, this process is more time consuming and expensive. Gas-jet milling of lyophilized proteins can cause contamination and inactivation. Thus abrasion-resistant mills using high-purity nitrogen and milling stabilizers such as human serum albumin and sorbitol are required [101]. Supercritical carbon dioxide is a good anti-solvent for precipitation because it is non-toxic, economical, and has a low critical temperature of 31.1°C for operation. However, due to its non-polar nature it is not readily miscible with water. Special nozzles have been employed to enhance the

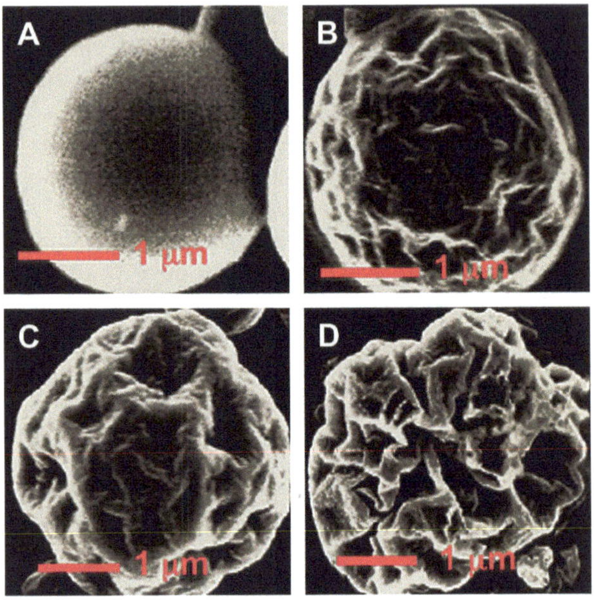

FIGURE 2.5 Scanning electron microscopy images of BSA particles produced by spray drying from aqueous solutions containing (A) 60 mg/mL; (B) 40 mg/mL; (C) 25 mg/mL and (D) 10 mg/mL BSA, respectively. *Reprinted from reference [95] with permission from Elsevier.*

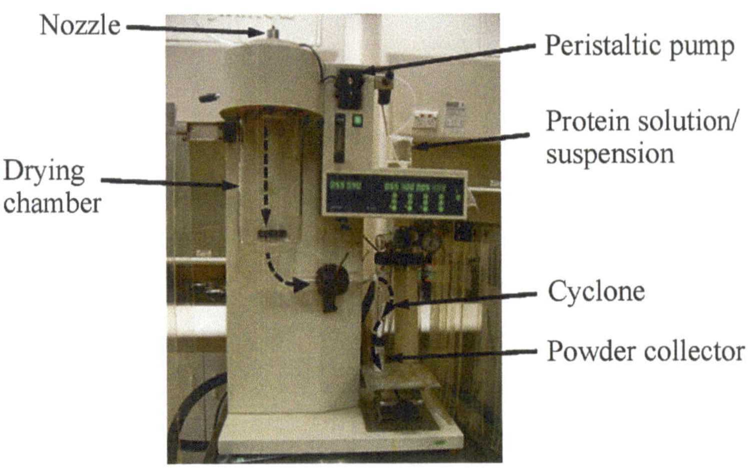

FIGURE 2.6 Laboratory-scale spray dryer. The drug solution/suspension is sprayed into the drying chamber through a nozzle. The droplets are dried as they travel through the drying chamber and cyclone. The dried particles are collected in the powder collector. Dashed arrows indicate the transit path of the droplets and particles.

mixing with aqueous protein solutions [104,105]. Alternatively, supercritical carbon dioxide modified with ethanol can be used as an anti-solvent [102,103]. Since carbon dioxide is acidic, the pH of the mixture should be controlled to avoid protein degradation. Excipients such as polyethylene glycol and self-assembling small organic molecules can be combined with proteins to form microparticles by precipitation [109–111].

2.5. PACKAGING AND STORAGE

The packaging and storage conditions of protein and peptide products are important for minimizing their degradation after manufacture. The relevant environmental factors include exposure to moisture and temperature, because both can vary during transport and storage. Of the aerosol formulations discussed above, DPIs are the most susceptible to moisture from the environment. For instance, the aggregation of salmon calcitonin [87] and anti-IgE monoclonal antibody [112] increased with increasing humidity for the co-spray dried powders containing the protein and mannitol. Thus the packaging of protein powders should employ excellent moisture barriers and/or desiccants. The temperature during transport and storage should also be controlled, to remain below the T_g of the protein and stabilizing excipients.

The robustness of the packaging is tested for developmental and regulatory purposes. There are specific stability study guidelines for MDI and DPI formulations due to their complexity. The studies should be performed on the product in the full packaging that is intended for marketing [113]. There are three types of storage conditions for testing:

(i) accelerated (40°C/75% RH) at 0, 1, 3, 6 months or beyond;
(ii) intermediate (30°C/60% RH) at 0, 3, 6, 9, 12 months; and
(iii) long-term (25°C/60% RH) at 0, 3, 6, 9, 12, 18, 24 months.

The products should be tested under more than one of the above conditions, because accelerated studies alone may not be predictive of the actual aerosol performance over the proposed shelf life [113]. The parameters with suitable tolerance criteria evaluated for MDIs and DPIs include the appearance/color of the device components and formulation, microbial contents, moisture content, drug content, degradation products and impurities, dose uniformity, particle size distribution and microscopic evaluation of the formulation [113]. In addition, the alcohol content (if present as a co-solvent), leak rate, valve delivery and presence of leachables are also tested on MDIs [113].

2.6. CONCLUSIONS

Pulmonary delivery of proteins is gaining increasing attention. Active research is being carried out in this area, but so far very few respirable proteins have been marketed and practically no absorption enhancers have been approved. Most of the *in vivo* efficacy and toxicity studies have been conducted on animals, and not yet on humans. As proteins and peptides are complex and fragile molecules there are formulation challenges that need to be overcome. Knowledge in this field is continually progressing and inhalation will become a convenient administration method for proteins and peptides, particularly for local delivery, in the future.

REFERENCES

[1] M.K. Campbell, Biochemistry, Harcourt Brace & Company, Orlando, USA, 1999.
[2] P.S. Bailey Jr., C.A. Bailey, Organic Chemistry: A Brief Survey of Concepts and Applications, Prentice-Hall International, Upper Saddle River, NJ, USA, 1995.
[3] D. Crommelin, E. van Winden, A. Mekking, Delivery of pharmaceutical proteins, in: M.E. Aulton (Ed.), Pharmaceutics: The Science of Dosage Form Design, Churchill Livingstone, Edinburgh, UK, 2002, pp. 544–553.
[4] R.J. Altiere, D.C. Thompson, Physiology and pharmacology of the airways, in: A.J. Hickey (Ed.), Inhalation Aerosols: Physical and Biological Basis for Therapy, Informa Healthcare USA, New York, USA, 2007.
[5] R.K. Wolff, Safety of inhaled proteins for therapeutic use, J. Aerosol Med. 11 (1998) 197–219.
[6] J.R. Wright, Clearance and recycling of pulmonary surfactant, Am. J. Physiol.-Lung Cell. Mol. Physiol. 259 (1990) L1–L12.
[7] W.C. Hinds, Aerosol Technology, John Willey & Sons, New York, USA, 1999.
[8] A. Hayem, A. Scharfman, A. Laine, J.J. Lafitte, B. Sablonniere, Proteases and antiproteases in brochoalveolar lavage, Bull. Eur. Physiopathol. Respir. (Suppl. 6) (1980) 247–260.
[9] A. Janoff, L. Raju, R. Dearing, Levels of elastase activity in bronchoalveolar lavage fluids of healthy smokers and nonsmokers, Am. Rev. Respir. Dis. 127 (1983) 540–544.
[10] S.F. Smith, A. Guz, N.T. Cooke, G.H. Burton, T.D. Tetley, Extracellular elastolytic activity in human lung lavage: A comparative study between smokers and non-smokers, Clin. Sci. 69 (1985) 17–27.
[11] J. Fujita, N.L. Nelson, D.M. Daughton, C.A. Dobry, J.R. Spurzem, S. Irino, et al., Evaluation of elastase and antielastase balance in patients with chronic bronchitis and pulmonary emphysema, Am. Rev. Respir. Dis. 142 (1990) 57–62.
[12] M. Orlowski, J. Orlowski, M. Lesser, K.H. Kilburn, Proteolytic enzymes in bronchopulmonary lavage fluids: Cathepsin B-like activity and prolyl endopeptidase, J. Lab. Clin. Med. 97 (1981) 467–476.
[13] D.B. Borson, Roles of neutral endopeptidases in airways, Am. J. Physiol.-Lung Cell. Mol. Physiol. 260 (1991) L212–L225.
[14] J.W. Ryan, Peptidase enzymes of the pulmonary vascular surface, Am. J. Physiol.-Lung Cell. Mol. Physiol. 257 (1989) L53–L60.
[15] A.R. Johnson, J. Ashton, W.W. Schultz, E.G. Erdös, Neutral metalloendopeptidase in human lung tissue and cultured cells. Am. Rev. Respir. Dis. 132 (1985) 564–568.
[16] D.A. Wall, A.T. Lanutti, High levels of exopeptidase activity are present in rat and canine bronchoalveolar lavage fluid, Int. J. Pharm. 97 (1993) 171–181.
[17] A. Janoff, Elastases and emphysema, Am. Rev. Respir. Dis. 132 (1985) 417–433.

[18] S.A. Adibi, E.L. Morse, Intestinal transport of dipeptides in man: Relative importance of hydrolysis and intact absorption, J. Clin. Invest. 50 (1971) 2266–2275.
[19] R.E. Stratford Jr., V.H.L. Lee, Aminopeptidase activity in homogenates of various absorptive mucosae m the albino rabbit: Implications in peptide delivery, Int. J. Pharm. 30 (1986) 73–82.
[20] G.H. Caughey, Peptide metabolism by lung and airway peptidases, J. Biopharm. Sci. 3 (1992) 233–240.
[21] J.S. Patton, S. Nagarajan, A. Clark, Pulmonary absorption and metabolism of peptides and proteins, in: R.N. Dalby, P.R. Byron, S.J. Farr (Eds.), Respiratory Drug Delivery VI, Interpharm Press, Bufflao Grove, IL, USA, 1998, pp. 17–24.
[22] S. Onoue, N. Hashimoto, S. Yamada, Dry powder inhalation systems for pulmonary delivery of therapeutic peptides and proteins, Expert Opin. Ther. Patents 18 (2008) 429–442.
[23] U.B. Kompella, V.H.L. Lee, (C) Means to enhance penetration: (4) Delivery systems for penetration enhancement of peptide and protein drugs: Design considerations, Adv. Drug Deliv. Rev. 8 (1992) 115–162.
[24] D. Deshpande, D. Toledo-Velasquez, L.Y. Wang, C.J. Malanga, J.K.H. Ma, Y. Rojanasakul, Receptor-mediated peptide delivery in pulmonary epithelial monolayers, Pharm. Res. 11 (1994) 1121–1126.
[25] D.A. Groneberg, M. Nickolaus, J. Springer, F. Döring, H. Daniel, F. Axel, Localization of the peptide transporter PEPT2 in the lung: Implications for pulmonary oligopeptide uptake, Am. J. Pathol. 158 (2001) 707–714.
[26] A.N.O. Dodoo, S. Bansal, D.J. Barlow, F.C. Bennet, R.C. Hider, A.B. Lansley, et al., Systematic investigations of the influence of molecular structure on the transport of peptides across cultured alveolar cell monolayers, Pharm. Res. 17 (2000) 7–14.
[27] R.H. Hastings, M. Grady, T. Sakuma, M.A. Matthay, Clearance of different-sized proteins from the alveolar space in humans and rabbits, J. Appl. Physiol. 73 (1992) 1310–1316.
[28] J.S. Patton, Mechanisms of macromolecule absorption by the lungs, Adv. Drug Deliv. Rev. 19 (1996) 3–36.
[29] K. Morimoto, H. Yamahara, V.H.L. Lee, K.-J. Kim, Transport of thyrotropin-releasing hormone across rat alveolar epithelial cell monolayers, Life Sci. 54 (1994) 2083–2092.
[30] A. Hussain, J.J. Arnold, M.A. Khan, F. Ahsan, Absorption enhancers in pulmonary protein delivery, J. Control. Release 94 (2004) 15–24.
[31] Y. Fukuda, T. Tsuji, T. Fujita, A. Yamamoto, S. Muranishi, Susceptibility of insulin to proteolysis in rat lung homogenate and its protection from proteolysis by various protease inhibitors, Biol. Pharm. Bull. 18 (1995) 891–894.
[32] K. Okumura, S. Iwakawa, T. Yoshida, T. Seki, F. Komada, Intratracheal delivery of insulin absorption from solution and aerosol by rat lung, Int. J. Pharm 88 (1992) 63–73.
[33] F. Johansson, E. Hjertberg, S. Eirefelt, A. Tronde, U.H. Bengtsson, Mechanisms for absorption enhancement of inhaled insulin by sodium taurocholate, Eur. J. Pharm. Sci. 17 (2002) 63–71.
[34] A. Yamamoto, S. Okumura, Y. Fukuda, M. Fukui, K. Takahashi, S. Muranishi, Improvement of the pulmonary absorption of ($Asu^{1,7}$)-eel calcitonin by various absorption enhancers and their pulmonary toxicity in rats, J. Pharm. Sci. 86 (1997) 1144–1147.
[35] S. Kobayashi, S. Kondo, K. Juni, Study on pulmonary delivery of salmon calcitonin in rats: Effects of protease inhibitors and absorption enhancers, Pharm. Res. 11 (1994) 1239–1243.
[36] S. Kobayashi, S. Kondo, K. Juni, Pulmonary delivery of salmon calcitonin dry powders containing absorption enhancers in rats, Pharm. Res. 13 (1996) 80–83.

[37] S.-A. Cryan, Carrier-based strategies for targeting protein and peptide drugs to the lungs, AAPS J. 7 (2005) E20–E41.

[38] F.-Y. Liu, Z. Shao, D.O. Kildsig, A.K. Mitra, Pulmonary delivery of free and liposomal insulin, Pharm. Res. 10 (1993) 228–232.

[39] G.V. Letsou, H.J. Safi, M.J. Reardon, M. Ergenoglu, Z. Li, C.N. Kolonaris, et al., Pharmacokinetics of liposomal aerosolized cyclosporine A for pulmonary immunosuppression, Ann. Thorac. Surg. 68 (1999) 2044–2048.

[40] C. Khanna, D.E. Hasz, J.S. Klausner, P.M. Anderson, Aerosol delivery of interleukin 2 liposomes is nontoxic and biologically effective: Canine studies, Clin. Cancer Res. 2 (1996) 721–734.

[41] G.D. Griffiths, G.J. Phillips, S.C. Bailey, Comparison of the quality of protection elicited by toxoid and peptide liposomal vaccine formulations against ricin as assessed by markers of inflammation, Vaccine 17 (1999) 2562–2568.

[42] Z. Shao, Y. Li, T. Chermak, A.K. Mitra, Cyclodextrins as mucosal absorption promoters of insulin. II. Effects of β-cyclodextrin derivatives on α-chymotryptic degradation and enteral absorption of insulin in rats, Pharm. Res. 11 (1994) 1174–1179.

[43] A. Hussain, T. Yang, A.-A. Zaghloul, F. Ahsan, Pulmonary absorption of insulin mediated by tetradecyl-β-maltoside and dimethyl-β-cyclodextrin, Pharm. Res. 20 (2003) 1551–1557.

[44] W.T. Cefalu, C.C. Balagtas, W.H. Lanschultz, R.A. Gelfand, Sustained efficacy and pulmonary safety of inhaled insulin during 2-years of outpatient therapy, Diabetes 49 (2000) A101.

[45] S. Dokka, D. Toledo, X. Shi, V. Castranova, Y. Rojanasakul, Oxygen radical-mediated pulmonary toxicity induced by some cationic liposomes, Pharm. Res. 17 (2000) 521–525.

[46] C.J. Bailey, A.H. Barnett, Why is Exubera being withdrawn? Br. Med. J. 335 (2007) 1156.

[47] L. Garcia-Contreras, H.D.C. Smyth, Liquid-spray or dry-powder systems for inhaled delivery of peptide and proteins? Am. J. Drug Deliv. 3 (2005) 29–45.

[48] M.C. Manning, K. Patel, R.T. Borchardt, Stability of protein pharmaceuticals, Pharm. Res. 6 (1989) 903–918.

[49] R.W. Niven, A.J. Hickey, Atomization and nebulizers, in: A.J. Hickey (Ed.), Inhalation Aerosols: Physical and Biological Basis for Therapy, Informa Healthcare USA, New York, USA, 2007.

[50] B.J. Greenspan, Ultrasonic and electrohydrodynamic methods for aerosol generation, in: A.J. Hickey (Ed.), Inhalation Aerosols: Physical and Biological Basis for Therapy, Informa Healthcare USA, New York, USA, 2007.

[51] D.C. Cipolla, A.R. Clark, H.-K. Chan, I. Gonda, S.J. Shire, Assessment of aerosol delivery systems for recombinant human deoxyribonuclease, STP Pharma. Sci. 4 (1994) 50–62.

[52] T. Arakawa, S.J. Prestrelski, W.C. Kenney, J.F. Carpenter, Factors affecting short-term and long-term stabilities of proteins, Adv. Drug Deliv. Rev. 10 (1993) 1–28.

[53] P.K. Gupta, A.L. Adjei, Therapeutic inhalation aerosols, in: A.L. Adjei, P.K. Gupta (Eds.), Inhalation Delivery of Therapeutic Peptides and Proteins, Marcel Dekker, New York, USA, 1997, pp. 185–234.

[54] R.W. Niven, A.Y. Ip, S.D. Mittelman, C. Farrar, T. Arakawa, S.J. Prestrelski, Protein nebulization: I. Stability of lactate dehydrogenase and recombinant granulocyte-colony stimulating factor to air-jet nebulizaion, Int. J. Pharm. 109 (1994) 17–26.

[55] H. Steckel, F. Eskander, K. Witthohn, Effect of excipients on the stability and aerosol performance of nebulised aviscumine, J. Aerosol Med. 16 (2003) 417–432.

[56] H. Steckel, F. Eskander, K. Witthohn, Effect of cryoprotectants on the stability and aerosol performance of nebulized aviscumine, a 57-kDa protein, Eur. J. Pharm. Biopharm. 56 (2003) 11–21.
[57] H. Steckel, F. Eskander, K. Witthohn, The effect of formulation variables on the stability of nebulised aviscumine, Int. J. Pharm. 257 (2003) 181–194.
[58] Pulmozyme® Consumer Medicine Information, Roche Products Pty Ltd, Dee Why, NSW, Australia, 2008.
[59] I. Gonda, Deoxyribonuclease inhalation, in: A.L. Adjei, P.K. Gupta (Eds.), Inhalation Delivery of Therapeutic Peptides and Proteins, Marcel Dekker, New York, USA, 1997, pp. 355–365.
[60] R. Rubsamen, Novel aerosol peptide drug delivery systems, in: A.L. Adjei, P.K. Gupta (Eds.), Inhalation Delivery of Therapeutic Peptides and Proteins, Marcel Dekker, New York, USA, 1997, pp. 703–731.
[61] S. Mudumba, M. Khossravi, D. Yim, T. Rossi, D. Pearce, M. Hughes, et al., Delivery of rhDNase by the AERx® pulmonary delivery system, in: R.N. Dalby, P.R. Byron, S. Farr, J. Peart (Eds.), Respiratory Drug Delivery VII, Serentec Press, Raleigh, NC, USA, 2000, pp. 329–332.
[62] R.R. Henry, S.R.D. Mudaliar, W.C. Howland III, N. Chu, D. Kim, B. An, et al., Inhaled insulin using the AERx Insulin Diabetes Management System in healthy and asthmatic subjects, Diabetes Care 26 (2003) 764–769.
[63] S. Farr, D. Reynolds, A. Nat, S. Srinivasan, M. Roach, S. Jensen, et al., Technical development of AERx® Diabetes Management System: Essential characteristics for diabetes treatment with pulmonary insulin, in: R.N. Dalby, P.R. Byron, J. Peart, S. Farr (Eds.), Respiratory Drug Delivery VIII, Davis Horwood International, Raleigh, NC, USA, 2002, pp. 51–60.
[64] S. Sangwan, J.M. Agosti, L.A. Bauer, B.A. Otulana, R.J. Morishige, D.C. Cipolla, et al., Aerosolized protein delivery in asthma: Gamma camera analysis of regional deposition and perfusion, J. Aerosol Med. 14 (2001) 185–195.
[65] T.S. Purewal, D.J.W. Grant, (Eds.), Metered Dose Inhaler Technology, Interpharm Press, Buffalo Grove, IL, USA, 1998.
[66] É.Á. Quinn, R.T. Forbes, A.C. Williams, M.J. Oliver, L. McKenzie, T.S. Purewal, Protein conformational stability in the hydrofluoroalkane propellants tetrafluoroethane and heptafluoropropane analysed by Fourier transform Raman spectroscopy, Int. J. Pharm. 186 (1999) 31–41.
[67] A.R. Brown, D.W. George, Tetrafluoroethane (HFC 134A) propellant-driven aerosols of proteins, Pharm. Res. 14 (1997) 1542–1547.
[68] R.O. Williams III, J. Liu, Formulation of a protein with propellant HFA 134a for aerosol delivery, Eur. J. Pharm. Sci. 7 (1998) 137–144.
[69] M. Hausmann, S. Dellweg, C. Osborn, L. Heinemann, A. Buchwald, R. Rosskamp, et al., Inhaled insulin as adjunctive therapy in subjects with type 2 diabetes failing oral agents: A controlled proof-of-concept study, Diabetes Obes. Metab. 8 (2006) 574–580.
[70] A.L. Adjei, P. Genova, Y. Zhu, F. Sexton (2006) Method of treating a systemic disease. Patent applicant: KOS Pharmaceuticals, Inc. Publication info: US2004042967 (A1) – 2004–03–04, and US7056494 (B2) – 2006–06–06.
[71] G. Brambilla, S. Berrill, R.J. Davies, D. Ganderton, S.C. George, D.A. Lewis, et al., Formulation of leuprolide as an HFA solution pMDI, J. Aerosol Med. 16 (2003) 209.
[72] N. Islam, E. Gladki, Dry powder inhalers (DPIs) – A review of device reliability and innovation, Int. J. Pharm. 360 (2008) 1–11.

[73] H.-K. Chan, I. Gonda, Solid state characterization of spray-dried powders of recombinant human deoxyribonuclease (rhDNase), J. Pharm. Sci. 87 (1998) 647–654.
[74] A.R. Clark, N. Dasovich, I. Gonda, H.-K. Chan, The balance between biochemical and physical stability for inhalation protein powders: rhDNase as an example, in: R.N. Dalby, P.R. Byron, S. Farr (Eds.), Resiratory Drug Delivery V, Interpharm Press, Buffalo Grove, IL, USA, 1996, pp. 167–174.
[75] J.D. Andya, Y.-F. Maa, H.R. Constantino, P.-A. Nguyen, N. Dasovich, T.D. Sweeney, et al., The effect of formulation excipients on protein stability and aerosol performance of spray-dried powders of a recombinant humanized anti-IgE monoclonal antibody, Pharm. Res. 16 (1999) 350–358.
[76] J.S. Patton, L. Foster, R.M. Platz (1999) Methods and compositions for pulmonary delivery of insulin. Patent Applicant: Nektar Therapeutics. Publication info: JP2006077032 (A) – 2006–03–23.
[77] C.P. Quan, S. Wu, N. Dasovich, C. Hsu, T. Patapoff, E. Canova-Davis, Susceptibility of rhDNase I to glycation in the dry-powder state, Anal. Chem. 71 (1999) 4445–4454.
[78] F. Frank, Long-term stabilization of biologicals, Biotechnol. 12 (1994) 253–256.
[79] K.-I. Izutsu, S. Yoshioka, T. Terao, Decreased protein-stabilizing effects of cryoprotectants due to crystallization, Pharm. Res. 10 (1993) 1232–1237.
[80] M.J. Pikal, K.M. Dellerman, M.L. Roy, R.M. Riggin, The effects of formulation variables on the stability of freeze-dried human growth hormone, Pharm. Res. 8 (1991) 427–436.
[81] J.F. Carpenter, S.J. Prestrelski, A. Dong, Application of infrared spectroscopy to development of stable lyophilized protein formulations, Eur. J. Pharm. Biopharm. 45 (1998) 231–238.
[82] M.J. Hageman, Sorption and solid-state stability of proteins, in: T.J. Ahern, M.C. Manning (Eds.), Stability of Protein Pharmaceuticals, Part A: Chemical and Physical Pathways of Protein Degradation, Plenum Press, New York, USA, 1992, pp. 273–309.
[83] F. Separovic, Y.H. Lam, X. Ke, H.-K. Chan, A solid-state NMR study of protein hydration and stability, Pharm. Res. 15 (1998) 1816–1821.
[84] H.-K. Chan, K.-L. Au-Yeung, I. Gonda, Development of a mathematical model for the water distribution in freeze-dried solids, Pharm. Res. 16 (1999) 660–665.
[85] Y. Roos, M. Karel, Plasticizing effect of water on thermal behavior and crystallization of amorphous food models, J. Food Sci. 56 (1991) 38–43.
[86] B.C. Hancock, S.L. Shamblin, G. Zografi, Molecular mobility of amorphous pharmaceutical solids below their glass transition temperatures, Pharm. Res. 12 (1995) 799–806.
[87] H.-K. Chan, A.R. Clark, J.C. Feeley, M.-C. Kuo, R. Lehrman, K. Pikal-Cleland, et al., Physical stability of salmon calcitonin spray-dried powders for inhalation, J. Pharm. Sci. 93 (2004) 792–804.
[88] N.Y.K. Chew, H.-K. Chan, Effect of humidity on the dispersion of dry powers, in: R.N. Dalby, P.R. Byron, S. Farr, J. Peart (Eds.), Respiratory Drug Delivery VII, Serentec Press, Raleigh, NC, USA, 2000, pp. 615–618.
[89] C. Yamashita, T. Nishibayashi, S. Akashi, H. Toguchi, M. Odomi, A novel formulation of dry powder for inhalation of peptides and proteins, in: R.N. Dalby, P.R. Byron, S. Farr (Eds.), Respiratory Drug Delivery V, Interpharm Press, Buffalo Grove, IL, USA, 1996, pp. 483–486.
[90] H.-K. Chan, A. Clark, I. Gonda, M. Mumenthaler, C. Hsu, Spray dried powders and powder blends of recombinant human deoxyribonuclease (rhDNase) for aerosol delivery, Pharm. Res. 14 (1997) 431–437.

[91] D.A. Edwards, J. Hanes, G. Caponetti, J. Hrkach, A. Ben-Jebria, M.L. Eskew, et al., Large porous particles for pulmonary drug delivery, Science 276 (1997) 1868–1872.
[92] Y.-F. Maa, P.-A. Nguyen, T.D. Sweeney, S.J. Shire, C.C. Hsu, Protein inhalation powders: Spray drying vs spray freeze drying, Pharm. Res. 16 (1999) 249–254.
[93] A.I. Bot, T.E. Tarara, D.J. Smith, S.R. Bot, C.M. Woods, J.G. Weers, Novel lipid-based hollow-porous microparticles as a platform for immunoglobulin delivery to the respiratory tract, Pharm. Res. 17 (2000) 275–283.
[94] H. Adi, D. Traini, H.-K. Chan, P.M. Young, The influence of drug morphology on the aerosolisation efficiency of dry powder inhaler formulations, J. Pharm. Sci. 97 (2008) 2780–2788.
[95] S. Adi, H. Adi, P. Tang, D. Traini, H.-K. Chan, P.M. Young, Micro-particle corrugation, adhesion and inhalation aerosol efficiency, Eur. J. Pharm. Sci. 35 (2008) 12–18.
[96] N.Y.K. Chew, H.K. Chan, Use of solid corrugated particles to enhance powder aerosol performance, Pharm. Res. 18 (2001) 1570–1577.
[97] N.Y.K. Chew, P. Tang, H.K. Chan, J.A. Raper, How much particle surface corrugation is sufficient to improve aerosol performance of powders? Pharm. Res. 22 (2005) 148–152.
[98] M. Mumenthaler, C.C. Hsu, R. Pearlman, Feasibility study on spray-drying protein pharmaceuticals: Recombinant human growth hormone and tissue-type plasminogen activator, Pharm. Res. 11 (1994) 12–20.
[99] Y.-F. Maa, P.-A.T. Nguyen, S.W. Hsu, Spray-drying of air–liquid interface sensitive recombinant human growth hormone, J. Pharm. Sci. 87 (1998) 152–159.
[100] R. Vanbever, J.D. Mintzes, J. Wang, J. Nice, D. Chen, R. Batycky, et al., Formulation and physical characeterization of large porous particles for inhalation, Pharm. Res. 16 (1999) 1735–1742.
[101] R.M. Platz, A. Ip, C.L. Whitham (1994) Process for preparing micronized polypeptide drugs. Patent Applicant: Stanford Res Inst Int. Publication info: S5354562 (A) – 1994–10–11.
[102] R.T. Bustami, H.-K. Chan, F. Dehghani, N.R. Foster, Generation of micro-particles of proteins for aerosol delivery using high pressure modified carbon dioxide, Pharm. Res. 17 (2000) 1360–1366.
[103] R.T. Bustami, H.-K. Chan, N.R. Foster, Aerosol delivery of protein powders processed by supercritical fluid technology, in: R.N. Dalby, P.R. Byron, S. Farr, J. Peart (Eds.), Respiratory Drug Delivery VII, Serentec Press, Raleigh, NC, USA, 2000, pp. 611–614.
[104] R.W. Sievers, S.P. Sellers, K.D. Kusek, G.S. Glark, B.J. Korte, Fine-particle formation using supercritical carbon dioxide-assisted aerosolization and bubble drying, in: Proceedings of the 218th ACS National Meeting, New Orleans, LA, USA, 1999.
[105] R. Sloan, H.E. Hollowood, G.O. Hupreys, W. Ashraf, Y. P, Supercritical fluid processing: Preparation of stable protein particles, in: Proceedings of the Fifth Meeting of Supercritical Fluids, Nice, France, 1998.
[106] W.J. Glover, E. Wan, J.S. Yun, J. Chen (2009) A process for making micro-sized protein particles. Patent Applicant: Nanomaterials Technology Pte Ltd. Publication info: WO2009020434 (A1) – 2009–02–12.
[107] H. Chiou, H.-K. Chan, R.K. Prud'homme, J.A. Raper, Evaluation on the use of the confined liquid impinging jets for the synthesis of nanodrug particles, Drug Dev. Ind. Pharm. 34 (2008) 59–64.
[108] H. Chiou, H.-K. Chan, D. Heng, R.K. Prud'homme, J.A. Raper, A novel production method for inhalable cyclosporine A powders by confined liquid impinging jet precipitation, J. Aerosol Sci. 39 (2008) 500–509.

[109] S.S. Steiner, A. Pfützner, B.R. Wilson, O. Harzer, L. Heinemann, K. Rave, Technosphere/Insulin: Proof of concept study with a new insulin formulation for pulmonary delivery, Exp. Clin. Endocrinol. Diabet. 110 (2002) 17–21.

[110] H. Lian, S.S. Steiner, R.D. Sofia, J.H. Woodhead, H.H. Wolf, H.S. White, et al., A self-complementary, self-assembling microsphere system: Application for intravenous delivery of the antiepileptic and neuroprotectant compound felbamate, J. Pharm. Sci. 89 (2000) 867–875.

[111] L.R. Brown, J. Rashba-Step, T.L. Scott, Y. Qin, P.W. Rulon, J. McGeehan, et al., Pulmonary delivery of novel insulin microspheres, in: R.N. Dalby, P.R. Byron, J. Peart, S. Farr (Eds.), Respiratory Drug Delivery VIII, Davis Horwood International, Raleigh, NC, USA, 2002, pp. 431–434.

[112] Y.-F. Maa, P.-A. Nguyen, J.D. Andya, N. Dasovich, T.D. Sweeney, S.J. Shire, et al., Effect of spray drying and subsequent processing conditions on residual moisture content and physical/biochemical stability of protein inhalation powders, Pharm. Res. 15 (1998) 768–775.

[113] Food and Drug Administration, Guidance for industry: Metered Dose Inhaler (MDI) and Dry Powder Inhaler (DPI) Drug Products. Chemistry, Manufacturing, and Controls Documentation, Food and Drug Administration, Rockville, MD, USA, 1998.

Chapter 3

Nasal Delivery of Peptides and Proteins with Chitosan and Related Mucoadhesive Polymers

Garnpimol C. Ritthidej
Faculty of Pharmaceutical Sciences, Department of Manufacturing Pharmacy, Chulalongkorn University, Bangkok, Thailand

Chapter Outline

Peptide and Protein Delivery. DOI: 10.1016/B978-0-12-384935-9.10003-3

3.1. INTRODUCTION

Advances in pharmaceutical biotechnology have led to numerous and large-scale production of proteins and peptides, several of which are therapeutic agents. Because of gastro-intestinal enzymatic degradation and first-pass hepatic metabolism, these active biopharmaceuticals are mainly administered by parenteral routes [1–2]. Nasal delivery is needle-free, non-invasive and painless; it does not require sterile preparation, can be self-administered and facilitates the rapid onset of both local and systemic drug action [3–5]. In order to maximize patient compliance and expand/extend any existing patent position, the nasal route of administration is the focus of much attention as a non-invasive alternative to the peroral route, and to invasive parenteral application. Moreover, the nose to brain pathway opens the door for targeting drugs to the central nervous system [2,4–5].

Similarly to other mucosal routes of drug delivery, the physical barrier of the nasal epithelium is the main limitation with respect to drug absorption, especially for large molecules such as proteins and peptides. Another important obstacle is the rapid mucociliary clearance which limits the time available for drug absorption. Enzymes (particularly proteases) that are present in the nasal mucosa and the mucus layer are of less importance (certainly when compared to oral delivery), but not negligible [6]. Another possible drawback is the relatively small amount of drug that can be nasally administered; this is related to the method and technique of administration which, if poor, may lead to drug absorption problems depending on the actual site of drug deposition. To achieve a stable, safe and effective nasal delivery of peptides and proteins, these limitations have to be overcome. Incorporating excipients in the formulation, and the design of novel nasal dosage systems are exciting approaches [2,7], but these areas still require further exploration to be of real clinical and commercial benefit.

3.2. PHYSIOLOGICAL OBSTACLES TO NASAL ABSORPTION

3.2.1. The Nasal Mucosa

The nasal mucosa facilitates a relatively high level of drug absorption per unit area. The mucosa is lined with extensive pseudostratified columnar epithelia containing ciliated cells and mucous-secreting goblet cells in the non-olfactory region. Although the volume of the nasal cavity is low (15–20 cm^3), the surface area for absorption is relatively large (150 cm^2) because of the numerous epithelial microvilli. Moreover, underlining the mucosa is a porous endothelial basement membrane which is richly supplied with blood vessels from both the external and internal carotid arteries [8]. A drug can permeate through the epithelial cell membrane either by the transcellular route, or by the paracellular route. The lipophilicity and size of the drug molecule play important roles in nasal permeation. Lipophilic drugs generally diffuse via the transcellular route, both passively and actively depending on the cell receptors engaged. In contrast, polar drugs with molecular weights below 1000 Da generally pass across the membrane through cell–cell tight junctions, which are less than 10 Å wide; much wider than the equivalent cell–cell tight junctions of the intestinal mucosa. However, since polypeptides are generally large, hydrophilic molecules (> 10 Å in size), only very small amounts can pass the nasal membrane, and this transport generally involves endocytosis [9].

3.2.2. Mucociliary Clearance

Each ciliated cell at the apical surface of the nasal epithelium has numerous cilia which are mobile and beat with a high frequency (around 1000 beats per minute). The beat consists of a rapid forward movement followed by a slow return. Covering the epithelium is a mucus layer. Nasal mucus is produced from submucosal seromucus glands and goblet cells. The main component of mucus is water, then mucin (*ca.* 2%) which is responsible for the viscous and elastic gel-like properties, and finally other components (< 3% in total). The pH of the mucus layer varies from 5.5 to 6.5 in adults. For mucociliary clearance, the mucus layer is propelled from the anterior towards the posterior part of the nasal cavity in a coordinated manner by the beating, synchronized cilia. This is generally called the methachronal wave, which thus gradually translocates particulates towards the nasopharynx for eventual discharge into the gastrointestinal tract. Moreover, this mucus layer, which comprises a lower sol layer and upper gel layer, is renewed every 15–20 mins, which concomitantly restricts the time available for drug absorption, which in turn further limits nasal absorption [1,10].

3.2.3. Enzymatic Degradation and Mucus Interaction

Although there is an attenuated degradation of proteinaceous drugs in the nasal mucosa compared to the gastrointestinal tract, enzymes are present in nasal secretions, especially proteases and peptidases, which naturally reduce the permeation of various peptide/protein drugs. Other enzymes include the cytochrome P450 family, carboxyl esterase, aldehyde dehydrogenase, epoxide hydrolase, DT-diaphorase and glutathione S-transferase. In addition, the components of the mucus may undergo electrostatic interaction with the drug molecule and thereby alter its absorption and activity. This is especially true for the heterogeneous mucin glycoproteins which, due to their numerous oligosaccharide chains, are highly negatively charged [4].

3.3. MECHANISMS OF MUCOADHESION

The low permeation of peptides and proteins through the nasal mucosa can generally be improved by either chemical modification or loosening ("widening") of the tight junctions. Chemical modification must of course be performed carefully so as to not alter the pharmacological action (or toxicity) of the drug. Insofar as nasal delivery technologies are concerned, modulation of the tight junctions is considered the more practical approach since it is a reversible process and applicable to all drugs, in theory. Lipoencapsulation in nanocarriers can also be beneficial to increasing nasal drug transport. However, methods for improving nasal absorption are primarily focused on overcoming mucociliary clearance. In principle, this problem can be overcome by various methods such as the inhibition of mucus secretion, the addition of mucolytic agents, calcium depletion, or the addition of mucoadhesive polymers to the peptide/protein formulation. The latter technique is the most extensively investigated, including the search for next generation mucoadhesive polymers and novel nasal delivery products [1].

The mechanisms of mucoadhesion fundamentally involve chemical interaction and interpenetration of the polymer (in this case, the polymer added to the drug formulation) with the mucin chains. In the former case, mucoadhesion results from either covalent bonding, which is irreversible and hence undesirable, or from non-covalent bonding involving hydrogen bonding, van der Waals interactions, or hydrophobic and electrostatic forces [11]. The key sites of mucin-facilitating mucoadhesive interactions with the polymer appear to be at the carbohydrate residues which either promote electrostatic-type interactions through terminal sialic acid or sulphonated residues, or hydrophobic interactions through clusters of fucose residues each of which possess a methyl group [12]. In the case of non-covalent polymer-mucin interactions, wetting and swelling of the polymer followed by the penetration of polymer molecules into tissue crevices/cell glycocalyx and interpenetration between the polymer and mucin chains leads to mucoadhesion.

Hydration plays an important role in mucoadhesion. In the nasal cavity, about 1.5 to 2 l of mucus is secreted daily, most of which is water. Complete hydration of the mucoadhesive polymer is necessary in order to produce adequate free and available polymer chains for interpenetration (with mucin). However, an excessive amount of water, either from a liquid formulation or from pathological conditions, weakens the polymers' mucoadhesiveness through hydrogen bonding with non-bound (bulk) water molecules; an optimum polymer concentration is therefore required to promote interaction and mucoadhesion [13]. Moreover, the pKa of the functional groups of a polymer and the pH of the nasal mucosa are very important since some polymers owe their mucoadhesiveness to ionic and electrostatic interactions, which naturally depend on the charge existing on the functional groups [14].

3.4. MUCOADHESIVE POLYMERS AND NASAL APPLICATION

Mucoadhesive polymers have been used with some notable success in extending the residence time of nasally administered polypeptides, as well as in inhibiting enzyme-degradation and improving absorption [15]. For reasons discussed above, mucoadhesive polymers should be able to form hydrogen bonds, possess charged groups (preferably anionic groups to interact with mucin), have a high MW and chain flexibility, and possess surface energy properties that favor spreading onto mucus layers [16]. Compared with synthetic polymers, the vast majority of natural polysaccharides are non-toxic, many being used widely in food products as thickeners and stabilizers. However, although some existing, natural, mucoadhesive polymers improve the nasal absorption of polypeptides impressively, new classes of polymers have been successively synthesized and examined. The chemical structures of some commonly used mucoadhesive polymers are shown in Figure 3.1.

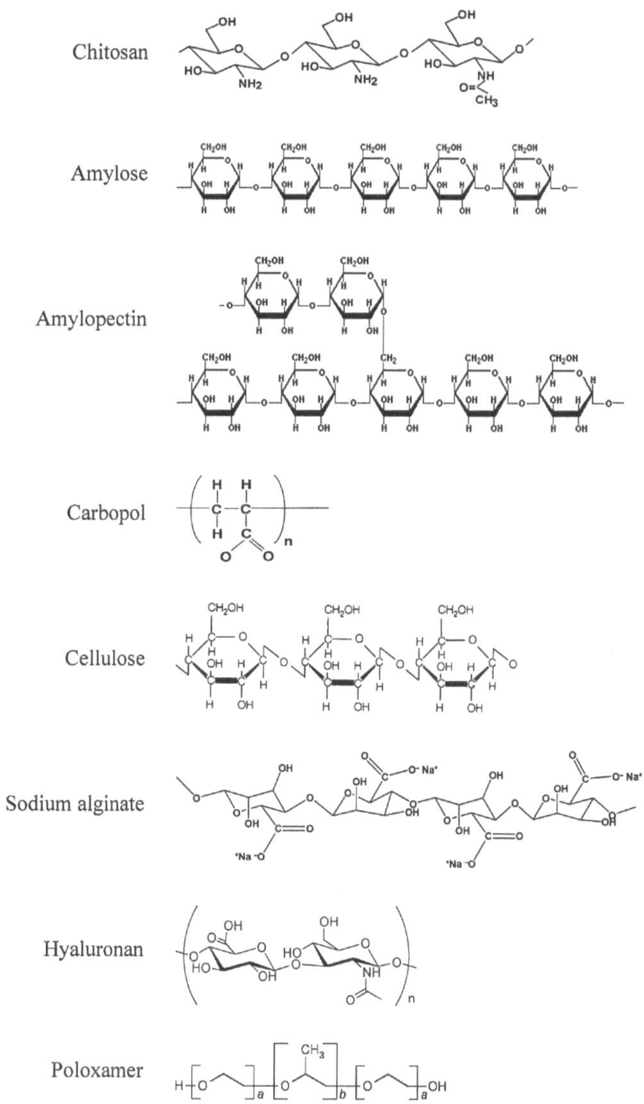

FIGURE 3.1 The chemical structures of some commonly used mucoadhesive polymers in nasal delivery.

3.4.1. Chitosan

Chitosan is a natural polysaccharide consisting of glucosamine and N-acetyl-glucosamine, and is available in a range of molecular weights and degrees of deacetylation. The amino group of chitosan has a pKa value of about 6.5; hence, chitosan is positively charged and is soluble in weakly acidic solutions with a charge density that is dependent on the pH and the degree of deacetylation. The chitosan is thought to promote the transport of polar drugs across the epithelial membrane by a combination of strong mucoadhesive properties and transient opening of the cell–cell tight junctions. Chitosan is also non-toxic and non-irritating to the nasal membrane [17]. Chitosan solutions can be easily prepared by dissolving the solid in dilute acidic solutions such as those of acetic, citric, lactic or glutamic acids. Chitosan salts, such as chitosan glutamate or chitosan hydrochloride can be easily dissolved in water. As with all other polymers, the higher the molecular weight of chitosan, the more viscous is the resultant solution (for equivalent % w/v).

Numerous reports have demonstrated an increased absorption of macromolecules and improved (*in vivo*) bioavailability across the nasal epithelial barrier after co-administration with chitosan; measured, for example, using peak plasma levels or the "area under the curve" (AUC) [18–20]. In addition, chitosan (free) base enhances the absorption of peptide in comparison with chitosan salts at lower pH [21]. Thus, the ability to enhance nasal absorption and mucoadhesion depends on the chitosan type, its molecular weight, degree of acetylation, concentration and the osmolarity of the surrounding medium [22,23].

For chitosan, a comparison of three studies suggests that the most effective formulation strategy for enhancing the nasal absorption of insulin is a chitosan powder (microsphere) delivery system, followed by a nanoparticulate chitosan formulation, and finally a chitosan solution formulation [24–26]. Compared to chitosan solution, the powder formulation delayed mucociliary clearance and the swelling of the microspheres dehydrated the epithelial cells, causing tight junction loosening as seen in Figure 3.2 [27]. Moreover, nasal powder formulations, with an optimal microsphere size of 10–30 μm provide greater stability than liquid forms, although it is noted that they may result in nasal irritation. Finally, chitosan solutions may drain more rapidly from the nasal passages and induce a strong flavor in the mouth.

Chitosan powders can be prepared by several methods, such as physical mixing, co-grinding, freeze-drying, spray drying, spray freeze-drying, emulsion solvent evaporation and precipitation in supercritical anti-solvents. Spray drying appears to be the most promising method due to controlled characteristics and narrow size distribution of the particles so obtained. In addition, spray drying is fast, simple and relatively energy-efficient. The physicochemical properties of spray dried chitosan microspheres are affected by formulation and process parameters. Examples of surface morphologies of chitosan

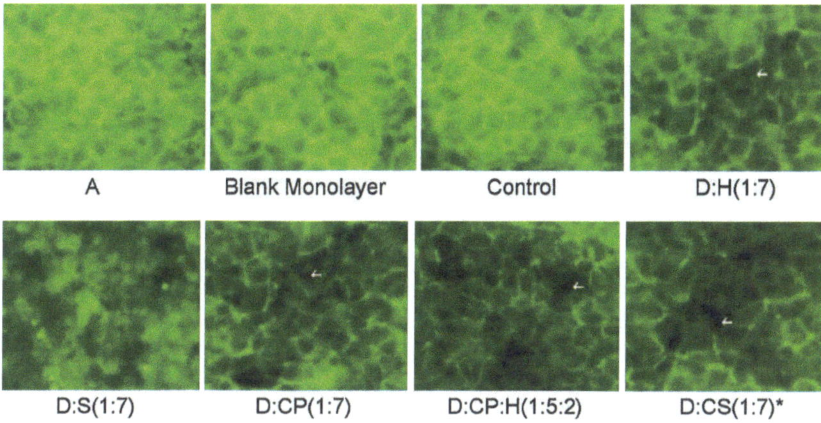

FIGURE 3.2 The opening of tight junction of nasal cell line RPMI 2650 by spray dried powder of 1:7 drug (D): mucoadhesive polymer after 5 hours exposure: H = hydroxylpropylmethylcellulose, CS = chitosan, CP = carbopol, * = after 3 hours exposure. Reprinted with permission from Harikarnpakdee S, Sutanthavibul N, Lipipun V and Ritthidej GC. Spray dried mucoadhesive microspheres: Preparation, and transport through nasal cell monolayer. *AAPS Pharm. Sci. Tech.* 2006;7:E79–E88. Copyright 2006, American Association of Pharmaceutical Scientists.

microspheres at various spray drying conditions with a model protein are shown in Figure 3.3 [28]. Analysis by sodium dodecyl sulfate polyacrylamide gel electrophoresis (SDS-PAGE) showed that the integrity of the primary protein structure remained intact following spray drying into chitosan powders. However, analysis of the protein secondary structure by circular dichroism showed that the α-helical structure, measured at wavelengths 208 and 222 nm, changed as a function of protein loading (Figure 3.4) [28].

The amine and hydroxyl groups of chitosan can be used to generate semi-synthetic polymers with (custom) modified physicochemical characteristics. In addition, the positively charged amine groups will bind negatively charged polymers, such as mucin (discussed above). Therefore, several chitosan derivatives and conjugates have been synthesized and semi-sythesized in order to improve the low aqueous solubility and mucoadhesiveness of chitosan; these include N-trimethyl chitosan (TMC) chloride, thiolated chitosan, thiolated carboxymethyl chitosan-*g*-β-cyclodextrin, methylpyrrolinone chitosan and chitosan/ethylene oxide-propylene oxide block copolymers [29–33]. These chitosan derivatives appear to be more effective in increasing nasal drug absorption, especially the well studied TMC derivatives, which in different dosage forms are known to improve nasal absorption by amounts that depend on the degree of quaternization [34]. Similarly, thiolated chitosan microspheres have been shown to enhance the nasal absorption of peptide drugs [35], and PEG grafted chitosan nanoparticles or nanocomplexes improve nasal insulin absorption *in vivo* [36–37].

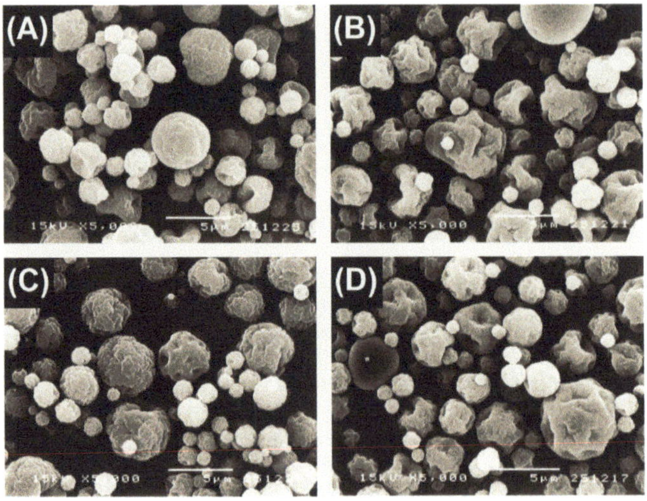

FIGURE 3.3 SEM of bovine serum albumin (BSA)-loaded chitosan microparticles, obtained by spray drying of (A) 1%; (B) 10%, BSA in LCS solution; (C) 10% BSA in HCS solution at the spray rate of 3 ml min^{-1} and the inlet air temperature of 120°C; (D) 10% BSA in LCS solution at the spray rate of 5 ml min^{-1} and the inlet air temperature of 100°C, respectively. LCS and HCS = low and high molecular weight of chitosan. *Reprinted from reference [28] with permission from Informa Healthcare.*

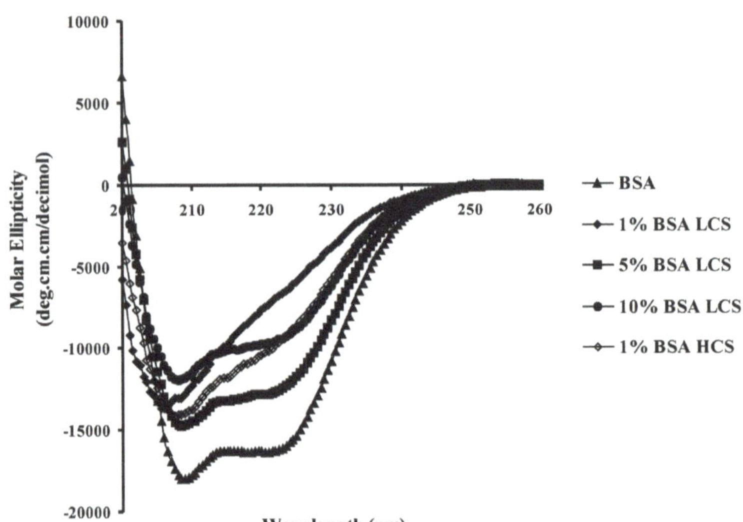

FIGURE 3.4 The circular dichroism spectra of unprocessed bovine serum albumin (BSA) and BSA recovered from chitosan microspheres with different % BSA loading: LCS and HCS = low and high molecular weight chitosan. *Reprinted from reference [28] with permission from Informa Healthcare.*

3.4.2. Starches

Starch is a carbohydrate which is composed of amylose and amylopectin, and is biocompatible, biodegradable and bioadhesive. Starches from different natural resources are widely used in foods and pharmaceuticals [38]. In addition, processed starches are used which have undergone modification at the hydroxyl groups to yield carboxy methyl starch, sodium starch glycolate, drum dried waxy corn starch, soluble starch and degradable starch. When placed in contact with aqueous media, starches undergo a high degree of swelling and form gel-like systems which strongly adhere to the nasal mucosa and significantly prolong the nasal residence time.

Mucoadhesive starch microspheres containing insulin reduced the plasma glucose in animals more effectively than insulin in dextran spheres, due to the increase in the drug residence time [39]. Nasal administration of insulin-degradable starch microspheres (DSM) also reduced blood glucose when compared to DSM alone or insulin solution [40]. Although starch undergoes a high degree of swelling, which is governed by the degree of cross-linking (which also controls peptide release), the starch polymer chains only interact weakly with mucin. It is hypothesized that absorption enhancement occurs via other mechanisms such as interaction with the cell lipid bilayers. Moreover, good tolerance was reported after starch microspheres were administered nasally to healthy human volunteers for a one week period [41]. Increasing the surface area via nanoparticulate formulation of starch was hypothesized to increase the insulin concentration gradient across the nasal epithelium and increase the release rate. Formulation of the nanoparticles with sodium glycocholate improved insulin absorption, producing a hypoglycemic effect over 6 hours post-administration [42].

3.4.3. Polyacrylic Acid and Polyacrylates

Polyacrylic acid is a polyelectrolyte which is soluble in aqueous media at neutral pH, due to the ionization of the pendent carboxyl side chains. Carbopols are polymers of acrylic acid that have been cross-linked with polyalkanyl esters or divinyl glycol. Carbopols readily absorb water, become hydrated and swell: their hydrophilic nature and cross-linked structure make them suitable for controlled drug delivery systems [15,43]. Moreover, these polymers are non-toxic and non-irritating. Polycarbophil or calcium polycarbophil is a calcium salt of polyacrylic acid cross-linked with divinyl chloride. The nomenclature of the many different commercial grades of the carbopols is various: Carbopol 934P is also referred to as "carboxy polymethylene", while Carbopol 980NF is referred to as "poly-acrylic acid", and 1342NF and 1382 as "carbomer". In general, the carbopols are known to be useful for nasal drug delivery, either in liquid, gel or powder formulations [15]. These polymers have yet to be investigated fully for nasal delivery of peptides and proteins, although carbopol gels are known to facilitate the systemic absorption of progesterone via the nasal route [44].

3.4.4. Celluloses

Celluloses are polysaccharides containing 8000–10000 glucose units linked by β-1,4 glucosidic bonds. The hydroxyl groups on the glucose residues are amenable to chemical modification, producing several derivatives: hydroxypropylmethylcellulose, microcrystalline cellulose, carboxymethylcellulose, hydroxypropylcellulose, methylcellulose, hydroxyethylcellolose and ethylcellulose. The mucoadhesive properties of the polymer depend on the type of derivative and the molecular weight. Although they have gel forming properties, some water soluble celluloses show only poor mucoadhesion. However, they are safe and are extensively used in pharmaceuticals, including in nasal formulations where they work to increase the residence time in the nasal mucosa [45]. Microcrystalline cellulose microspheres, alone or with hydroxypropylcellulose, were observed to enhance the absorption of various peptides and proteins such as insulin [46], calcitonin and leuprolide [45]. Increased nasal bioavailability of human parathyroid hormone and insulin by ethylcellulose and a mucolytic agent, N-acetyl-L-cysteine (NAC) has also been reported [47]. As for chitosan, the improvement in peptide/protein absorption via nasal delivery can be envisaged using different dosage forms such as gels, liquids, sprays and powders.

3.4.5. Other Adhesive Macromolecules

Alginate: Sodium alginate, the sodium salt of alginic acid, is the purified carbohydrate product extracted from brown seaweed by the use of a dilute alkali. In water, this polymer is slowly soluble forming a viscous, colloidal solution. Alginates have a poor mucoadhesive property, although this is improved by conjugation with cysteine [48]. However, the alginate polymers are not popular for the nasal delivery of peptide drugs.

Gelatin: Gelatin is a large, fibrillar protein with varying molecular weight (*ca.* 75 kDa). Gelatin A is an acid hydrolytic product of collagen, while gelatin B is generated from alkaline treatment of collagen. Gelatin forms a thermally reversible gel with water, the gelation temperature of which is around 35°C, dependent on the pH of the solution. Gelatin microspheres have been shown to enhance the absorption of calcitonin [49]. Aminated gelatin is cationized gelatin with ethylene diamine, such that the cationic charges can interact with negatively charged mucin on the nasal mucosa. Aminated gelatin, in both liquid and powder dosage forms, was reported to enhance the absorption of fluorescein isothiocyanate (FITC)-dextran and insulin, compared to native gelatin, with only negligible nasal toxicity [50]. Sperminated gelatin also improved the absorption of insulin by modulation of the epithelial tight junctions [51].

Poly-L-arginine: Poly-L-arginine is a cationic polymer and may be classified as a cell penetrating peptide for short peptide chains (see Chapter 10). Here, we will consider high molecular weight cationic peptide chains which have mucoadhesive properties, achieved by electrostatic interaction with the negatively charged mucins. Accordingly, poly-L-arginine has been shown to substantially enhance the absorption of insulin [52] and recombinant human granulocyte colony-stimulating factor (rhG-CSF) in rats [53].

Hyaluronans: Hyaluronic acid ester is an abundant mucopolysaccharide found in the synovial fluid and extracellular matrices of vertebrates. Solutions of hyaluronan enhanced the absorption of vasopressin [54]. Hyaluronan microspheres also significantly increased the nasal absorption of insulin [55].

Lectins: Lectins are a heterogenous family of proteins that bind sugars (they are not glycoproteins). On binding the epithelial cell surface via, for example, glycolipids, they are internalized and are considered to be a promising, alternative route for bioadhesion [56], although they have not yet been investigated for the nasal delivery of peptides/proteins.

Penetratins: The D- and L-forms of the penetratins are cell penetrating peptides which have been studied for their ability to increase the nasal absorption of lactate dehydrogenase and insulin. They are reported to be more effective than the L- or D-forms of octaarginine, yielding a pharmacological availability and bioavailability of nasally administered insulin of ~77% and 51% (cf. subcutaneous administration), respectively, yet without cell toxicity [57].

3.5. THERMAL AND pH SENSITIVE POLYMERS

To overcome the difficulty of administering gel formulations to the nose, this class of mucoadhesive polymers has been developed to form gels *in situ* in the nasal cavity, consequently enhancing the residence time and nasal absorption of therapeutic drugs. These polymers exhibit a low viscosity under standard, ambient conditions but gel upon nasal administration as a result of a change in pH and/or temperature within the local nasal mucosa. Examples of pH-sensitive polymers include chitosan–thioglycolic acid conjugates, which gel at physiological pH through the formation of inter- and intra-molecular disulfide bonds [58]. Examples of thermoreversible polymers include the Pluronics™, particularly Pluronic F127, which is one of the most extensively investigated temperature-responsive materials. Pluronics, also known as poloxamers, are triblock copolymers of poly(ethylene oxide)-poly(propylene oxide)-poly(ethylene oxide). *In situ* gel formulations of Pluronic F127, administered in liquid form, have been used as controlled release depots for delivery of plasmid DNA in combination with bioadhesive polymers such as polycarbophil [59]. Because the Pluronics are also soluble in water and offer non-denaturing conditions for protein formulation (e.g. by simple mixing) it is somewhat surprising that their use for nasal delivery has not been studied more extensively. Novel thermosensitive gels were generated by cross-linking chitosan and poloxamer with glutaraldehyde (the reaction was stopped with glycine). It was demonstrated that these gels produce a prolonged hyperglycemic effect in diabetic rats following nasal delivery of insulin [60]. The potential value of simple mixtures of polymers has been demonstrated with N-[(2-hydroxy-3-trimethylammonium) propyl] chitosan chloride, poly(ethylene glycol) (PEG) and (small amounts) of α-β-glycerophosphate. Insulin formulated with this mixture dropped or sprayed into nasal cavity of rats was shown to lower blood glucose levels following sustained release of insulin from the thermosetting gel, which decreased the nasal mucociliary clearance rate [61].

As alluded to above, combinations of polymers with different physicochemical properties, produced either through synthesis or simple mixing, yield novel, multi-functional delivery systems (e.g. mucoadhesive, thermosetting, etc.), which synergistically enhance the nasal absorption of peptide drugs. Most of the combinations are in the form of powders to prolong adhesion to nasal mucosa; for example, amioca starch and Carbopol 974P [62], drum-dried waxy maize starch or maltodextrin DE8 and Carbopol 974P [63], chitosan and hyaluronan [64], chitosan containing N-acetyl-L-cysteine [65], and starch and carbopol mixture with calcium ions [66]. In each case, these combinations of polymers have been examined with respect to their potential for nasal drug delivery, generally using insulin as the model peptide, and the studies have shown that they are well tolerated on the basis of histological and toxicological evaluation.

3.6. CHARACTERIZATION AND EVALUATION OF MUCOADHESIVE POLYMERS

Several *in vitro*, *ex vivo* and *in vivo* methods have been used to characterize and evaluate the mucoadhesive and absorption-enhancing properties of polymers in nasal delivery. These methods provide an excellent means of analyzing the bioadhesion of a polymer, or the efficacy of the drug, sometimes under specific or simplified conditions.

3.6.1. Viscosity

The viscosity of the polymers mentioned above depends on their molecular weight. Higher molecular weight and longer polymer chain lengths produce high viscosity, and consequently a greater adhesion to the nasal mucosa. However, highly viscous preparations are more difficult to administer, and this viscosity may retard the release of drug beyond the clearance time in the nasal cavity [67]. Viscosity can be determined by several methods: capillary viscometers, orifice viscometers, high temperature and high shear rate viscometer, rotational viscometers, falling ball viscometer, vibrational viscometers and ultrasonic viscometers [68]. In the drug development field, the capillary and rotational viscometers are normally used for Newtonian and non-Newtonian types of liquids, respectively.

3.6.2. Mucoadhesion

Various methods have been used to determine mucoadhesion, both *in vitro, ex vivo* and *in vivo*. For *in vitro* tests, the measurement of tensile strength and shear strength are widely used. Other methods, such as the viscometric method and thumb test are less studied. *In vivo* evaluation is based on measurement of the residence time of bioadhesives at the application site, historically performed using radioisotopes [69]. Ionic polymers bind the nasal mucosa more effectively than polymers without charge. Permanently charged (anionic) sulfate groups are more effective than carboxylic groups whose charge depends on the surrounding pH. It is generally accepted that polycarbophil and sodium carboxymethylcellulose adhere more strongly to mucus than hydroxypropylmethylcellulose, methylcellulose or pectin [70]. An extensive study of the mucoadhesive potential of various polymers by Grabovac *et al.* [71] generated a series which followed, in order of decreasing mucoadhesive potential: chitosan-4-thiobuthylamidine > Carbopol 980 > Carbopol 974P > polycarbophil, dependent on pH and drying method; while the neutral sodium salts of the polyacrylates were most mucoadhesive and the cellulose derivatives, polyvinylpirrolidone and PEGs had very little mucoadhesion (in contrast to previous data).

3.6.3. Swelling

Hydration of the above polymers leads to swelling, which: (1) extends and expands the polymer chains for interpenetration with mucin, and (2) controls drug release from the polymer matrix, for example, by Fickian diffusion. Application of the polymers as a dry powder therefore draws up water from the nasal mucosa, resulting in a transient dehydration of the epithelial cells, which in turn is believed to loosen the cell—cell tight junctions and increase drug permeation. Normally, swelling can be determined by the weight change during water absorption or the increase in sample volume. The swelling of different mucoadhesive microspheres in pH 6.8 phosphate buffer is ranked: carbopol > chitosan > hydroxypropylmethylcellulose, as shown in Figure 3.5 [27].

3.6.4. Release

The rate and profile of release of a drug from a particular dosage form is important to characterize in order to determine the amount available for absorption. Because the turnover of mucus layer in the nose takes about 15—20 minutes, prolonged release of drug for several hours is impractical, even with formulations containing a mucoadhesive polymer. Generally, the drug release rate from a powder is slower than from a gel preparation, since there is a lag time for the ingress of water; the rate of release may be dependent on the formulation parameters in addition to the pore tortuosity and evolution within the polymer matrix.

No standard *in vitro* method has yet been developed for nasal formulations. Various apparatus have been used to assess drug release from mucoadhesive powders, most commonly a modification to the US Pharmacopoeial rotating basket (i.e. the paddle or wire basket method) [28]. Studies of drug release and

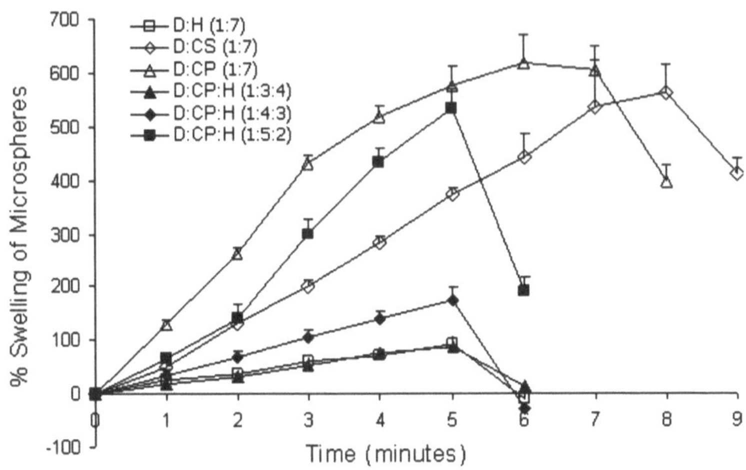

FIGURE 3.5 The profiles of percentage swelling with time of spray dried microspheres with different ratios of drug to polymer; D = drug; H = hydroxypropylmethylcellulose; CS = chitosan; CP = carbopol. Reprinted with permission from Harikarnpakdee S, Sutanthavibul N, Lipipun V and Ritthidej GC. Spray dried mucoadhesive microspheres: preparation, and transport through nasal cell monolayer. *AAPS Pharm. Sci. Tech.* 2006;7:E79–E88. *Copyright 2006, American Association of Pharmaceutical Scientists.*

subsequent passage across a membrane will use a Franz diffusion cell set-up [27]. Some studies have suffered from the limitations of dispersing powdered particulates directly in an excess amount of release medium, which does not simulate the conditions of the nasal mucosa.

3.6.5. Permeation Studies

Nasal drug permeation and absorption can be investigated by the use of *in vitro* and *in vivo* techniques. *In vivo* studies represent the most critical test for any nasal drug application or formulation [72]. However, there are, of course, species differences in the nasal anatomy and physiology, which makes extrapolation of animal data to humans complicated. In addition, a large number of animals and quantities of drug are required. By contrast, *in vitro* permeability studies using human tissue can be standardized and performed more rapidly with simpler procedures. However, there is only a limited supply of human tissue and so, alternatively, various animal tissues are used; porcine mucosa in particular closely resembles human tissue in its anatomy, physiology, histological and biochemical aspects [73]. Moreover, the ease of maintaining sufficient viability and integrity of the isolated porcine cavity mucosa makes it suitable for permeation studies using Franz cells.

Use of primary and immortalized cell lines of human nasal epithelial cells in culture is promising and a number of cell culture systems have been developed:

Primary cell culture: Human nasal epithelial tissue can be obtained by atraumatic or traumatic techniques, enabling a supply of tissue for cell isolation. The main disadvantages are the limited number of cells that can be harvested from one donor, the relatively short-term culture of the cells, heterogeneity within cultures and between cultures, contamination with pathogens, and difficulties in culturing itself. Primary respiratory epithelial cells can be serially passaged up to two or three times [74–75].

Immortalized cell lines: Cell lines are derived from carcinomas of epithelial origin and offer to the researcher an extended lifespan, improved proliferation, homogeneity and relative ease of culture. However, important biochemical and morphological properties of the cells (and therefore the cellular response to drugs and formulations) are changed after immortalization; in the case of nasal studies this includes the lack of beating cilia. The most popular nasal cells currently used are RPMI 2650 cells, derived from a human nasal anaplastic squamous cell carcinoma, though Calu-3 cells derived from a human bronchial epithelial adenocarcinoma are also used [76–80]. The RPMI 2650 cell line forms clusters of round and slightly flattened cells, as shown in Figure 3.2. RPMI 2650 cells do not express goblet and ciliated cells and, in some culture conditions, can be induced to spread [77,78]. Although presenting peripheral rings of occludin and claudin, tight junctions between the cells in culture are not clearly expressed. Drug permeation across confluent monolayers of RPMI 2650 cells in either submersed or air–liquid surface cultures has been reported [27,80], and the cell line is also thought to be useful for metabolism studies. In contrast, Calu-3 cells grow as a confluent sheet and form polarized monolayers with tight junctions and a uniform mucus layer. These cells can also be useful in studies screening the nasal and lung permeation potential of drugs and drug candidates [79].

3.7. NASAL (CILIARY) TOXICITY

Studies of toxicity associated with nasal drug administration have to date examined markers of inflammation that are indicative of local irritation of the nasal mucosa, mucociliary clearance and epithelial damage. Impairment of the nasal mucociliary clearance system can have clinical consequences, since mucociliary transport is one of the most important local defense mechanisms of the respiratory tract. The ciliary beat frequency (CBF) has been widely studied as a central parameter of tissue/cell function [81]. The consequences of interference with mucociliary clearance may, for example, cause local infection and subsequent inflammation. The method for classification of assessing the toxic potential of components in nasal drug formulations is based mainly on the recovery of the CBF after removal of the test compound/formulation. There are three categories: cilio-friendly, in which the CFB recovers to 75% or more of its initial frequency; cilio-inhibiting, where recovery is between 25% and 75%; and ciliostatic, where the CFB recovery is 25% or less [82].

Most conventional mucoadhesive polymers are deemed safe, especially when used in liquid form, though this safety is dependent on the concentration of polymer applied. Mucociliary transport rates for human nasal tissue treated *ex vivo* and *in vivo* with chitosan solutions showed a transient inhibitory effect that was related to the volume applied and the molecular weight of the chitosan (higher MW chitosans being more inhibitory). However, cell histology and clearance rates were not affected by once daily application of a 0.25% chitosan solution over 7 days [83]. Similarly, N-trimethyl chitosans were marginally cytotoxic and ciliotoxic [29].

Mild to moderate cilio-inhibition was also observed for Carbopol 971P and carboxymethylcellulose, but as for chitosan, these effects were transient and reversible. Mild nasal mucosal inflammation and CBF inhibition could be observed following application of carboxymethylcellulose [84]. However, Carbopol 971P appeared to cause severe concentration- and time-dependent inhibitory effects on the CBF, with one particular report suggesting that this polymer alone may not be suitable for nasal delivery [85]. In contrast, Carbopol 974P, formulated with drum-dried waxy maize, was well tolerated by rabbit nasal mucosa following administration as the powder, and measured as the release of marker compounds versus the negative control [86]. Finally, it is interesting to note in the review of Dimova *et al.* that cilio-inhibitory effects may arise not as a consequence of the drug or polymer but of the preservative (or other additive) in the nasal formulation [87]. In conclusion, there are now standard measurements of nasal epithelial toxicity and this should make the selection of excipients for the development of safe nasal drug formulations more straightforward.

3.8. NOSE TO BRAIN PATHWAY: FUTURE CONSIDERATIONS

The olfactory region is situated in the roof of the nasal cavity, and only covers about 10% of the total nasal area. Its epithelium consists of specialized olfactory cells, supporting cells, serous and mucosal glands. In the upper nasal passage, the dendritic processes of the olfactory neurons can internalize a drug via endocytosis and translocate the drug to the olfactory bulb in the brain by axonal transport. In addition to this transcellular pathway paracellular (extracellular) pathways exist, wherein drugs first cross between the olfactory neurons or the trigeminal nerve before being transported to the olfactory bulb or trigeminal region. The potential of these mechanisms in the context of nasal drug delivery to the central nervous systems are the subject of ongoing debate [88–91].

Numerous studies in animal models and in man have demonstrated the possibility of polypeptide uptake into the cerebrospinal fluid and the brain tissue [92–95]. For example, via nasal administration, insulin-like growth factor (IGF-1) was shown to bypass the blood–brain barrier and reach the central nervous system (CNS) directly from the nasal cavity [92]. *In vitro* studies with bovine olfactory epithelium and Franz diffusion cells were used to optimize chitosan formulations of nerve growth factor. *In vivo* studies for the nasal administration of the optimized formulations then showed that there was a *ca.* 14-fold increase in the bioavailability of nerve growth factor, compared to administration of the peptide alone [93]. Intranasal administration was studied for a radiolabeled peptide, exendin(9-39), an antagonist of full length exendin, a member of the glucagon-like peptide-1 family. Distribution patterns showed that the peptide had been transported primarily to the olfactory bulb, but also to the hippocampus, cerebellum and brain stem, probably via the cerebrospinal fluid; relatively little entered the bloodstream [94]. One study hypothesized that antibodies for treatment of airborne infections would be best delivered to the nasal cavity instead of systemic administration, and accordingly developed spray-dried microparticulate formulations with albumin and disaccharide excipients. Aerosolization using the UniDose dry powder nasal device (Bespak Ltd., UK) was studied, showing that 55% of the dose was deposited at the nasal vestibule, with 45% deposited in the turbinates, olfactory region and nasal-pharynx [95].

It is clear from the above discussions that the nasal delivery of peptides and proteins is at a reasonably advanced stage in terms of research. Proof of principle has been demonstrated for the systemic uptake of nasally administered polypeptides and there are documented toxicological methods for selection of a wide range of well characterized excipients. In relation to this, it is anticipated that in our aging population, Alzheimer's disease, dementia, Parkinson's disease and other CNS diseases will become more prevalent. The field is therefore set to remain an exciting area of research, in which the nose to brain pathway is exploited for non-invasive delivery of peptide neurotransmitters.

REFERENCES

[1] M.I. Ugwoke, R.U. Agu, N. Verbeke, R. Kinget, Nasal mucoadhesive drug delivery: Background, applications, trends and future perspectives, Adv. Drug Deliv. Rev. 57 (2005) 1640–1655.

[2] L. Illum, Nasal drug delivery: New developments and strategies, Drug Discov. Today 7 (2002) 1184–1189.

[3] H.R. Costantino, L. Illum, G. Brandt, P.H. Johnson, S.C. Quay, Intranasal delivery: Physicochemical and therapeutic aspects, Int. J. Pharm. 337 (2007) 1–24.

[4] P.H. Johnson, S.C. Quay, Advances in nasal drug delivery through tight junction technology, Expert Opin. Drug Deliv. 2 (2005) 281–298.

[5] P. Arora, S. Sharma, S. Garg, Permeability issues in nasal drug delivery, Drug Discov. Today 7 (2002) 967–975.

[6] A.-L. Cornaz, P. Buri, Nasal mucosa as an absorption barrier, Eur. J. Pharm. Biopharm. 40 (1994) 261–270.

[7] C.R. Behl, H.K. Pimplaskar, A.P. Sileno, W.J. Xia, W.J. Gries, J.C. de Meireles, et al., Optimization of systemic nasal drug delivery with pharmaceutical excipients, Adv. Drug Deliv. Rev. 29 (1998) 117–133.

[8] N. Migand, M. Pedersen, M.H. Nielsen, Morphology of the upper airway epithelium, in: D.F. Proctor, I.B. Andersen (Eds.), The Nose – Upper Airway Physiology and the Atmospheric Environment, Elsevier Science Publishing Co., New York, 1982, pp. 71–97.

[9] S.C. Chen, K. Eiting, K. Cui, A.K. Leonard, D. Morris, C.Y. Li, et al., Therapeutic utility of a novel tight junction modulating peptide for enhancing intranasal drug delivery, J. Pharm. Sci. 95 (2006) 1364–1371.
[10] E. Marttin, N. Schipper, J. Verhoef, W.H.M. Merkus, Nasal mucociliary clearance as a factor in nasal drug delivery, Adv. Drug Deliv. Rev. 29 (1998) 13–38.
[11] M.A. Longer, J.R. Robinson, Fundamental aspects of bioadhesion, Pharm. Int. 5 (1986) 114–117.
[12] P. Roussel, G. Lamblin, M. Lhermitte, N. Houdret, J.J. Lafitte, J.M. Perini, et al., The complexity of mucins, Biochimie. 70 (1998) 1471–1482.
[13] P.O. Glantz, T. Arnebrant, T. Nylander, R.E. Baier, Bioadhesion – A phenomenon with multiple dimensions, Acta Odontol. Scand. 57 (1999) 238–241.
[14] H. Park, J.R. Robinson, Physicochemical properties of water insoluble polymers important to mucin/epithelial adhesion, J. Control. Release 2 (1985) 47–57.
[15] L. Jiang, L. Gao, X. Wang, L. Tang, J. Ma, The application of mucoadhesive polymers in nasal drug delivery, Drug Devel. Ind. Pharm. 36 (2010) 323–336.
[16] N.A. Peppas, P.A. Buri, Surface, interfacial and molecular aspects of polymers bioadhesion on soft tissue, J. Control. Release 2 (1985) 257–275.
[17] L. Illum, N.F. Farraj, S.S. Davis, Chitosan as a novel nasal delivery system for peptide drugs, Pharm. Res. 11 (1994) 1186–1189.
[18] T. Cerchiara, B. Luppi, F. Bigucci, V. Zecchi, Chitosan salts as nasal sustained delivery systems for peptidic drugs, J. Pharm. Pharmacol. 55 (2003) 1623–1627.
[19] M. Hinchcliffe, I. Jabbal-Gill, A. Smith, Effect of chitosan on the intranasal absorption of salmon calcitonin in sheep, J. Pharm. Pharmacol. 57 (2005) 681–687.
[20] L. Illum, P. Watts, A.N. Fisher, I. Jabbal-Gill, S.S. Davis, Novel chitosan based delivery systems for nasal administration of an LHRH-analogue, STP Pharma. 10 (2000) 89–94.
[21] P. Tengamnuay, A. Sahamethapat, A. Sailasuta, A. Mitra, Chitosans as nasal absorption enhancers of peptides: Comparison between free amine chitosans and soluble salts, Int. J. Pharm. 197 (2000) 53–67.
[22] D. Mei, S. Mao, W. Sun, Y. Wang, T. Kissel, Effect of chitosan structure properties and molecular weight on the intranasal absorption of tetramethylpyrazine phosphate in rats, Eur. J. Pharm. Biopharm. 70 (2008) 874–881.
[23] S. Yu, Y. Zhao, F. Wu, X. Zhang, W. Lu, H. Zhang, et al., Nasal insulin delivery in the chitosan solution: *In vitro and in vivo* studies, Int. J. Pharm. 81 (2004) 11–23.
[24] A.M. Dyer, M. Hinchcliffe, P. Watts, J. Castile, I. Jabbal-Gill, R. Nankervis, et al., Nasal delivery of insulin using novel chitosan based formulations: A comparative study in two animal models between simple chitosan formulations and chitosan nanoparticles, Pharm. Res. 19 (2002) 998–1008.
[25] J. Varshosaz, H. Sadrai, R. Alinagari, Nasal delivery of insulin using chitosan microspheres, J. Microencap. 21 (2004) 761–774.
[26] L. Illum, Nanoparticulate systems for nasal delivery of drugs: A real improvement over simple systems? J. Pharm. Sci. 96 (2007) 473–483.
[27] S. Harikarnpakdee, N. Sutanthavibul, V. Lipipun, G.C. Ritthidej, Spray dried mucoadhesive microspheres: Preparation, and transport through nasal cell monolayer, AAPS Pharm. Sci. Tech. 7 (2006) E79–E88.
[28] C. Kusonwiriyawong, W. Pichayakorn, V. Lipipun, G.C. Ritthidej, Retained integrity of protein encapsulated in spray-dried chitosan microparticles, J. Microencap. 26 (2009) 111–121.

[29] M.W. Thanou, J.C. Verhoef, S.G. Romeijn, J.F. Nagelkerke, F.W. Merkus, H.E. Junginger, Effects of N-trimethyl chitosan chloride, a novel absorption enhancer, on Caco-2 intestinal epithelia and the ciliary beat frequency of chicken embryo trachea, Int. J. Pharm. 185 (1999) 73–82.

[30] D. Sakloetsakun, G. Perera, J. Hombach, G. Millotti, A. Bernkop-Schnürch, The impact of vehicles on the mucoadhesive properties of orally administrated nano particles: A case study with chitosan-4-thiobutylamidine conjugate, AAPS Pharm Sci Tech. 11 (2010) 1185–1192.

[31] M. Prabaharan, S. Gong, Novel thiolated carboxymethyl chitosan-*g*-β-cyclodextrin as mucoadhesive hydrophobic drug delivery carriers, Carbohyd. Polym. 73 (2008) 117–125.

[32] E. Gavini, G. Rassu, C. Muzzarelli, M. Cossu, P. Giunchedi, Spray dried microspheres based on methylpyrrolinone chitosan as new carrier for nasal administration of metoclopramide, Eur. J. Pharm. Biopharm. 68 (2008) 245–253.

[33] P. Calvo, C. Remunan-Lopez, J.L. Vila-Jato, M.J. Alonso, Chitosan and chitosan/ethylene oxide-propylene oxide block copolymer nanoparticles as novel carries for proteins and vaccines, Pharm. Res. 14 (1997) 1431–1436.

[34] J.H. Hamman, M. Stander, A.F. Kotze, Effect of the degree of quaternisation of N-trimethyl chitosan chloride on absorption enhancement: *In vivo* evaluation in rat nasal epithelia, Int. J. Pharm. 232 (2002) 235–242.

[35] A.H. Krauland, D. Guggi, A. Bernkop-Schnurch, Thiolated chitosan microparticles: A vehicle for nasal peptide drug delivery, Int. J. Pharm. 307 (2006) 270–277.

[36] X. Zhang, H. Zhang, Z. Wu, Z. Wang, H. Niuand, C. Li, Nasal absorption enhancement of insulin using PEG-grafted chitosan nanoparticles, Eur. J. Pharm. Biopharm. 68 (2008) 526–534.

[37] A. Jindapattanakit, P. Peungvicha, A. Sailasuta, T. Kissel, V.B. Junyaprasert, Nasal absorption and local tissue reaction of insulin nanocomplexes of trimethyl chitosan derivatives in rats, J. Pharm. Pharmacol. 62 (2010) 583–591.

[38] R.F. Tester, J. Karkalas, X. Qi, Starch – composition, fine structure and architecture, J. Cereal Sci. 39 (2004) 151–165.

[39] P. Edman, E. Bjork, L. Ryden, Microspheres as a nasal delivery system for peptide drugs, J. Control. Release 21 (1992) 165–172.

[40] E. Björk, P. Edman, Degradable starch microspheres as a nasal delivery system for insulin, Int. J. Pharm. 47 (1988) 233–238.

[41] L. Illum, A.N. Fisher, I. Jabbal-Gill, S.S. Davis, Bioadhesive starch microspheres and absorption enhancing agents act synergistically to enhance the nasal absorption of polypeptides, Int. J. Pharm. 222 (2001) 109–119.

[42] A.K. Jain, R.K. Khar, F.J. Ahmed, P.V. Diwan, Effective insulin delivery using starch nanoparticles as a potential trans-nasal mucoadhesive carrier, Eur. J. Pharm. Biopharm. 69 (2008) 426–435.

[43] H. Blanco-Fuente, S. Anguiano, F.J. Otero-Espinar, J. Balnco-Mendez, *In vitro* bioadhesion of carbopol hydrogels, Int. J. Pharm. 142 (1996) 169–174.

[44] G. Rathnam, N. Narayanan, R. Ilavarasan, Carbopol-based gels for nasal delivery of progesterone, AAPS Pharm. Sci. Tech. 9 (2008) 1078–1082.

[45] Y. Suzuki, Y. Makino, Mucosal drug delivery using cellulose derivatives as a functional polymer, J. Control. Release 62 (1999) 101–107.

[46] D. Teshima, A. Yamauchi, K. Makino, Y. Kataoka, Y. Arita, H. Nawata, et al., Nasal glucagon delivery using microcrystalline cellulose in healthy volunteers, Int. J. Pharm. 233 (2002) 61–66.

[47] T. Matsuyama, T. Morita, Y. Horikiri, H. Yamahara, H. Yoshino, Influence of fillers in powder formulations containing N-acetyl-l-cysteine on nasal peptide absorption, J. Control. Release 120 (2007) 88–94.

[48] A. Bernkop-Schnurch, C.E. Kast, M.F. Richter, Improvement in the mucoadhesive properties of alginate by the covalent attachment of cysteine, J. Control. Release 71 (2001) 277–285.
[49] K. Morimoto, H. Katsumata, T. Yabuta, K. Iwanaga, M. Kakemi, Y. Tabata, et al., Evaluation of gelatin microspheres for nasal and intramuscular administrations of salmon calcitonin, Eur. J. Pharm. Sci. 13 (2001) 179–185.
[50] J. Wang, S. Sakai, Y. Deguchi, D. Bi, Y. Tabata, K. Morimoto, Aminated gelatin as a nasal absorption enhancer for peptide drugs: Evaluation of absorption enhancing effect and nasal mucosa perturbation in rats, J. Pharm. Pharmacol. 54 (2002) 181–188.
[51] T. Seki, H. Kanbayashi, S. Chono, Y. Tabata, K. Morimoto, Effects of a sperminated gelatin on the nasal absorption of insulin, Int. J. Pharm. 338 (2007) 213–218.
[52] K. Ohtake, T. Maeno, H. Ueda, H. Natsume, Y. Moritomo, Poly-l-Argenine predominantly increases the paracellular permeability of hydrophilic macromolecules across rabbit nasal epithelium, *in vivo*. Pharm. Res. 20 (2003) 153–160.
[53] M. Miyamoto, H. Natsume, I. Satoh, K. Ohtake, M. Yamaguchi, D. Kobayashi, et al., Effect of poly-l-arginine on the nasal absorption of FITC-dextran of different molecular weights and recombinant human granulocyte colony-stimulating factor (rhG-CSF) in rats, Int. J. Pharm. 226 (2001) 127–138.
[54] K. Morimoto, H. Yamaguchi, Y. Iwakura, K. Morisaka, Y. Ohashi, Y. Nakai, Effects of viscous hyaluronate sodium solutions on the nasal absorption of vasopressin and an analogue, Pharm. Res. 8 (1991) 471–474.
[55] L. Illum, N.F. Farraj, A.N. Fisher, I. Jabbal-Gill, M. Miglietta, L.M. Benedetti, Hyaluronic acid ester microspheres as a nasal delivery system for insulin, J. Control. Release 29 (1994) 133–141.
[56] C.-M. Lehr, Lectin-mediated drug delivery: The second generation of bioadhesives, J. Control. Release 65 (2000) 19–29.
[57] E.-S. Khafagy, M. Morishita, K. Isowa, J. Imai, K. Takayama, Effect of cell penetrating peptides on the nasal absorption of insulin, J. Control. Release 133 (2009) 103–108.
[58] M.D. Hornof, C.E. Kast, A. Bernkop-Schnurch, *In vitro* evaluation of the viscoelastic properties of chitosan–thioglycolic acid conjugates, Eur. J. Pharm. Biopharm. 55 (2003) 185–190.
[59] S. Park, Y.K. Oh, H. Yoon, J.M. Kim, C.K. Kim, *In situ* gelling and mucoadhesive polymer vehicles for controlled intranasal delivery of plasmid DNA, J. Biomed. Mater. Res. 59 (2002) 144–151.
[60] T.-W. Chung, D.-Z. Liu, J.-S. Yang, Effect of interpenetration of thermo-sensitive gels by crosslinking of chitosan on nasal delivery of insulin: *In vitro* characterization and *in vivo* study, Carbohyd. Polym. 82 (2010) 316–322.
[61] J. Wu, W. Wei, L.-Y. Wang, Z.-G. Su, G.-H. Ma, A thermosensitive hydrogel based on quaternized chitosan and poly(ethylene glycol) for nasal drug delivery system, Biomaterials 28 (2007) 2220–2232.
[62] E. Adriaens, D. Ameye, M.M. Dhondt, P. Foreman, J.P. Remon, Evaluation of the mucosal irritation potency of co-spray dried Amioca/poly(acrylic acid) and Amioca/Carbopol 974P mixtures, J. Control. Release 88 (2003) 393–399.
[63] C. Callens, J.P. Remon, Evaluation of starch–maltodextrin–Carbopol 974 P mixtures for the nasal delivery of insulin in rabbits, J. Control. Release 66 (2000) 215–220.
[64] S.T. Lim, G.P. Martin, D.J. Berry, M.B. Brown, Preparation and evaluation of the *in vitro* drug release properties and mucoadhesion of novel microspheres of hyaluronic acid and chitosan, J. Control. Release 66 (2000) 281–292.

[65] X. Wang, C. Zheng, Z. Wu, T. Teng, X. Zhang, C. Li, Chitosan-NAC nanoparticles as a vehicle for nasal absorption enhancement of insulin, J. Biomed. Mater. Res. B. 88 (2009) 150–161.

[66] E. Pringels, C. Vervaet, B. Verbeeck, P. Foreman, J.P. Remon, The addition of calcium ions to starch/carbopol mixture enhances the nasal bioavailability of insulin, Eur. J. Pharm. Biopharm. 68 (2008) 201–206.

[67] C. Callens, J. Ceulemans, A. Ludwig, P. Foreman, J.P. Remon, Rheological study on mucoadhesivity of some nasal powder formulations, Eur. J. Pharm. Biopharm. 55 (2003) 323–328.

[68] D.S. Viswanath, T.K. Ghosh, D.H.L. Prasad, N.V.K. Dutt, K.Y. Rani, Viscosity of Liquids: Theory, Estimation, Experiment, and Data, Springer, Dordrecht, The Netherlands, 2007.

[69] A. Ahuja, R.K. Khar, J. Ali, Mucoadhesive drug delivery systems, Drug Dev. Ind. Pharm. 23 (1997) 489–515.

[70] K.V.R. Rao, P. Buri, A novel *in situ* method to test polymers and coated microparticles for bioadhesion, Int. J. Pharm. 52 (1989) 265–270.

[71] V. Grabovac, D. Guggi, A. Bernkop-Schurch, Comparison of the mucoadhesive properties of various polymers, Adv. Drug Deliv. Rev. 57 (2005) 1713–1723.

[72] C.-P. Lee, R.L.A. de Vrueh, P.L. Smith, Selection of development candidates based on *in vitro* permeability measurements, Adv. Drug Deliv. Rev. 23 (1997) 47–62.

[73] C. Wadell, E. Björk, O. Camber, Nasal drug delivery – evaluation of an *in vitro* model using porcine nasal mucosa, Eur. J. Pharm. Sci. 7 (1999) 197–203.

[74] S.H. Randell, L. Walstad, U.E. Schwab, B.R. Grubb, J.R. Yankaskas, Isolation and culture of airway epithelial cells from chronically infected human lungs, In vitro Cell Dev. Biol. – Anim. 37 (2001) 480–489.

[75] D.C. Gruenert, W.E. Finkbeiner, J.H. Widdicombe, Culture and transformation of human airway epithelial cells, Am. J. Physiol. 268 (1995) L347–L360.

[76] M.C. Schmidt, H. Peter, S.R. Lang, G. Ditzinger, H.P. Merkle, *In vitro* cell models to study nasal mucosal permeability and metabolism, Adv. Drug Deliv. Rev. 29 (1998) 51–79.

[77] U. Werner, T. Kissel, Development of a human nasal epithelial cell culture model and its suitability for transport and metabolism studies under *in vitro* conditions, Pharm. Res. 12 (1995) 565–571.

[78] A. De Fraissinette, R. Brun, H. Felix, J. Vonderscher, A. Rummelt, Evaluation of the human cell line RPMI 2650 as an *in vitro* nasal model, Rhinology 33 (1995) 194–198.

[79] K.A. Foster, M.L. Avery, M. Yazdanian, K.L. Audus, Characterization of the Calu-3 cell line as a tool to screen pulmonary drug delivery, Int. J. Pharm. 208 (2000) 1–11.

[80] A. Wengst, S. Reichl, RPMI2650 epithelial model and three dimensional reconstructed human nasal mucosa as *in vitro* models for nasal permeation studies, Eur. J. Pharm. Biopharm. 74 (2010) 290–297.

[81] R.U. Agu, M. Jorissen, R. Kinget, N. Verbeke, P. Augustijns, Alternatives to *in vivo* nasal toxicological screening for nasally-administered drugs, STP Pharm. Sci. 12 (2002) 13–22.

[82] P. Markus, S.G. Romeijn, J.C. Verhoef, F.W. Merkus, P.F. Schonwenburg, Classification of cilio-inhibiting effects of nasal drops, Laryngoscope 111 (2001) 595–602.

[83] T.J. Aspden, J.D.T. Mason, N.S. Jones, J. Lowe, O. Skaugrud, L. Illum, Chitosan as a nasal delivery system: The effect of chitosan solutions on *in vitro* and *in vivo* mucociliary transport rates in human turbinates and volunteers, J. Pharm. Sci. 86 (1997) 509–513.

[84] M.I. Ugwoke, R.U. Agu, M. Jorissen, P. Augustijns, R. Sciot, N. Verbeke, et al., Toxicological investigations of the effects of carboxymethylcellulose on ciliary beat frequency of

human nasal epithelial cells in primary suspension culture and *in vivo* on rabbit nasal mucosa, Int. J. Pharm. 205 (2000) 43–51.

[85] M.I. Ugwoke, R.U. Agu, M. Jorissen, P. Augustijns, R. Sciot, N. Verbeke, et al., Nasal toxicological investigations of Carbopol 971P formulation of apomorphine: Effects on ciliary beat frequency of human nasal primary cell culture and *in vivo* on rabbit nasal mucosa, Eur. J. Pharm. Sci. 9 (2000) 387–396.

[86] C. Callens, E. Adriaens, K. Dierckens, J.P. Remon, Toxicological evaluation of a bioadhesive nasal powder containing a starch and carbopol 974P on rabbit nasal mucosa and slug mucosa, J. Control. Release 76 (2001) 81–91.

[87] S. Dimova, M.E. Brewster, M. Noppe, M. Jorissen, P. Augustijns, The use of human nasal *in vitro* cell systems during drug discovery and development, Toxicol. Vitro 19 (2005) 107–122.

[88] T.K. Vyas, A. Shahiwala, S. Marathe, A. Misra, Intranasal drug delivery for brain targeting, Curr. Drug Deliv. 2 (2005) 165–175.

[89] C.L. Graff, G.M. Pollack, Nasal drug administration: Potential for targeted central nervous system delivery, J. Pharm. Sci. 94 (2005) 1187–1195.

[90] S.V. Dhuria, L.R. Hanson, W.H. Prey II, Intranasal delivery to the central nervous system: Mechanisms and experimental considerations, J. Pharm. Sci. 99 (2010) 1654–1673.

[91] S. Horvat, A. Feher, H. Wolburg, P. Sipos, S. Veszelka, A. Toth, et al., Sodium hyaluronate as a mucoadhesive component in nasal formulation enhances delivery of molecules to brain tissue, Eur. J. Pharm. Biopharm. 72 (2009) 252–259.

[92] R.G. Thorne, G.J. Pronk, V. Padmanabhan, W.H. Frey, Delivery of insulin-like growth factor-I to the rat brain and spinal cord along olfactory and trigeminal pathways following intranasal administration, Neuroscience 127 (2004) 481–496.

[93] S.R.K. Vaka, S.M. Sammeta, L.B. Day, S.N. Murthy, Delivery of nerve growth factor to brain via intranasal administration and enhancement of brain uptake, J. Pharm. Sci. 98 (2009) 3640–3646.

[94] W.A. Banks, M.J. During, M.L. Niehoff, Brain uptake of the glucagon-like peptide-1 antagonist exendin(9–39) after intranasal administration, J. Pharmacol. Exp. Ther. 309 (2004) 469–475.

[95] R.S. Kaye, T.S. Purwal, O.H. Alpar, Development and testing of particulate formulation for nasal delivery of antibodies, J. Control. Release 135 (2009) 127–135.

Transdermal Delivery of Peptides and Proteins

Anushree Herwadkar and Ajay K. Banga
College of Pharmacy and Health Sciences, Mercer University, Atlanta, GA-30341, USA

Chapter Outline

4.1. INTRODUCTION

Proteins and peptides are being increasingly used as therapeutic agents. However, these molecules cannot be administered orally as they are subjected to proteolytic degradation in the gastrointestinal tract. Parenteral administration is possible, but is associated with poor patient compliance. Transdermal delivery offers a potential alternative, overcoming setbacks associated with oral as well as injectable formulations. The passive delivery of proteins through the skin has to date been impracticable due to their macromolecular and hydrophilic nature, which limits their transport through the lipophilic skin barrier (stratum corneum). Hence, several enhancement techniques have been investigated to assist delivery of these molecules through skin. Amongst these, the use of chemical enhancers, iontophoresis, phonophoresis and microporation

Peptide and Protein Delivery. DOI: 10.1016/B978-0-12-384935-9.10004-5

have shown considerable promise. This chapter will focus on the enhancement techniques that aid the transport of proteins and peptides through skin, so advancing their delivery.

4.2. SKIN STRUCTURE

The skin is the largest organ of the human body and functions as a barrier, protecting the internal structures of the body from the environment. Structurally, skin primarily consists of the epidermis, dermis and the underlying subcutaneous fat. The epidermis forms the outermost layer, being exposed to the environment on one side and separated from the dermal tissue by a basement membrane on the other. From this basal membrane to the surface of the skin, the epidermis can be differentiated into several strata, namely: stratum germanitivum or stratum basale, stratum spinosum, stratum granulosum, stratum lucidum and stratum corneum [1]. The lower strata: stratum basale, stratum spinosum, stratum spinosum and stratum lucidum are together known as the viable epidermis.

The stratum corneum, on the other hand, is made up of dead cells, and hence forms the non-viable portion of the epidermis. Epidermal tissue is entirely avascular and is primarily composed of epithelial cells, termed keratinocytes, embedded in a lipid matrix. In addition to this, there are melanocytes which produce the pigment, melanin. Merkel cells and Langerhans cells, which are responsible for skin's immune response, are sparsely distributed throughout this tissue [2]. Keratinocytes are formed in the stratum basale and undergo continuous differentiation as they move from this layer towards the surface of the skin. As these cells move away from the dermis, they are deprived of oxygen and nourishment and by the time they reach the stratum corneum, they are transformed into completely keratinized, flattened, dead cells called corneocytes. The conversion of basal cell keratinocytes to corneocytes takes 28 days, including 14 days residence time of the corneocytes in the stratum corneum [2]. Corneocytes are then subjected to a process called desquamation, where one cell layer of stratum corneum is sloughed off per day [2]. The stratum corneum is composed of around 15–25 layers of corneocytes, stacked one above another [3]. Intercellular spaces between these cells are filled with lipid, composed mainly of cholesterol, ceramides and free fatty acids. Because of the hexagonal shape and close packing of the corneocytes, the stratum corneum stands as a rate-limiting barrier to the transdermal permeation of molecules.

The dermis is composed of collagen and elastic fibers interspersed in proteoglycan ground substance. This composition is responsible for imparting plasticity to the skin, and makes the dermal tissue relatively hydrophilic as compared to the stratum corneum, and hence polar molecules such as proteins and peptides can permeate relatively easily through it. Skin appendages, such as hair follicles, sebaceous glands and sweat glands originate in the dermis.

These appendages account for only 0.1% of the skin area [2,4], but they play an important role in the transdermal delivery of macromolecules by acting as shunt pathways which bypass the stratum corneum barrier. Appendageal transport is also favored during iontophoresis, where molecules tend to travel via the shortest pathway through skin under the influence of an electric current [5]. The dermal tissue is innervated with nerves and supplied with blood and lymphatic vessels. In addition, spindle shaped fibroblasts and mast cells are also present in the ground substance.

The thickness of human skin varies between individuals and also varies between different anatomical sites on an individual. On average, skin thickness ranges from 2–5 mm, with the stratum corneum being 10–15 μm thick. The epidermis ranges between 0.06 mm on eyelids to 0.8 mm on the palms and soles, and the dermis is 3–5 mm thick. In addition, the underlying subcutaneous fat is several millimeters thick, and provides insulation and mechanical support to the skin [1].

4.3. RATIONALE AND CHALLENGES FOR TRANSDERMAL DELIVERY OF PEPTIDES AND PROTEINS

4.3.1. Rationale

Proteins and peptides are being used widely as therapeutic agents to treat a variety of indications. Several marketed products include: insulin to treat diabetes mellitus, calcitonin for osteoporosis, monoclonal antibodies such as rituximab (Rituxan®) to treat non-Hodgkin's lymphoma or gemtuzumab (Mylotarg®) to treat leukemia, interferons, epidermal growth factor and vasopressin to treat diabetes insipidus. Besides these, a large number of proteins are under investigation to treat further disorders. However, proteinaceous agents are denatured in the acidic conditions of the stomach and degraded by proteolytic enzymes in the gastro-intestinal tract (GIT), and hence oral delivery has not been a feasible route of administration. Delivery via nasal or ocular routes has also been challenging because of enzymatic activity in these mucosal routes [6]. Proteins and peptides are therefore almost exclusively administered by the parenteral route. Although parenteral administration serves the purpose, it has several shortcomings. Apart from being invasive, administration has to be done under the supervision of a health care professional. This reduces patient compliance and adds to the cost. Also, repeated administration is necessary as many proteins and peptides have a very short half-life. In such a situation, the transdermal route offers a continuous, non-invasive and hence painless administration. The skin is easily accessible and delivery can be controlled and terminated when required. Additionally, since proteins and peptides are potent molecules with short half-lives, continuous delivery by the transdermal route would help avoid repeated administration and improve patient compliance [7].

4.3.2. Challenges

Transdermal delivery therefore seems to offer several advantages over conventional parenteral administration. However, several barriers have limited the widespread use of this route. Since the skin functions to restrict the passage of exogenous substances into the body, achieving the penetration of these macromolecules through it is a challenging task.

4.3.2.1. Penetration Barrier

As discussed in section 4.2, the skin acts as a barrier to all exogenous materials and restricts their entry into the body. Most of the barrier function of the skin is achieved by the stratum corneum [8]. This presents a highly organized, brick-and-mortar type of structure where keratinocytes are embedded in a lipid matrix. Permeation of drugs through this lipid matrix (the intercellular route of permeation) is considered to be a major pathway for molecules to cross the stratum corneum [9]. Due to its lipid content, the permeability of the stratum corneum to hydrophilic molecules is limited, so highly hydrophilic molecules, like proteins and peptides, are unable to partition into the lipoidal stratum corneum layers and hence cannot passively permeate through skin [8]. Also, the diffusivity of a molecule decreases exponentially with its molecular volume and hence its molecular weight. Macromolecules such as proteins hence have an inherently low diffusivity, which further limits their ability to permeate into and through skin [8]. In general, moderately lipophilic, low molecular weight compounds are ideal candidates for transdermal permeation. Thus, proteins and peptides being hydrophilic macromolecules with ionic character cannot undergo passive permeation through skin [6].

4.3.2.2. Enzymatic Barrier

The epidermal enzymatic barrier of the skin is another factor that limits the delivery of peptides and proteins through the skin. Even though the enzymatic activity of skin is considerably lower than that in liver and GIT, it may affect topical and transdermal delivery of drugs [10]. Skin metabolism is comprised of primary and secondary metabolic reactions such as oxidation, reduction, hydrolysis and conjugation reactions. However, the delivery of peptides and proteins is mainly limited by proteolytic enzymes. The epidermal and dermal layers of the skin contain a variety of endopeptidases, which cleave peptide bonds within the peptide structure, and exopeptidases which cleave the N- or C-terminal peptide bonds. Some of the endopeptidases identified in the skin include collagenases, elastases, fibrinolysins, casenolytic enzymes and enzymes of the kallikrein–kinin system. Among the exopeptidases, amino-peptidases are relatively abundant while carboxypeptidases are present in lower amounts [11].

These enzymes are located in the subcellular compartments within the skin and hence proteins and peptides may or may not encounter them during permeation. Thus, studies carried out with skin homogenates may not be very good predictors of skin metabolism, as all the subcellular compartments would be ruptured by the homogenization process, which exposes and sometimes even inactivates enzymes [11]. Also, *in vitro* studies may sometimes overestimate metabolism, as there is increased enzyme activity due to lack of removal of enzymes by systemic circulation. Use of animal models may also be inaccurate as proteolytic activity may be higher in animal skin compared to human skin [2].

4.4. TECHNIQUES ENABLING TRANSDERMAL DELIVERY OF PEPTIDES AND PROTEINS

In order to overcome the formidable barrier function of the skin, physical and chemical penetration enhancement techniques are being widely investigated. These techniques will be discussed in detail in this section.

4.4.1. Formulation Approaches

Formulation approaches to the transdermal delivery of peptide and protein therapeutics include the following:

- Encapsulation technologies;
- Prodrugs;
- Protease inhibitors.

4.4.1.1. Encapsulation Technologies

Encapsulation of a protein or peptide into a colloidal carrier is an attractive approoach for enhancing its transdermal delivery. Carrier systems such as conventional and specialized types of liposomes, and different types of nanoparticles are being widely investigated to assist peptide delivery into and through the skin. Encapsulation into liposomal structures is believed to help in enhancing localized delivery and skin targeting (cf. Chapter 2, page 27; Chapter 6, pages 116–117). Liposomes increase deposition of molecules in the stratum corneum and upper skin layers, but do not necessarily enhance the permeation of molecules through skin [12]. Limited penetration of these vesicular structures in the skin layers, and interaction with stratum corneum lipids causing drug release are the proposed mechanisms of action for liposomal topical drug delivery [1]. Liposomal formulations have been used to assist the delivery of hydrophilic molecules and macromolecules into skin. Liposomes have been effective for cutaneous delivery of vaccines and have shown enhanced skin delivery of proteins such as interferon-α (INF-α) [13]. Special types of liposomal structures, such as ethosomes, niosomes and transfersomes, have also been developed. Ethosomes have around 30% ethanol incorporated in their structure. Their high alcohol content is believed to fluidize both vesicular and stratum corneum lipids, thereby helping drug molecules to penetrate into deeper skin layers.

Niosomes are vesicles composed of non-ionic surfactants and have been used for a variety of cosmetic applications. Transferosomes are also known as deformable or elastic liposomes and in addition to the phospholipid, they contain around 10 to 24 wt % of surfactant and 3–10% of ethanol. The surfactant imparts a highly flexible nature to these vesicles, which enables them to permeate through channels that are one-tenth of their own diameter [1]. Cevc *et al.* compared a transferosome formulation to liposomes and micellar structures for the transdermal delivery of insulin [14]. While insulin could not be delivered passively, or by liposomes and micelles, transdermal permeation was observed with the transfersome formulation. Encapsulation technology thus seems to be advantageous and promising for the delivery of peptides and proteins. However, scale-up of these formulations for therapeutic use needs to be optimized.

4.4.1.2. Prodrugs

Proteins and peptides are hydrophilic molecules and hence not inherently suitable for transdermal delivery. In order to present these molecules as better candidates, lipophilic prodrugs can be made by conjugating proteins with lipophilic moieties. Lipophilic prodrugs would help increase retention in the stratum corneum and improve the overall permeation through skin. Foldvari *et al.* showed that the permeation of INF-α was enhanced 2.5- to 5-fold when delivered as an acyl derivative [13]. They also demonstrated the dependence of permeation on fatty acid chain length. As the alkyl chain length increased from 12 to 16, dermal and transdermal delivery was enhanced, but further increase in chain length decreased delivery. During or after its passage into and through the skin, the prodrug is converted back into its active form by enzymatic or chemical release. Since the prodrug approach leads to the formation of a new chemical entity, additional chemical, biological and toxicological testing may be required. However, in studies carried out by Hashimoto *et al.* and Murashini *et al.* for delivery of insulin and thyrotropin hormone respectively, 60–80% of the parent peptide active was retained after derivatization with fatty acids [15,16].

4.4.1.3. Protease Inhibitors

Protease inhibitors help in bypassing the enzymatic barrier of the skin and hence can assist in the transdermal delivery of proteins and peptides. The mechanism of action behind inhibiting enzymatic activity varies for different protease inhibitors. Enzyme activity is inhibited by tight binding or covalent linkage of the inhibitor to active sites on the enzyme. Inhibitors may also act by chelation of metal ions essential for activity of the proteolytic enzyme [11].

4.4.2. Chemical Permeation Enhancers

Certain chemicals, when applied to the skin, have the property of increasing permeation of drugs into it. Chemical permeation enhancement has been widely used in assisting the transdermal delivery of small molecules, and has also been investigated for enhancing delivery of peptides through skin.

A variety of molecules belonging to various chemical classes have shown efficacy as permeation enhancers. Some well-known and widely used enhancers include alcohols, polyalcohols, esters, fatty acids, pyrrolidones, sulfoxides, amines, amides, surfactants and phospholipids. The mechanism of permeation enhancement differs for different molecules. For example, fatty acids act by fluidizing stratum corneum lipid bilayers, while alcohols act by extracting stratum corneum lipids. Polar co-solvents such as propylene glycol help in solubilizing the enhancers in the stratum corneum layers and have been shown to have a synergistic action with enhancers like fatty acids [8]. When the organized lipid bilayers of the stratum corneum are disturbed by any of the above means, the skin barrier is compromised and enhanced transdermal delivery is possible.

Permeation enhancers have been tested for their ability to enhance transdermal and topical delivery of peptide and protein drugs. However, since proteins are relatively large, hydrophilic molecules, higher amounts of chemical enhancers have to be used to achieve effective delivery. This may not be feasible, since most enhancers would have an irritating effect if used in quantities above a certain threshold level. Compatibility of the protein/peptide with the enhancer could be another issue, and it needs to be ensured that the chemical enhancer does not degrade or alter the peptide or protein structure [17].

Chemical enhancers have shown a 2–3-fold enhancement in transdermal delivery of smaller peptides. N-decylmethyl sulfoxide, a non-ionic surfactant, has been shown to increase the permeability of the amino acids tyrosine and phenylalanine, diphenylalanine and the pentapeptide enkephalin across hairless mouse skin [6]. Linolinic acid and menthone have assisted the permeation of insulin both *ex vivo* and *in vivo*. Pretreatment of rat skin with these chemical enhancers improved permeation (*ex vivo*) compared to passive delivery. Also, a synergistic enhancement in permeation was observed when chemical enhancer pretreatment was combined with iontophoresis. *In vivo* studies also indicated that higher plasma insulin concentrations were achieved by a combination of linolinic acid pretreatment with iontophoresis [18]. Similarly, the effect of the permeation enhancer ethanol, and a combination of ethanol and cineole on the permeation of the thyrotropin releasing hormone analog (M-TRH) across human epidermal membrane was studied. An enhancement ratio of around 3 to 4.7 was observed when donor solutions containing 50% ethanol, as well as 3% cineole and 47% ethanol, were used [19]. However, in spite of these promising results, the efficacy of these molecules in enhancing the permeation of larger peptides and proteins remains limited.

4.4.3. Microporation

Microporation refers to the creation of micron sized channels through the skin to assist transdermal delivery of molecules. These channels bypass the stratum corneum barrier and reach into the epidermis layer. However, they do not reach the dermis where nerves and blood vessels are located. Hence, in comparison to hypodermic needles, delivery by microporation is non-invasive. Also, the microchannels are typically large in dimension compared to any drug molecule. Since macromolecules like proteins have dimensions in the nano range, they can be easily delivered through these microchannels into and across skin. Microporation is thus painless, non-invasive and yet an effective and elegant method to achieve enhanced delivery of molecules through the skin [20,21].

Mechanical and thermal microporation are being widely investigated. Several academic and industrial groups are actively involved in the development of mechanical microneedles, including 3M, Corium, Becton Dickinson, the Georgia Institute of Technology and Zosano Pharma. Microneedles are created using microfabrication technology, which involves techniques such as micromachining and microelectromechanical systems to create micron-sized patterns and structures [22]. Microneedles have been created from a variety of materials including silicon, metals, glass and polymers, as well as from sugars such as maltose which can dissolve when inserted in skin. Microneedles of dimensions varying from 300–900 μm have also been fabricated from commercially available 30G hypodermic needles, by a method where the needles were placed at predetermined lengths through holes created in a polyetheretherketone mold and then cut and glued at the back of the mold (Figure 4.1) [23].

Microneedles can have a variety of geometries, and can be solid or hollow in structure. Microneedle mediated delivery can be achieved in several different ways. The "poke and patch" method involves piercing the skin with microneedles followed by application of the drug formulation. In the "coat and poke" method, microneedles are coated with drug formulation and then inserted in the skin (Figure 4.2). The active ingredient can also be encapsulated within biodegradable polymer microneedles. In addition, the drug formulation can be directly injected into the skin with the help of hollow microneedles [24].

Microneedles have been used effectively to help transdermal delivery of macromolecules. The authors have studied the *in vitro* transdermal delivery of a relatively large protein (immunoglobulin G) assisted by mechanical microporation [25]. Characterization of the biodegradable maltose microneedles used in this study showed that the needles created microchannels in hairless rat skin, which could be easily visualized using techniques such as methylene blue staining and histology. Delivery increased with an increase in the concentration of immunoglobulin G, an increase in the number of microneedle arrays and an increase in the length of microneedles [25].

Thermal microporation involves the application of rapid and controlled thermal pulses with the help of small resistive elements to microscopic sites on

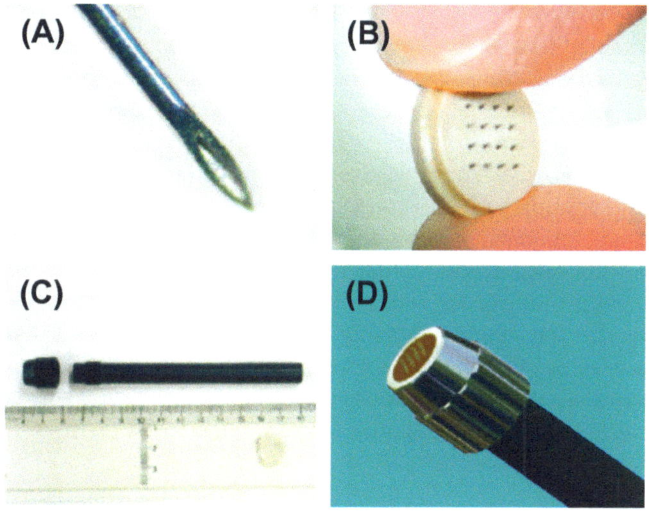

FIGURE 4.1 Microneedle array assembled from commercially available 30G hypodermic needles. (A) tip of a 30G hypodermic needle; (B) the assembled microneedle array; (C) microneedle array applicator in comparison with a ruler; (D) applicator at an angled view. *Reprinted from reference [23] with permission from Elsevier.*

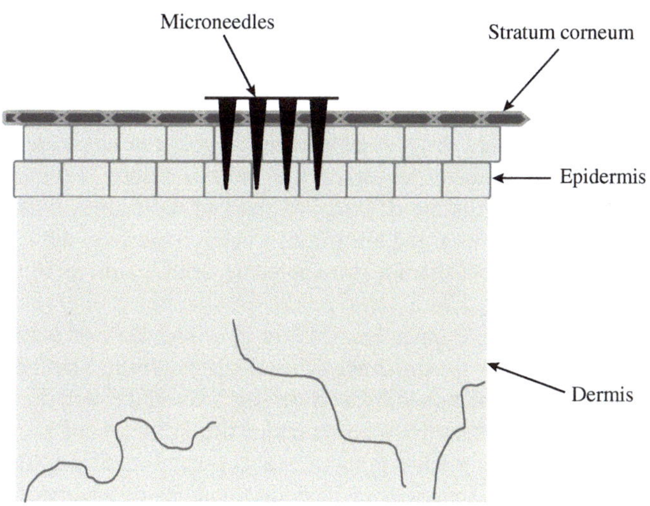

FIGURE 4.2 A schematic of microneedles applied on skin. Microneedles selectively penetrate the stratum corneum and a part of the epidermis, thus allowing painless but efficient drug delivery.

the skin. This resistive heating causes flash vaporization of stratum corneum cells, thus creating micropores in the skin. The PassPort™ system, by Altea Therapeutics, was used to assist the transdermal delivery of interferon alpha-2b (INFα2b) *in vivo* in the hairless rat model. An array of micropores was created on rat abdomen skin (2 cm^2, 72 micropores/cm^2) by the thermal ablation of the stratum corneum. INFα2b could not permeate the skin by passive delivery or iontophoresis alone. However, *in vivo* transdermal delivery of INFα2b was achieved following thermal ablation, which was further enhanced by cathodal iontophoresis [26].

Microporation has shown promise in assisting the transdermal delivery of peptides and proteins in both *in vitro* and *in vivo* models [20]. Pharmacologically relevant amounts of desmopressin were delivered across skin using Macroflux® transdermal microprojection technology (Zosano Pharma Inc.) in an *in vivo* guinea pig model with bioavailability as high as 85% [27]. Microneedles are also considered promising for vaccination purposes. Microporation can direct vaccines to antigen-presenting cells in the skin which will further elicit an immune response [21]. Microneedle arrays dry coated with ovalbumin using a novel gas-jet drying approach were found to be highly effective for vaccination [28].

4.4.4. Iontophoresis

Iontophoresis is a non-invasive, electrically assisted technique in which a physiologically acceptable amount of electric current (up to 0.5 mA/cm^2) is used to facilitate transdermal delivery of charged and neutral molecules [10,29]. Unlike other physical enhancement techniques, which tend to disrupt the skin barrier in promoting transdermal flux, iontophoresis acts on the drug molecule itself [30]. These molecules are propelled into the deeper layers of the skin under the influence of an electric current. Depending on the physicochemical nature of the drug, mechanisms involving electrorepulsion or electro-osmosis predominate to drive the molecule into and across skin. Electrorepulsion facilitates the delivery of charged molecules; when they are placed under a similarly charged electrode, repulsion between the like charges of the electrode and the therapeutic molecule tend to drive the molecule through the skin (Figure 4.3).

Electro-osmosis on the other hand refers to a bulk flow of water from an anode to a cathode under the influence of an electric current. The movement of positively charged ions like sodium is especially favored through the negatively charged skin [31]. As these ions move under the influence of a current, the water of hydration associated with them is also transported from the anode to the cathode. This water, in turn, can carry drug solutes of therapeutic interest, and thus deliver them into the skin. Also, since ions tend to take the path of least resistance when subjected to an electric current, it was believed that the appendageal pathway predominates during iontophoresis [5,32]. This hypothesis is supported by confocal laser scanning microscopy studies and vibrating probe electrode techniques [32]. In addition to the appendageal route, polar regions in the intercellular pathway may also aid transdermal iontophoretic transport [32].

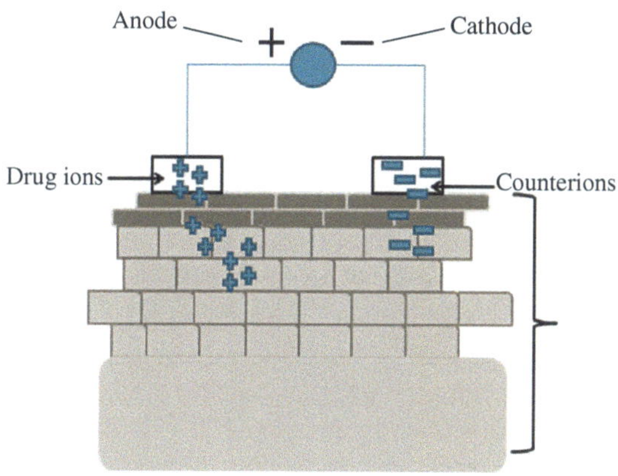

FIGURE 4.3 A schematic representing iontophoresis carried out on skin. In this figure, a positively charged drug molecule is placed under the positively charged electrode (anode). Repulsion between like charges drives the molecule through the skin (bracketed).

Iontophoresis helps in active transdermal transport of large, hydrophilic molecules like peptides and proteins which cannot undergo passive permeation. Iontophoretic delivery can be controlled because the amount of drug that is delivered using this technique is directly proportional to the quantity of charge applied [10]. Delivery can be tailored by changing the amount as well as the type of current. It is hypothesized that continuous direct current may be favored in acute conditions, while pulsed direct current may be suited for use in treating chronic skin conditions [32]. In addition to the amount of current, several other factors also play a role in iontophoretic delivery of peptides. The peptide charge is important and those with a high charge to mass ratio can be delivered with high iontophoretic efficiency. The secondary and tertiary structures of (large) peptides also play an important role in determining the overall charge of the peptide. The secondary structure determines the environment of the charged groups in the peptide structure, which may intensify or weaken the overall charge on the peptide and consequently assist or challenge iontophoresis [29].

In addition, it is important that the peptide remains charged throughout its transit across the skin. Since the pH of skin ranges from around 4 at the surface to 7.4 in the dermis region, peptides with isoelectric points below 4 and above 7.4 are ideal candidates for transdermal iontophoresis [30]. The molecular weight of the peptide, along with its size, shape, structure and volume, also plays a critical role in determining its mobility through skin. Mobility is inversely proportional to the size of the peptide which is again determined by its secondary, tertiary and (where appropriate) quaternary structure [29]. It has been hypothesized by quantitative 3-dimensional structure-permeability relationships that the hydrophilicity of the peptide structure assists iontophoresis, while bulky hydrophobic structures reduce iontophoretic efficiency [10].

A potential problem associated with the transdermal delivery of peptides is their metabolic breakdown in the skin by cutaneous peptidases. However, the increased flux associated with iontophoresis may help avoid this breakdown [29]. Also, since the drug is being delivered iontophoretically, similarly charged ions in the formulation and oppositely charged ions in the skin compete with the drug to carry current [33]. Thus, to ensure efficient delivery it is essential to minimize these extraneous ions.

Thus, iontophoresis appears to be a promising technology for the delivery of peptides which are hydrophilic and charged at physiological pHs. Iontophoresis has enabled the topical/transdermal delivery of several therapeutic small peptides, such as insulin [34], calcitonin [35,36], cyclosporine [37], arginine vasopressin [38], nafarelin [39,40], leutenizing hormone releasing hormone (LHRH) [40] and thyrotropin-releasing hormone [41]. In general, local delivery of small peptides (~10 kDa) is therapeutically feasible [10,30]. However, additional enhancement techniques are required for systemic delivery of larger peptides and proteins [29]. Combining iontophoresis with microneedle-mediated delivery was reported by the authors to be the most effective way to achieve transdermal delivery of a ~13 kDa protein, daniplestim (an engineered cytokine), *in vivo* across hairless rat skin [42]. Although iontophoresis is associated with a transient redness and erythema of the skin, these mild symptoms are generally well tolerated and severe effects have not been reported [10].

4.4.5. Sonophoresis

Sonophoresis is the use of ultrasound to drive molecules into and across skin [43]. Ultrasound is typically used for medical and diagnostic applications, and low-frequency sonophoresis (20 to 100 KHz) [44] is now gaining importance as a transdermal permeation enhancement technique. Sonophoresis has been successful in the topical delivery of small molecules such as lidocaine, to induce local anesthesia. Since this technique is non-invasive and painless, it is being investigated for transdermal delivery of macromolecules.

Sonophoresis equipment consists of a piezoelectric crystal and a coupling medium. Ultrasound is produced when electrical energy is converted into mechanical energy (oscillations) by a piezoelectric crystal, thereby generating acoustic waves. The mechanisms behind sonophoretic enhancement are not very clear, though acoustic cavitation is believed to have a major role. Application of ultrasound leads to the formation of microbubbles in the coupling medium. Collapse of these bubbles at the surface of stratum corneum gives rise to shock waves which render the skin permeable [45]. Additionally, acoustic microjets may be produced, which act by impacting or even penetrating the stratum corneum surface [46]. Other mechanisms such as thermal effects [47] and radiation pressures [30] have also been proposed to have some effect in sonophoretic permeation enhancement.

Important parameters to be monitored during sonophoresis are frequency, intensity, pulse length, application time and mode. Ultrasonic frequency depends on the size of the piezoelectric crystal and is inversely proportional to cavitation effects. Therefore, transdermal permeation enhancement is higher when sonophoresis is used at low frequencies rather than at high frequencies. Intensity is related to acoustic energy and in most cases sonophoresis is operated at intensities of 0.5 to 2 W/cm^2. Sonophoresis can be carried out in a continuous or discontinuous (pulsed) mode [48]. Sonophoresis can also be started simultaneously with drug application or a sonophoretic pretreatment can be carried out, after which the drug would be applied to the permeabilized skin [44].

The application of sonophoresis at higher acoustic intensities – in the range of 2–50 W/cm^2 – may lead to a phenomenon called sonomacroporation. It has been reported that at higher intensities, micron sized pathways were created through the stratum corneum and a part of the epidermis. These channels permitted the transdermal delivery of a range of poly-L-lysines having molecular weight up to 51 kDa [49]. *In vitro* delivery of peptides and proteins has been feasible by sonophoresis. Transdermal delivery of insulin (~6 kDa), interferon γ (~17 kDa) and erythropoetin (~48 kDa) has been reported [43]. Heparin and low molecular weight heparin have also been delivered into and across skin by sonophoresis [50].

The equipment in current use includes the SonoPrep™ System and cymbal array transducers [51]. The SonoPrep System has been approved by the US Food and Drug Administration (FDA) for topical delivery of lidocaine to induce local anaesthesia. Wearable ultrasound cymbal transducers are also in development. Smith *et al.* have demonstrated the enhanced *in vivo* delivery of insulin by use of these cymbal arrays (Figure 4.4) [51]. Low frequency ultrasound applied for

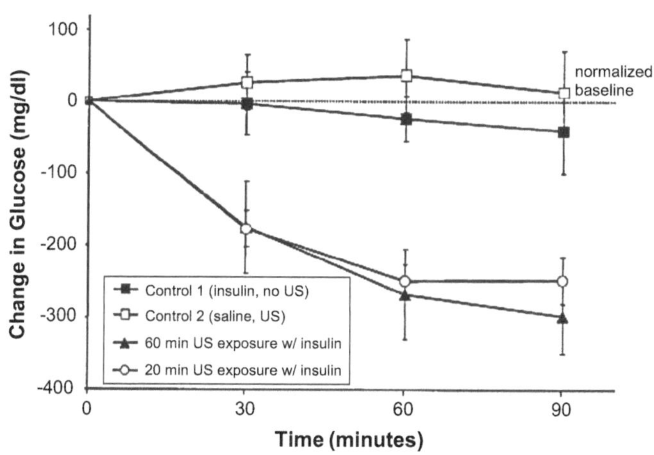

FIGURE 4.4 Changes in blood glucose levels of rats with insulin treatment using cymbal ultrasound arrays (I_{SPTP} = 100 mW/cm^2) over 90 min. The two controls studied (insulin delivery without ultrasound and ultrasound application with saline) did not show deviations of more than 40 mg/dL from the normalized baseline. A decrease in blood glucose level up to −296.7 mg/dL was observed at 90 min for rats treated with ultrasound for 60 min. Similar results were obtained for a 20 min ultrasound exposure group. *Reprinted from reference [51] with permission from Elsevier.*

short durations is generally well tolerated and patient compliant. However, at higher intensity reactions, such as utricaria or even second degree burns are possible and patients should be monitored for the same [44].

4.4.6. Other Techniques

Other techniques such as electroporation and laser ablation are being investigated for protein and peptide delivery. Also, a technology called ViaDerm™, based on radiofrequency ablation is being developed by TransPharma Medical. This approach consists of a microarray of electrodes to which current is applied in the radiofrequency range. Microchannels are created in the skin by localized heating and cell ablation, which enable transdermal delivery [10,20]. Another technique is the use of jet injectors which use high velocity to propel powders or liquids into skin. In addition to this, suction blister ablation [10], photomechanical waves, controlled heat assisted drug delivery (CHADD) and microscissioning are also being explored for transdermal delivery of peptides and proteins [17].

Electroporation: Electroporation involves exposing the skin to very high voltages (~10−1000 V) for very short durations of time (1−100 ms). Electroporation was traditionally used for introducing DNA material into cells, before its use in transdermal applications. It causes transient structural changes to the upper skin layers, thereby creating aqueous pathways across the skin. Enhancement in the permeation of molecules into and across skin occurs by several mechanisms, such as improved diffusion, electrophoretic movement and electro-osmosis. This technique has been effective in assisting and enhancing transdermal delivery of peptides and proteins [10,17,52]. Application of electroporation prior to iontophoresis was shown to enhance the flux of luteinizing hormone releasing hormone (LHRH) across the isolated perfused porcine skin flap model. LHRH delivery increased with an increase in the number of electroporative pulses [53]. However, patient acceptability might be an issue for this technique and very few clinical studies have been conducted so far.

Laser Ablation: Laser ablation involves the creation of microchannels in skin by using a high energy laser beam. Laser irradiation is absorbed by the skin tissue in the form of vibrational heating. In the irradiated area, this superheating causes water in the skin to boil, leading to a rapid evaporative local explosion, which further leads to creation of microchannels. Since tissue and water vaporization is very rapid, damage to unexposed surrounding skin is avoided. Optimum wavelengths for laser ablation are 2940 nm and 2790 nm, which are absorbed by tissue proteins and tissue water, respectively [17]. Currently, the Er:YAG (erbium: yttrium-aluminium-garnet) laser operating at 2940 nm is being used for cosmetic treatments. It finds applications in reconstructive surgery for treating rhytides, scars, photodamage and melasma [10]. Lee *et al.* have demonstrated that Er:YAG laser pretreatment was useful in assisting transdermal delivery of a variety of polypeptides and vaccines [7].

Issues such as safety and recovery of skin post-treatment still remain to be studied in detail. Previously, laser ablation was not very cost effective and was mostly limited to hospital settings. However, the handheld devices now available have lead to an increase in the scope and application of this technique. P.L.E.A.S.E. technology by Pantec Biosolutions consists of one such handheld device using the Er:YAG laser. This device is capable of creating microchannels in skin (~150–200 μm in diameter). Similarly, the Epicure Easytouch system by Norwood Abbey uses laser energy to induce dermal anaesthesia in 5 minutes post application of a 4% lidocaine cream [20].

4.5. CONCLUSIONS

The transdermal route offers several advantages, but is currently not feasible for delivery of hydrophilic macromolecules like peptides and proteins. Several techniques, such as chemical penetration enhancers, iontophoresis and sonophoresis have been developed to bypass the stratum corneum barrier of skin and assist in transdermal and topical delivery. Iontophoresis in particular holds promise for the topical delivery of peptides. Protease inhibitors help bypass the enzymatic barrier faced by proteins and peptides. Microporation would appear to be a very effective and promising technique in assisting transdermal delivery of macromolecules and is gaining increasing attention.

REFERENCES

[1] A.C. Williams, Transdermal and Topical Drug Delivery from Theory to Clinical Practice, Pharmaceutical Press, London, UK, 1993, pp. 1–14.

[2] A.K. Banga, Electrically Assisted Transdermal and Topical Drug Delivery, Taylor & Francis, London, UK, 1998, pp. 2–4.

[3] K.A. Walters, M.S. Roberts, The structure and function of skin, in: K.A. Walters (Ed.), Dermatological and Transdermal Formulations, Informa Healthcare USA Inc., New York, 2007, pp. 1–41.

[4] R. Alvarez-Roman, A. Naik, Y.N. Kalia, R.H. Guy, H. Fessi, Skin penetration and distribution of polymeric nanoparticles, J. Control. Release 99 (2004) 53–62.

[5] C. Cullander, R.H. Guy, Sites of Iontophoretic current flow into the skin: Identification and characterization with the vibrating probe electrode, J. Invest. Dermatol. 97 (1991) 55–64.

[6] H.K. Choi, G.L. Flynn, G.L. Amidon, Transdermal delivery of bioactive peptides: The effect of n-decylmethyl sulfoxide, pH, and inhibitors on enkephalin metabolism and transport, Pharm. Res. 7 (1990) 1099–1106.

[7] W.R. Lee, T.L. Pan, P.W. Wang, R.Z. Zhuo, C.M. Huang, J.Y. Fang, Erbium-YAG laser enhances transdermal peptide delivery and skin vaccination, J. Control. Release 128 (2008) 200–208.

[8] A. Naik, Y.N. Kalia, R.H. Guy, Transdermal drug delivery: Overcoming the skins barrier function, Pharmaceut. Sci. Tech. Today 3 (2000) 318–326.

[9] H.A. Benson, Transdermal Drug Delivery: Penetration Enhancement Techniques, Curr. Drug Deliv. 2 (2005) 23–33.

[10] H.A. Benson, S. Namjoshi, Proteins and peptides: Strategies for delivery to and across skin, J. Pharm. Sci. 97 (2008) 3591–3610.

[11] A.K. Banga, Therapeutic Peptides and Proteins: Formulation, Processing, and Delivery Systems, 2nd Ed., CRC Press, Taylor and Francis Group, Boca Raton, USA, 2006.

[12] G.M. El Maghraby, B.W. Barry, A.C. Williams, Liposomes and skin: From drug delivery to model membranes, Eur. J. Pharm. Sci. 34 (2008) 203–232.

[13] M. Foldvari, M.E. Baca-Estrada, Z. He, J. Hu, S. Attah-Poku, M. King, Dermal and transdermal delivery of protein pharmaceuticals: Lipid-based delivery systems for interferon alpha, Biotechnol. Appl. Biochem. 30 (1999) 129–137.

[14] G. Cevc, D. Gebauer, J. Stieber, A. Schätzlein, G. Blume, Ultraflexible vesicles, Transfersomes, have an extremely low pore penetration resistance and transport therapeutic amounts of insulin across the intact mammalian skin, Biochim. Biophys. Acta 1368 (1998) 201–215.

[15] S. Muranishi, A. Sakai, K. Yamada, M. Murakami, K. Takada, Y. Kiso, Lipophilic peptides: Synthesis of lauroyl thyrotropin-releasing hormone and its biological activity, Pharm. Res. 8 (1991) 649–652.

[16] M. Hashimoto, K. Takada, Y. Kiso, S. Muranishi, Synthesis of palmitoyl derivatives of insulin and their biological activities, Pharm. Res. 6 (1989) 171–176.

[17] Y.B. Schuetz, A. Naik, R.H. Guy, Y.N. Kalia, Emerging strategies for the transdermal delivery of peptide and protein drugs, Expert Opin. Drug Deliv. 2 (2005) 533–548.

[18] O. Pillai, R. Panchagnula, Transdermal delivery of insulin from poloxamer gel: *Ex vivo* and *in vivo* skin permeation studies in rat using iontophoresis and chemical enhancers, J. Control. Release 89 (2003) 127–140.

[19] B.M. Magnusson, P. Runn, Effect of penetration enhancers on the permeation of the thyrotropin releasing hormone analogue pGlu-3-methyl-His-Pro amide through human epidermis, Int. J. Pharm. 178 (1999) 149–159.

[20] A.K. Banga, Microporation applications for enhancing drug delivery, Expert Opin. Drug Deliv. 6 (2009) 343–354.

[21] R.K. Sivamani, D. Liepmann, H.I. Maibach, Microneedles and transdermal applications, Expert Opin. Drug Deliv. 4 (2007) 19–25.

[22] S. Henry, D.V. McAllister, M.G. Allen, M.R. Prausnitz, Microfabricated microneedles: A novel approach to transdermal drug delivery, J. Pharm. Sci. 87 (2000) 922–925.

[23] F.J. Verbaan, S.M. Bal, D.J. van den Berg, W.H. Groenink, H. Verpoorten, R. Luttge, et al., Assembled microneedle arrays enhance the transport of compounds varying over a large range of molecular weight across human dermatomed skin, J. Control. Release 117 (2007) 238–245.

[24] H.S. Gill, M.R. Prausnitz, Coated microneedles for transdermal delivery, J. Control. Release 117 (2007) 227–237.

[25] G. Li, A. Badkar, S. Nema, C.S. Kolli, A.K. Banga, *In vitro* transdermal delivery of therapeutic antibodies using maltose microneedles, Int. J. Pharm. 368 (2009) 109–115.

[26] A.V. Badkar, A.M. Smith, J.A. Eppstein, A.K. Banga, Transdermal delivery of interferon alpha-2B using microporation and iontophoresis in hairless rats, Pharm. Res. 24 (2007) 1389–1395.

[27] M. Cormier, B. Johnson, M. Ameri, K. Nyam, L. Libiran, D.D. Zhang, et al., Transdermal delivery of desmopressin using a coated microneedle array patch system, J. Control. Release 97 (2004) 503–511.

[28] X. Chen, T.W. Prow, M.L. Crichton, D.W. Jenkins, M.S. Roberts, I.H. Frazer, et al., Dry-coated microprojection array patches for targeted delivery of immunotherapeutics to the skin, J. Control. Release 139 (2009) 212–220.

[29] P.G. Green, Iontophoretic delivery of peptide drugs, J. Control. Release 41 (1996) 33–48.

[30] A.K. Banga, Theme section: Transdermal delivery of proteins, Pharm. Res. 24 (2007) 1357–1359.

[31] R.H. Guy, Y.N. Kalia, M.B. Delgado-Charro, V. Merino, A. Lopez, D. Marro, Iontophoresis: Electrorepulsion and electro-osmosis, J. Control. Release 64 (2000) 129–132.

[32] R. Panchagnula, O. Pillai, V.B. Nair, P. Ramarao, Transdermal iontophoresis revisited, Curr. Opin. Chem. Biol. 4 (2000) 468–473.

[33] O. Pillai, S.D. Borkute, N. Sivaprasad, R. Panchagnula, Transdermal iontophoresis of insulin. II. Physicochemical considerations, Int. J. Pharm. 254 (2003) 271–280.

[34] N. Kanikkannan, J. Singh, P. Ramarao, Transdermal iontophoretic delivery of bovine insulin and monomeric human insulin analogue, J. Control. Release 59 (1999) 99–105.

[35] S.L. Chang, G.A. Hofmann, L. Zhang, L.J. Deftos, A.K. Banga, Transdermal iontophoretic delivery of salmon calcitonin, Int. J. Pharm. 200 (2000) 107–113.

[36] A. Chaturvedula, D.P. Joshi, C. Anderson, R.L. Morris, W.L. Sembrowich, A.K. Banga, *In vivo* iontophoretic delivery and pharmacokinetics of salmon calcitonin, Int. J. Pharm. 297 (2005) 190–196.

[37] R.R. Boinpally, S.L. Zhou, G. Devraj, P.K. Anne, S. Poondru, B.R. Jasti, Iontophoresis of lecithin vesicles of cyclosporin A, Int. J. Pharm. 274 (2004) 185–190.

[38] B. Vinod, R.P. Nair, Effects of iontophoresis and fatty acids on permeation of Arginine Vasopressin through rat skin, Pharmacol. Res. 47 (2003) 563–569.

[39] M.B. Delgadocharro, A.M. Rodriguezbayon, R.H. Guy, Iontophoresis of nafarelin: Effects of current density and concentration on electrotransport *in vitro*, J. Control. Release 35 (1995) 35–40.

[40] J. Raiman, M. Koljonen, K. Huikko, R. Kostiainen, J. Hirvonen, Delivery and stability of LHRH and Nafarelin in human skin: The effect of constant/pulsed iontophoresis, Eur. J. Pharm. Sci. 21 (2004) 371–377.

[41] A. Jadoul, C. Hanchard, S. Thysman, V. Preat, Quantification and localization of fentanyl and TRH delivered by iontophoresis in the skin, Int. J. Pharm. 120 (1995) 221–228.

[42] S. Katikaneni, G. Li, A. Badkar, A.K. Banga, Transdermal delivery of a ~13 kDa protein – an *in vivo* comparison of physical enhancement methods, J. Drug Target. 18 (2010) 141–147.

[43] S. Mitragotri, D. Blankschtein, R. Langer, Ultrasound-mediated transdermal protein delivery, Science 269 (1995) 850–853.

[44] S. Mitragotri, J. Kost, Low-frequency sonophoresis: A review, Adv. Drug Deliv. Rev. 56 (2004) 569–601.

[45] N.B. Smith, Perspectives on transdermal ultrasound mediated drug delivery, Int. J. Nanomedicine 2 (2007) 585–594.

[46] A. Tezel, S. Mitragotri, Interactions of inertial cavitation bubbles with stratum corneum lipid bilayers during low-frequency sonophoresis, Biophys. J. 85 (2003) 3502–3512.

[47] K.Y. Ng, Enhancing transdermal drug transport with low-frequency ultrasound, Drug Discov. Today 9 (2004) 913.

[48] L. Machet, A. Boucaud, Phonophoresis: Efficiency, mechanisms and skin tolerance, Int. J. Pharm. 243 (2002) 1–15.

[49] L.J. Weimann, J. Wu, Transdermal delivery of poly-l-lysine by sonomacroporation, Ultrasound Med. Biol. 28 (2002) 1173–1180.

[50] S. Mitragotri, J. Kost, Transdermal delivery of heparin and low-molecular weight heparin using low-frequency ultrasound, Pharm. Res. 18 (2001) 1151–1156.

[51] N.B. Smith, S. Lee, K.K. Shung, Ultrasound-mediated transdermal *in vivo* transport of insulin with low-profile cymbal arrays, Ultrasound Med. Biol. 29 (2003) 1205–1210.

[52] A.K. Banga, M.R. Prausnitz, Assessing the potential of skin electroporation for the delivery of protein- and gene-based drugs, Trends Biotechnol. 16 (1998) 408–412.
[53] J.E. Riviere, N.A. Monteiroriviere, R.A. Rogers, D. Bommannan, J.A. Tamada, R.O. Potts, Pulsatile transdermal delivery of LHRH using electroporation: Drug delivery and skin toxicology, J. Control. Release 36 (1995) 229–233.

Chapter 5

Ocular Delivery of Peptides and Proteins

Suresh P. Vyas, Rishi Paliwal and Shivani Rai Paliwal

Drug Delivery Research Laboratory, Department of Pharmaceutical Sciences, Dr. H. S. Gour Vishwavidyalaya, Sagar, M.P., India

Chapter Outline

5.1. INTRODUCTION

Protein and peptide delivery via the ocular route is an interesting but delicate approach for pharmaceutical technology. A good understanding of the physiological parameters of the eye is necessary before designing an ocular product. For example, low drug contact time and poor ocular bioavailability due to the drainage of solution/suspensions, tear production (lacrimation) and turnover and consequent dilution are key problems [1,2]. The main challenge in ocular

Peptide and Protein Delivery. DOI: 10.1016/B978-0-12-384935-9.10005-7

delivery is to circumvent the protective barriers of the eye so that the therapeutic molecule can penetrate into the bio-milieu quantities sufficient to treat ophthalmic diseases or to exert its pharmacological action [3].

Although conventional drug delivery systems such as solutions, suspensions, gels, ointments and inserts have been investigated for controlled ocular delivery, they suffer from problems such as poor drainage of instilled solutions, tear turnover, poor corneal permeability, nasolacrimal drainage, systemic absorption and blurred vision [4]. Advanced drug delivery systems have been developed with the intention of optimizing and controlling delivery of ocular therapeutics to the target sites, either by increasing its penetration across the mucosa or by prolonging the contact time of the carrier with the ocular surface, and have shown promising results [1,5,6].

To date, most of the proteins and peptides that have been delivered to the eye have been for the treatment of local ocular disorders. Although the ocular route is not the preferred route for *systemic* delivery of proteins due to limitations such as the poor permeability through a membrane of these large, hydrophilic molecules, metabolism (enzymatic degradation) and low capacity for transport, some significant efforts have been made in this area. These include the ocular delivery of insulin, and also the use of various nanocarriers for controlled and/or targeted delivery [7,8]. The prodrug approach has also been developed to overcome the poor membrane permeability of peptides. The present chapter deals with aspects related to proteins and peptide delivery to or through the ocular route for treatment of local and systemic disorders.

5.2. THE PHYSIOLOGY OF THE EYE AND CHALLENGES TO THE DELIVERY OF BIOMACROMOLECULES

The human eye, from the anterior to the posterior segment, consists of vitreous humor, the ciliary body, lens, cornea, conjunctiva, aqueous humor, iris, choroid, retina and sclera. The shape of human eye is roughly spherical with a diameter of nearly 23 mm. It has complicated arrays of delicate mechanisms behind its visible portions, which work in concert to transmit an image of the seen object to the brain. The extent and quality of light entering into the eye is regulated and filtered by the pupil, which dilates and contracts as required.

Functionally, the structural components of the eyeball can be divided into three layers:

(i) the outermost coat which comprises the clear, transparent cornea and white, opaque sclera;
(ii) the middle layer which comprises the iris anteriorly, the choroid posteriorly, and the intermediate ciliary body;
(iii) the inner layer which possesses the retina, which is an extension of the central nervous system [9].

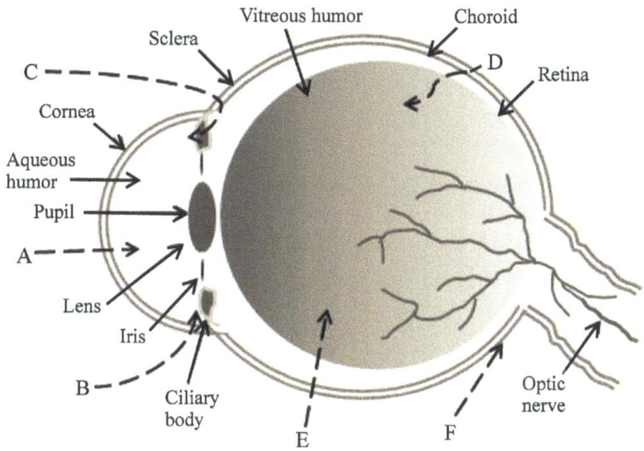

FIGURE 5.1 Structure of the eye showing the different routes of ocular drug delivery; (A) topical administration with trans-corneal permeation; (B) topical administration with non-corneal permeation across the conjunctiva and sclera; (C) drug distribution from the blood through the blood–aqueous barrier into the anterior chamber; (D) drug distribution from the blood through the blood–retina barrier into the posterior chamber; (E) intra-vitreal drug administration route; (F) sub-tenon injection.

The fluid systems, *viz* the aqueous humor and vitreous humor, play an important and decisive role in maintenance of homeostasis of the eye. The cornea, an optically transparent tissue having a diameter of 11.7 mm and thickness 0.5–0.7 mm, performs as the principal refractive element of the eye. Figure 5.1 shows the important components of eye and different routes for ocular drug delivery.

The most common eye disorders include age-related macular degeneration, diabetic macular edema, cataract, proliferative vitreoretinopathy, uveitis, cytomegalovirus and glaucoma [10]. The corneal epithelium is the main barrier to drug absorption into eye. The stratified corneal epithelium acts as a protective barrier against invasion of foreign molecules and also as a barrier to ion transport. The corneal epithelium consists of a basal layer of columnar cells, two to three layers of wing cells and one or two outermost layers of squamous, polygonal shaped, superficial cells. In the healthy corneal epithelium, intercellular tight junctions (zonula occludens) completely surround the most superficial cells, which restrict the passage of peptides and proteins. Absorption therefore relies on transcellular passage or strategies which can modulate (e.g. transiently open) the tight junctions. In contrast, the intercellular spaces between the wing cells and basal cells are wider and permit the paracellular diffusion of large molecules. The paracellular route would naturally be favored since this type of transport may occur irrespective of the charge or size of the peptide (to a certain limit). In general, the negatively charge membrane of the corneal epithelium offers greater resistance to negatively charged compounds than it does to positively charged ones.

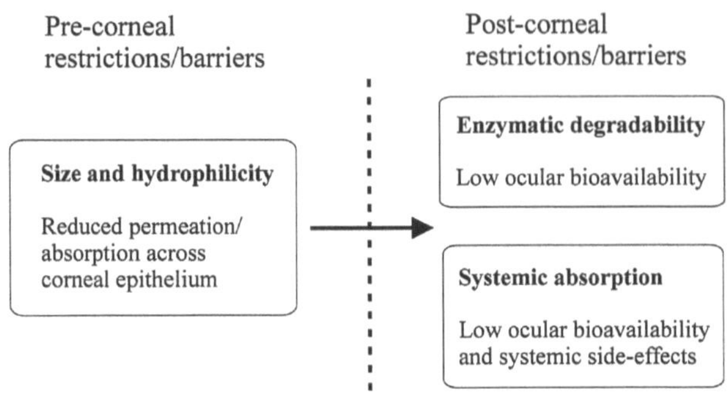

FIGURE 5.2 Schematic presentation of the different barriers for ocular delivery of proteins and peptides.

Ocular enzymes, specifically peptidases, effectively form an "enzymatic barrier" which effectively limits the ocular absorption of peptides, due to their metabolism and clearance. Some endopeptidases, such as plasmin and collagenase, exopeptidases and aminopeptidases, reside in the ocular fluids and tissues [11]. In general, the level of endopeptidase is low unless the eye is inflamed or injured, but the level is of little concern relative to the stability of topically applied doses of peptides. For example, it was reported that about 90% of leucine enkephalin, and almost 100% of methionine enkephalin (pentapeptides) were recovered in the rabbit corneal epithelium in hydrolyzed forms within 5 minutes of instillation [12]. It has therefore been suggested that aminopeptidase activity must be inhibited in order to increase ocular peptide absorption.

Three important barriers affect the ocular pharmacokinetics of locally delivered drug molecules:

(i) loss of drug from the ocular surface;
(ii) the lacrimal–fluid barrier; and
(iii) the blood–ocular barrier [13].

Figure 5.2 outlines some of the barriers in ocular delivery of proteins.

5.3. CONSIDERATIONS AND OPPORTUNITIES FOR PROTEIN/PEPTIDE OCULAR DELIVERY

The topical delivery of therapeutic molecules is considered to be the best option for treatment of most ocular disorders [4], but in some cases systemic delivery of molecules has also been demonstrated. Thus, a factor which must be taken into consideration during ocular absorption of topically applied peptides and proteins is their loss to the systemic circulation [14]. This can occur as a result of contact of the instilled solution with the conjunctival and the nasal mucosae, both of which have been shown to participate in the systemic absorption of ocularly applied drugs. Several peptides and proteins, notably cyclosporine, various growth factors, interferons and interleukins, have already been investigated as potential therapeutic agents in uveitis, wound healing, herpes simplex infections [15], and the induction of local immune responses in response to foreign agents. The possibility that ocularly applied peptides are subsequently absorbed into the systemic circulation and elicit various side-effects must be addressed, since this is another complicating factor in ocular peptide delivery for local eye treatment.

5.3.1. Treatment of Local Eye Disorders

Controlled local delivery of proteins and peptides to ocular sites can be achieved by suitable manipulation of their physicochemical properties without affecting their biological activity. Several peptides have been demonstrated as therapeutic agents in a number of ocular disorders including dry eye disease, age related macular degeneration or proliferative diabetic retinopathy, etc. Table 5.1 lists common disorders and the respective therapeutic peptides which can be delivered through ocular routes. However, adverse physicochemical properties and enzymatic degradation of these peptides within the ocular environment, as discussed above, may render them less effective. Sustained release of peptides, after loading them on a carrier system such as a liposome or biodegradable nanoparticle may limit some of these problems. A further problem in peptide delivery to posterior ocular sites is the impermeability of the eye to even small molecules which restricts the complete transport of therapeutic molecules to the desired site. Table 5.2 summarizes the reported literature related to the ocular delivery of proteins and peptides.

5.3.2. Systemic Absorption of Proteins and Peptides via the Ocular Route

Putative advantages of the ocular route include: the delivery of precise doses of peptide or protein; the relative ease and low cost of formulating and administering eye drops (compared to injection); the relatively rapid rate of systemic absorption (compared to oral delivery for example); the relative insensitivity of the eye tissues towards immunological reactions compared to other tissues such as the lung and gut; the absence of first pass metabolism through the hepatic circulation as occurs for oral delivery platforms; an apparently good tolerance without (ocular) side-effects, exemplified by a report following long-term (three months) daily administration of the peptide hormone insulin [16].

Systemic delivery of insulin via the ocular route is especially challenging with regard to the requirement for reproducible delivery. The formulation of insulin eye drops tends to result in low bioavailability, even for viscous aqueous solutions, oily solutions and emulsions, because they can be easily

TABLE 5.1 List of disorders/indications where therapeutic peptides could be delivered through ocular route

Disorder/Indication	Therapeutic peptide
Antiallergic, antiinflammatory	ACTH
Analgesic	β Endorphin, Leu-enkephalin
Antiscarring agent in glaucoma filtration surgery	Integrin-binding peptide
Attenuate miotic response	Somatostatin
Choroidal or retinal neovascularization	Octreotide, Urokinase derived peptide, Cyclic integrin-binding peptide
Corneal epithelial wound	Insulin-like growth factor derived peptide Substance P derived peptide
Diabetes mellitus	Insulin
Diabetes insipidus	Vasopressin
Diagnosis of thyroid cancer	TSH
Dry eye disease	Cyclosporine A
Hypoglycemic crisis	Glucagon
Immunostimulant	Met-enkephalin
Induction of uterine contractions	Oxytocin
Induction of vitreous detachment in vitretomy	Integrin-binding peptide
Paget's disease	Calcitonin
Secretion of insulin	VIP
Uveal melanoma and retinal blastoma	Apoptosis inducing peptide

drained from the eye by the lachrymal system. Chiou *et al.* administered insulin-containing eye drops to rabbits [17]. Although only a small hypoglycemic response to the insulin eye drops could be observed when no surfactant was added to the formulation, a dose-dependent decrease in blood glucose values was observed when saponin (1%) was added to the eye drop solution. These results demonstrated that the ocular tissue was not impermeable to insulin, and that the amount of insulin taken up by the rabbit eye was sufficient to cause a significant decrease in blood glucose. Both of these observations support the hypothesis that human diabetic patients could

TABLE 5.2 Reported literature related to ocular delivery of proteins and peptides

Protein/ peptide	Delivery strategies	Concluding remarks	Ref.
Insulin	Penetration enhancer	The insulin bioavailability was 5.7 to 12.6% with polyoxyethylene-9-lauryl ether, 4.9 to 7.9% with GC, 3.6 to 7.8% with Na taurocholate and 8.2 to 8.3% with Na deoxycholate, as compared to 0.7 to 1.3% in the absence of absorption promoters.	[18]
Cyclosporine A	Azone penetration enhancer	Cyclosporine-treated grafts contained significantly fewer infiltrating T-lymphocytes than did the drug/solvent-treated allografts, indicating that the topical application of cyclosporine actively inhibited the entry of T-cells into the grafts.	[38]
IgG protein	Transscleral delivery	IgG protein delivered to the retina and choroid in an optimum concentration for the treatment of chorio-retinal disorders with negligible systemic absorption.	[43]
Vancomycin (peptide)	PLGA microparticles	PLGA microparticles loaded with peptide drug showed high and prolonged concentration of vancomycin and increased level of AUC (2-fold) as compared to aqueous solutions.	[44]
Ganciclovir (GCV)	Prodrug	Glycine-valine-GCV is the effective and lead candidate for the treatment of Human Cytomegalovirus (HCMV).	[45]
Vasoactive intestinal peptide (VIP)	Liposome	Treatment of ocular inflammation by modulation of macrophage and T-cell activation of the immune system.	[46]
VIP	Liposomes	For the treatment of endotoxin induced uveitis (EIU), liposomal delivery increased VIP efficiency and bioavailability.	[47]
Ganciclovir (GCV)	Prodrug	Diester GCV prodrugs demonstrated excellent chemical stability, high aqueous solubility and markedly enhanced antiviral potency against the herpes viruses without any increase in cytotoxicity.	[48]

someday be treated with insulin-containing eye drops. There are, of course, other criteria that must be met before clinical application can be realized:

(i) in particular, the assessment of any potential ocular toxicity of insulin eye drops when administered several times daily;
(ii) demonstration in humans that the ocular route will permit enough insulin to be absorbed to significantly lower blood glucose levels;
(iii) demonstration in humans that insulin absorption from eye drops will be sufficiently reproducible and reliable to allow precise calculation of appropriate dosage size and concentration.

Yamamoto *et al.* also reported that eye drops can deliver insulin in rabbits and further suggested that an ocular insert would be another feasible approach to prolong and thus enhance the ocular delivery of insulin [18]. An ocular insert for the delivery of insulin using Gelform® as a drug carrier to deliver insulin with the aid of Brij-78 as an absorption enhancer has also been demonstrated [19]. The conjunctival cul-de-sac has also been targeted as a potential route for insulin delivery [20]. Permeation enhancers such as BL-9, Brij-78 and alkyl-polysaccharides have been found to be safe, and to stimulate the systemic absorption of insulin. These approaches will be discussed in the following sections.

5.4. PEPTIDE TRANSPORT SYSTEMS IN THE EYE

Epithelial cells express numerous nutrient transporters and receptors on their membrane surface which aid in the movement of various vitamins and amino acids across the cell membrane [21], and the corneal epithelium is no exception. Peptide transporter systems in the eye have gained attention in recent years as potentially being useful in targeted ocular drug delivery. These proton coupled transporters help in the translocation of di- and tripeptides across the epithelium [22]. The transporters are mainly classified into PepT1, PepT2 and peptide/histidine transporters (PHT1 and PHT2), and many drug molecules are known to be substrates for these transporters [23]. The presence of an oligopeptide transporter on rabbit cornea has been confirmed. Other peptide-derived drugs including the β-lactam antibiotics, renin inhibitors and angiotensin converting enzyme (ACE) inhibitors are known to be substrates for PepT1 and PepT2. The expression of PHT1 in bovine and human retinal pigment epithelium cells (BRPE and HRPE cells, respectively), ARPE-19 cells (a human RPE cell line), and bovine and human neural retina cells has been demonstrated [24]. However, whereas PepT2 and PHT2 expression was reported in bovine and human retina, PepT1 was not detected. It was also concluded that glycylsarcosine uptake studies did not demonstrate any significant functional activity of PHT1 on plasma membranes of RPE.

The mechanism of model dipeptide (glycylsarcosine) transport across the blood–ocular barriers following systemic administration has been investigated, including carrier mediated uptake of glycylsarcosine across the blood–ocular barrier, and a dependence on time and concentration was discovered [25]. Peptide prodrugs such as valine-ACV (where ACV = 6-(L-alpha-aminoadipy1)-L-cysteinyl-D-valine) and valine-valine-ACV exhibited higher concentrations of ACV in the aqueous humor following systemic administration, as compared to the parent drug [26]. This confirmed that peptide prodrugs can be taken up from the systemic circulation into the eye via carrier mediated transport mechanisms. It can be concluded that drugs with poor ocular bioavailability can be suitably modified by rational design so that they can be recognized and taken up by peptide transporters for enhanced ocular bioavailability.

5.5. STRATEGIES FOR THE OCULAR DELIVERY OF PROTEINS AND PEPTIDES

5.5.1. Prodrug Approach

In contrast to the uptake of peptide prodrugs from the systemic circulation into the eye, discussed above, the same strategy (the synthesis of prodrugs) can be used to change the physicochemical properties of a drug in order to improve its permeation across the cornea, hence enhancing ocular bioavailability by a different mechanism (cf. Chapter 4, page 74). The concept of prodrugs was first introduced to ophthalmology about 30 years ago with the introduction of dipivefrin, which is a prodrug of epinephrine used to treat glaucoma [27]. Since then, numerous other prodrugs have been designed to improve the ophthalmic bioavailability of various drug molecules, prolong their duration of action, improve their formulation properties or reduce systemic side-effects. It is expected that drugs developed for ophthalmic use should have good chemical stability to allow formulation into topical ready-to-use aqueous eye drops, which is the most practical and commonly used administration route for treatment of ophthalmic diseases. This is often a challenge for the development of ophthalmic prodrugs that are intended to be rapidly converted to the active drug after absorption. Therefore, only those prodrugs that show good chemical stability combined with a sufficiently high enzymatic lability can be easily developed without resorting to multi-vial reconstitutable products.

The major goal in designing prodrugs is to overcome various physicochemical, biopharmaceutical and/or pharmacokinetic problems that may be associated with the parent drug molecules, which would otherwise limit their clinical use. The most common barriers in ophthalmic drug formulation and delivery that may be overcome by a prodrug are:

(i) a low aqueous solubility, which prevents the development of aqueous eyedrops;
(ii) a low lipid solubility, which results in low corneal permeation and low ophthalmic bioavailability;

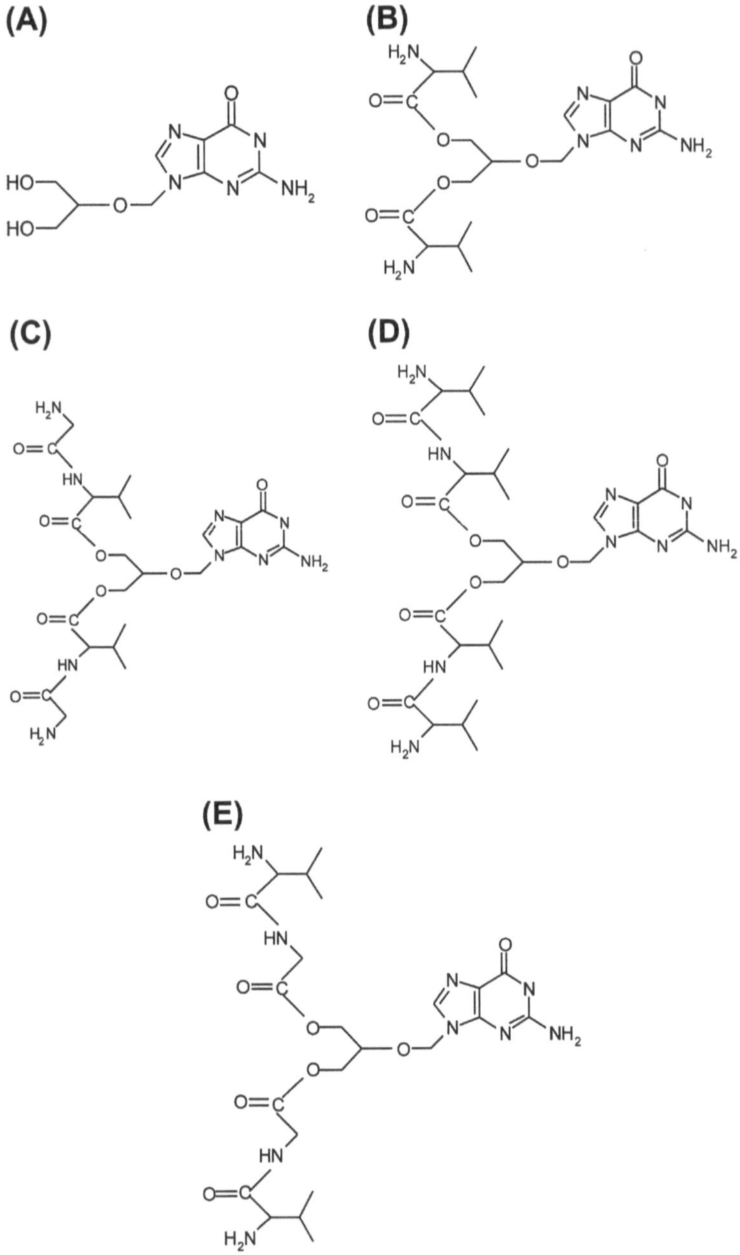

FIGURE 5.3 Chemical structure of peptides and their chemical modifications that have been used for the treatment of ocular disorders: (A) Ganciclovir; (B) Val-GCV diester; (C) Gly-Val-GCV diester; (D) Val-Val-GCV diester; (E) Val-Gly-GCV diester.

(iii) a short duration of action due to rapid drug elimination from the site of action, which necessitates frequent administration and often leads to poor patient compliance;
(iv) systemic side-effects, due to low corneal and high systemic absorption, which may lead to safety concerns and also poor patient compliance.

Figure 5.3 shows the chemical structures of peptide prodrugs developed for ocular delivery.

5.5.2. Mucoadhesive Particulate Carriers

The cornea and conjunctiva have a net negative charge and hence mucoadhesive cationic polymers might interact intimately with these extraocular structures, putatively increasing the concentration and residence time of polymer-associated drug. Among the mucoadhesive polymers, chitosan has attracted a great deal of attention because of its unique properties – such as acceptable biocompatibility, a biodegradable backbone and an ability to enhance the paracellular transport of drugs – possibly through a transient loosening of the tight junctions [28,29]. The potential of chitosan nanoparticles as a new vehicle for the improvement of the delivery of the hydrophobic, cyclic peptide cyclosporine A (CsA) to the ocular mucosa has been investigated by De Campos *et al.* [30]. They concluded that, following topical instillation of CsA-loaded chitosan nanoparticles to rabbits, it was possible to achieve therapeutic concentrations in external ocular tissues (i.e. the cornea and conjunctiva) for at least 48 h, while maintaining negligible or undetectable CsA levels in the inner ocular structures (i.e. the iris, ciliary body and aqueous humour), blood and plasma.

5.5.3. Engineered Nanoconstructs

Nanocarriers, which are now widely accepted for controlled and targeted drug delivery, are emerging as a new platform for the ocular delivery of proteins and peptides. Nanocarriers also offer selective targeting along with sustained release of molecules at the desired site. Liposomes, niosomes, biodegradable nanoparticles, solid-lipid nanoparticles, dendrimers, etc., are some of the examples [31]. These nanoconstructs provide protection to encapsulated peptide drugs from enzymatic degradation, and also from loss due to tear turnover by maintaining a sustained drug release over longer periods of time. In addition, mucoadhesive polymers, if incorporated with the drug in the nanocarrier complex, will also allow the nanocarriers to adhere to the corneal epithelium (*cf.* the discussion of chitosan in the above section).

The effectiveness of liposomes in aiding the ocular absorption of entrapped insulin in normal rabbits has been reported [31]. Administration of insulin entrapped in positively charged liposomes to normal rabbits produced a substantial reduction in blood glucose concentration 90 to 120 min after the administration of the formulation. Pleyer *et al.* investigated the ability of liposomes to deliver the immunosuppressive agent cyclosporine A (CsA) to the cornea, anterior sclera, aqueous and vitreous humor in rabbit eyes [31a]. They tested liposome-encapsulated CsA (CsA-LIP) or olive oil drops containing an equivalent concentration of CsA (CsA-DR) against "collagen shields" soaked for 30 min in the liposome preparation (CsA-LIP-CS), both *in vitro* and *in vivo*. CsA-CS-LIP yielded significantly higher levels of CsA in the aqueous and vitreous humor and in sclera compared to CsA-DR, evincing the requirement for the nanocarrier system.

CsA-loaded, solid, lipid nanoparticles (SLNs) for topical ophthalmic applications have also been investigated [32]. SLNs were prepared by using a high shear homogenization and ultrasound method, with Compritol 888 ATO (a wax for hot-melt coating and prolonged-release), Poloxamer 188 (a PEO–PPO–PEO triblock copolymer) and Tween 80 (polyethylene glycol sorbitan monooleate). These SLNs were then investigated for cellular uptake into rabbit corneal epithelial (RCE) cells and evaluated for potential cytotoxicity. CsA release from the SLNs was found to be enzyme (lipase/co-lipase complex) dependent. In the subsequent studies, it was observed that the topical ophthalmic efficacy of CsA was enhanced remarkably via administration of SLNs with a particle size of 225.9 ± 5.5 nm and a negative surface charge [33]. An aqueous humor drug level of up to 50.53 ng/ml was achieved without any serious irritation in the rabbit eye.

Similarly, CsA levels in ocular tissues and fluids after topical administration of poly-epsilon-caprolactone (PCL)/benzalkonium chloride (BKC) nanospheres, and hyaluronic acid (HA) coated PCL/BKC nanospheres into healthy rabbit corneas has been reported [34]. The CsA loaded PCL/BKC and HA coated PCL/BKC nanospheres were found to achieve high levels of CsA in the cornea, 10- to 15-fold higher than could be achieved with CsA that had been solubilized in castor oil. It can be concluded that the nanosphere formulation and HA coating both played an important role in delivering high levels of CsA into the cornea.

A novel formulation of vasoactive intestinal peptide (VIP) based on the incorporation of VIP-loaded rhodamine-conjugated liposomes (VIP-Rh-Lip) within an HA gel, for the treatment of endotoxin-induced uveitis (EIU), has been reported [35]. It was observed that interactions between the HA chains and liposomes resulted in an increased viscosity and reinforced elasticity of the gel. Retention of the liposomes by the HA gel was confirmed by *in vitro* and *in vivo* studies. It was further noted that the severity of the inflammatory response profoundly influenced the stability of the liposomal system, thereby resulting in the delayed release of VIP, which is desired for the treatment of uveitis. Hence, it was concluded that the HA-gel-containing VIP-Rh-Lip served as an efficient strategy for the sustained delivery of VIP in both the ocular and local lymph node tissues for better immunosupressor activity of VIP.

5.5.4. Penetration Enhancers

Penetration enhancers promote the penetration of drugs through the corneal barrier, and change the integrity of the epithelial cell layer. Frequently used penetration enhancers in ocular formulations include cyclodextrin, dimethylsulphoxide (DMSO), ethylenediaminetetraacetic acid (EDTA), sodium glycocholate and related cholates, Tween 20 (a non-ionic polysorbate surfactant), Brij 35 (polyoxyethylene lauryl ether), saponins and bile salts [36]. Generally, penetration enhancers such as EDTA and cholates transiently loosen the tight junctions between adjacent cells of the corneal epithelium. Thus, penetration enhancers, when applied topically to the eye, have been successfully applied to the delivery of protein and peptides through the corneal epithelium [37].

Several studies have reported the improved delivery of peptides and proteins into the systemic circulation via the ocular route [17,38,39]. Yamamoto *et al.* studied the effect of several penetration enhancers on the bioavailability of insulin administered via the ocular route [18]. They concluded that improved insulin delivery was achieved by enhancers, in the order of polyoxyethylene-9-lauryl ether $>$ sodium deoxycholate $>$ sodium glycocholate $\approx$ sodium taurocholate. Similarly, Newton *et al.* [38] reported that when cyclosporine was topically applied in conjunction with the novel cutaneous penetration enhancer, Azone® (1-dodecylazacycloheptan-2-one) to allografted rabbit eyes, a clinically significant concentration of cyclosporine could be measured in the treated corneas, but little or no cyclosporine could be found in the aqueous humor or blood. Hence, it was proposed that cyclosporine delivered in conjunction with Azone may result in suppression in the severity and incidence of graft rejection. However, the use of a penetration enhancer is always associated with the risk of toxicity, which becomes more prominent in case of the eye [40]. Therefore, toxicological parameters should be thoroughly addressed before selection of a penetration enhancer for ophthalmological products.

5.5.5. Miscellaneous Approaches

Apart from the strategies discussed above, other vehicle or delivery platforms for ocular delivery of proteins and peptides have been reported in the literature. For example, the efficacy of the TAT (the trans-activating transcription factor from the human immunodeficiency virus) protein transduction domain (residues 48–60) has been investigated as a carrier by which to deliver acidic fibroblast growth factor (aFGF) to the rat retina [41]. The TAT-conjugated aFGF-His peptide (TAT-aFGF-His) exhibited efficient penetration to the retina after topical administration. Immunohistochemical staining with anti-His antibody revealed that TAT-aFGF-His proteins were readily found in the retina, mainly in the ganglion cell layer (GCL), after 30 min and remained detectable for at least 8 h after administration. In contrast, His-positive proteins were undetectable in the retina after topical administration of aFGF-His, indicating that aFGF-His cannot penetrate the ocular barrier alone. Hence, it can be concluded that conjugation of TAT to aFGF-His can markedly improve the ability of aFGF-His to penetrate the ocular barrier without impairing its biological function for combating retinal disease. The reader should also cross reference with Chapter 9, which discusses in detail the use of TAT-peptide-mediated peptide/protein delivery.

Intravitreal injections can cause several ocular complications, including vitreous hemorrhage, endophthalmitis, retinal detachment and cataract, and, clearly, repeated injections can multiply the risk of these complications. Bevacizumab (a recombinant antibody) is used for the treatment of several different ocular diseases, but is delivered by intravitreal injection. In order to improve and prolong its ocular bioavailability after intravitreal administration, liposomal Bevacizumab, as a novel drug delivery system, has been described and compared with conventional formulations on the market [42]. The mean concentration of free Bevacizumab in the vitreous humor of eyes that received liposomal Bevacizumab was compared with eyes injected with soluble Bevacizumab, and shown to be nearly 2-fold (48 versus 28 μg/ml) and 5-fold (16 versus 3.3 μg/ml) higher at days 28 and 42, respectively. In contrast, the mean concentration of free Bevacizumab in the aqueous humor of both injected eyes (i.e. for the liposomal and soluble formulations) was almost equivalent at the same time intervals. Hence, the liposomal formulation provided a beneficial effect in prolonging the residency of Bevacizumab in the vitreous humor.

5.6. FUTURE PERSPECTIVES

Emerging trends in the development of ocular biotechnology products, especially for proteins and peptides, include the design of more specific delivery strategies intended to achieve therapeutic responses with minimal doses and controlled ocular pharmacokinetics. Prodrug design, the use of penetration enhancers and novel nanoparticulate delivery systems are some of the approaches that have been established more recently with several peptide molecules. However, a successful marketed product based on the above discussed strategies has yet to emerge, and will require a thorough assessment of issues regarding the potential toxicology of the delivery system.

REFERENCES

[1] S. Wadhwa, R. Paliwal, S.R. Paliwal, S.P. Vyas, Nanocarriers in ocular drug delivery: An update review, Curr. Pharm. Des. 15 (2009) 2724–2750.
[2] V.H.L. Lee, Ophthalmic delivery of peptides and proteins, Pharm. Technol. 11 (1987) 26–38.
[3] J.W. Shell, Ophthalmic drug delivery systems, Surv. Ophthalmol. 29 (1984) 117–128.
[4] V.H. Lee, J.R. Robinson, Topical ocular drug delivery: Recent developments and future challenges, J. Ocul. Pharmacol. 2 (1986) 67–108.

[5] S.P. Vyas, N. Mysore, V. Jaitely, N. Venkatesan, Discoidal niosome based controlled ocular delivery of timolol maleate, Pharmazie 53 (1998) 466–469.
[6] S.P. Vyas, S. Ramchandraiah, C.P. Jain, S.K. Jain, Polymeric pseudolatices bearing pilocarpine for controlled ocular delivery, J. Microencapsul. 9 (1992) 347–355.
[7] S.K. Sahoo, F. Dilnawaz, S. Krishnakumar, Nanotechnology in ocular drug delivery, Drug Discov. Today 13 (2008) 144–151.
[8] J. Vandervoort, A. Ludwig, Ocular drug delivery: Nanomedicine applications, Nanomedicine 2 (2007) 11–21.
[9] C.W. Oyster, The Human Eye: Structure and Function, Sinauer Associates, Sunderland, MA, USA, 1999.
[10] G.H. Cassel, M.D. Billig, H.G. Randall, M.F. Goldberg, The Eye Book: A Complete Guide to Eye Disorders and Health, Johns Hopkins Press, Maryland, USA, 1998.
[11] R.E. Stratford, V.H.L. Lee, Ocular aminopeptidase activity and distribution in the albino rabbit, Curr. Eye Res. 4 (1995) 995–1000.
[12] R.J. Erb, L.M. Her, A. Abdallah, A.K. Mitra, Pharmacodynamics and biophasic drug levels of methionine enkephalin, Pharm. Res. 8 (1991) 930–932.
[13] S. Duvvuri, S. Majumdar, A.K. Mitra, Drug delivery to the retina: Challenges and opportunities, Expert Opin. Biol. Ther. 3 (2003) 45–56.
[14] R.D. Schoenwald, Ocular drug delivery. Pharmacokinetic considerations, Clin. Pharmacokinet. 18 (1990) 255–269.
[15] G.C. Chiou, Systemic delivery of polypeptide drugs through ocular route, J. Ocul. Pharmacol. 10 (1994) 93–99.
[16] V.H. Lee, New directions in the optimization of ocular drug delivery, J. Ocul. Pharmacol. 6 (1990) 157–164.
[17] G.C. Chiou, Y.C. Ching, Improvement of systemic absorption of insulin through the eyes with absorption enhancers, J. Pharm. Sci. 78 (1989) 815–818.
[18] A. Yamamoto, A.M. Luo, S. Dodda-Kashi, V.H. Lee, The ocular route for systemic insulin delivery in the albino rabbit, J. Pharmacol. Exp. Ther. 249 (1989) 249–255.
[19] Y.C. Lee, P. Simamora, S.H. Yalkowsky, Effect of Brij-78 on systemic delivery of insulin from an ocular device, J. Pharm. Sci. 86 (1997) 430–433.
[20] E. Hayakawa, D.S. Chien, K. Inagaki, A. Yamamoto, W. Wang, V.H. Lee, Conjunctival penetration of insulin and peptide drugs in the albino rabbit, Pharm. Res. 9 (1992) 769–775.
[21] R. Gaudana, J. Jwala, S.H. Boddu, A.K. Mitra, Recent perspectives in ocular drug delivery, Pharm. Res. 26 (2009) 1197–1216.
[22] V. Ganapathyand, F.H. Leibach, Peptide transport in intestinal and renal brush border membrane vesicles, Life Sci. 30 (1982) 2137–2146.
[23] R.S. Talluri, S.K. Samanta, R. Gaudana, A.K. Mitra, Synthesis, metabolism and cellular permeability of enzymatically stable dipeptide prodrugs of acyclovir, Int. J. Pharm. 361 (2008) 118–124.
[24] S.M. Ocheltree, R.F. Keep, H. Shen, D. Yang, B.A. Hughes, D.E. Smith, Preliminary investigation into the expression of proton-coupled oligopeptide transporters in neural retina and retinal pigment epithelium (RPE): Lack of functional activity in RPE plasma membranes, Pharm. Res. 20 (2003) 1364–1372.
[25] H. Atluri, B.S. Anand, J. Patel, A.K. Mitra, Mechanism of a model dipeptide transport across blood–ocular barriers following systemic administration, Exp. Eye Res. 78 (2004) 815–822.
[26] C. Dias, Y. Nashed, H. Atluri, A. Mitra, Ocular penetration of acyclovir and its peptide prodrugs valacyclovir and val-valacyclovir following systemic administration in rabbits: An evaluation using ocular microdialysis and lc-ms, Curr. Eye Res. 25 (2002) 243–252.

[27] A. Hussain, J.E. Truelove, Prodrug approaches to enhancement of physicochemical properties of drugs IV: Novel epinephrine prodrug, J. Pharm. Sci. 65 (1976) 1510–1512.

[28] S. Wadhwa, R. Paliwal, S.R. Paliwal, S.P. Vyas, Chitosan and its role in ocular therapeutics, Mini Rev. Med. Chem. 9 (2009) 1639–1647.

[29] S. Wadhwa, R. Paliwal, S.R. Paliwal, S.P. Vyas, Hyaluronic acid modified chitosan nanoparticles for effective management of glaucoma: Development, characterization, and evaluation, J. Drug Target. 18 (2010) 292–302.

[30] A.M. De Campos, A. Sánchez, M.J. Alonso, Chitosan nanoparticles: A new vehicle for the improvement of the delivery of drugs to the ocular surface. Application to cyclosporin A, Int. J. Pharm. 224 (2001) 159–168.

[31] R. Srinivasan, S.K. Jain, Insulin delivery through the ocular route, Drug Deliv. 5 (1998) 53–55.

[31a] U. Pleyer, B. Elkins, D. Rückert, S. Lutz, J. Grammer, J. Chou, K.H. Schmidt, B.J. Mondino, Ocular absorption of cyclosporine A from liposomes incorporated into collagen shields, Curr. Eye Res. 13 (1994) 177–181.

[32] E.H. Gokce, G. Sandri, M.C. Bonferoni, S. Rossi, F. Ferrari, T. Güneri, et al., Cyclosporine A loaded SLNs: Evaluation of cellular uptake and corneal cytotoxicity, Int. J. Pharm. 364 (2008) 76–86.

[33] E.H. Gökçe, G. Sandri, S. Eğrilmez, M.C. Bonferoni, T. Güneri, C. Caramella, Cyclosporine a-loaded solid lipid nanoparticles: Ocular tolerance and *in vivo* drug release in rabbit eyes, Curr. Eye Res. 34 (2009) 996–1003.

[34] I. Yenice, M.C. Mocan, E. Palaska, A. Bochot, E. Bilensoy, I. Vural, et al., Hyaluronic acid coated poly-epsilon-caprolactone nanospheres deliver high concentrations of cyclosporine A into the cornea, Exp. Eye Res. 87 (2008) 162–167.

[35] L. Lajavardi, S. Camelo, F. Agnely, W. Luo, B. Goldenberg, M.C. Naud, et al., New formulation of vasoactive intestinal peptide using liposomes in hyaluronic acid gel for uveitis, J. Control. Release 139 (2009) 22–30.

[36] S.P. Vyas, R.K. Khar, Ocular drug delivery systems, in: Controlled Drug Delivery: Concepts and Advances, Vallabh Prakashan, Delhi, India, 2008.

[37] V.H.L. Lee, L.W. Carson, K.A. Takemoto, Macromolecular drug absorption in the albino rabbit eye, Int. J. Pharm. 29 (1986) 43–51.

[38] C. Newton, B.M. Gebhardt, H.E. Kaufman, Topically applied cyclosporine in zone prolongs corneal allograft survival, Invest. Ophthalmol. Vis. Sci. 29 (1988) 208–215.

[39] K. Morimoto, T. Nakai, K. Morisaka, Evaluation of permeability enhancement of hydrophilic compounds and macromolecular compounds by bile salts through rabbit corneas *in vitro*, J. Pharm. Pharmacol. 39 (1987) 124–126.

[40] P. Chetoni, S. Burgalassi, D. Monti, M.F. Saettone, Ocular toxicity of some corneal penetration enhancers evaluated by electrophysiology measurements on isolated rabbit corneas, Toxicol. Vitro 17 (2003) 497–504.

[41] Y. Wang, H. Lin, S. Lin, J. Qu, J. Xiao, Y. Huang, et al., Cell-penetrating peptide TAT-mediated delivery of acidic FGF to retina and protection against ischemia-reperfusion injury in rats, J. Cell. Mol. Med. 14 (2010) 1998–2005.

[42] M. Abrishami, S.Z. Ganavati, D. Soroush, M. Rouhbakhsh, M.R. Jaafari, B. Malaekeh-Nikouei, Preparation, characterization, and *in vivo* evaluation of nanoliposomes-encapsulated bevacizumab (avastin) for intravitreal administration, Retin.-J. Retin. Vitr. Dis. 29 (2009) 699–703.

[43] J. Ambati, E.S. Gragoudas, J.W. Miller, T.T. You, K. Miyamoto, F.C. Delori, et al., Transscleral delivery of bioactive protein to the choroid and retina, Invest. Ophthalmol. Vis. Sci. 41 (2000) 1186–1191.

[44] E. Gavini, P. Chetoni, M. Cossu, M.G. Alvarez, M.F. Saettone, P. Giunchedi, PLGA microspheres for the ocular delivery of a peptide drug, vancomycin using emulsification/spray-drying as the preparation method: *In vitro/in vivo* studies, Eur. J. Pharm. Biopharm. 57 (2004) 207–212.

[45] S. Majumdar, V. Kansara, A.K. Mitra, Vitreal pharmacokinetics of dipeptide monoester prodrugs of ganciclovir, J. Ocul. Pharmacol. Ther. 22 (2006) 231–241.

[46] L. Lajavardi, A. Bochot, S. Camelo, B. Goldenberg, M.C. Naud, F. Behar-Cohen, et al., Downregulation of endotoxin-induced uveitis by intravitreal injection of vasoactive intestinal Peptide encapsulated in liposomes, Invest. Ophthalmol. Vis. Sci. 48 (2007) 3230–3238.

[47] S. Camelo, L. Lajavardi, A. Bochot, B. Goldenberg, M.C. Naud, E. Fattal, et al., Ocular and systemic bio-distribution of rhodamine-conjugated liposomes loaded with VIP injected into the vitreous of Lewis rats, Mol. Vis. 13 (2007) 2263–2274.

[48] K. Patel, S. Trivedi, S. Luo, X. Zhu, D. Pal, E.R. Kern, et al., Synthesis, physicochemical properties and antiviral activities of ester prodrugs of ganciclovir, Int. J. Pharm. 305 (2005) 75–89.

Chapter 6

Brain Delivery of Peptides and Proteins

Christine Dufès

Strathclyde Institute of Pharmacy and Biomedical Sciences, University of Strathclyde, 161 Cathedral Street, Glasgow G4 0RE, United Kingdom.

Chapter Outline

6.1. INTRODUCTION

The last three decades have brought considerable advances in our understanding of the mechanisms involved in the development of brain diseases. As a result, numerous therapeutic proteins and peptides have been identified as showing great promise for the treatment of brain cancer and neurodegenerative diseases. However, their delivery to their intended site of action from the blood is hampered by the presence of the blood–brain barrier (BBB), a vital

Peptide and Protein Delivery. DOI: 10.1016/B978-0-12-384935-9.10006-9

element in the regulation of the internal environment of the brain and the spinal cord. In order to overcome this problem, different approaches have been developed to enable drug delivery to the site of action following systemic administration. This chapter combines an overview of the major CNS therapeutic peptides, highlighting their general characteristics and potential, with a description of the peptide transport to the CNS, highlighting the structural and physiological features of the BBB. It will then review the various strategies that are applicable for enhancing the delivery of protein/peptide delivery to the CNS.

6.2. PEPTIDES AS DRUGS FOR THE TREATMENT OF CNS RELATED DISORDERS

A wide variety of peptides and proteins have been identified as promising therapeutic agents for the treatment of various brain pathologies (for review see [1,2]). Peptides and proteins are present in the whole nervous system, with unique distribution patterns. They can exert numerous biological actions in the brain, such as regulation of the internal environment of the brain and of the cerebral blood flow, modulation of the permeability of the blood–brain barrier (BBB) to nutrients, neurotransmission and neuromodulation, various roles in the immune system, and hormonal regulation in the endocrine system. They are also involved in temperature control, food and water intake, cardiovascular, gastrointestinal and respiratory control, memory and affective states and potently modulate nerve development and regeneration [1–3].

The multiplicity of biological actions of peptides suggests that these agents may be used as pharmaceuticals in the treatment of a variety of disorders of the brain and spinal cord, as indeed is borne out in the examples below. However, as with any potential neuropharmaceutical, peptides must be able to undergo transport into brain from the blood. The drugs currently used for disorders of the central nervous system (CNS) such as diazepam (anxiety), flurazepam (insomnia), amitryptiline (depression), codeine (headache) and chlorpromazine (schizophrenia) are most often lipid-soluble drugs that readily cross the BBB following oral administration, which is not the case for most peptides and proteins with therapeutic potential.

6.2.1. CNS Tumors

Malignant tumors of the CNS represent an illness where therapeutic approaches so far have only been able to achieve a short extension in the lifespan of the patient. Brain tumors correspond to primary or metastatic tumors arising from peripheral cancers such as those of the lung, breast or gastrointestinal tract. Malignant gliomas represent 13–22% of the brain cancers. Despite surgery, external beam radiation therapy and systemic chemotherapy, these tumors tend to recur within centimeters of their original location. Regardless of the treatment, the median survival time is less than 1 year [4].

Operating on the inductive and effective phases of an anti-tumor immune response and uncovering pivotal functions that may reduce cancer cell growth, Interleukin-18 (IL-18) appears to be an attractive candidate for the sustained local adjuvant immunotherapeutic treatment of brain gliomas. IL-18 is a molecule made up of a single peptide chain of 18.3–18.4 kDa. In addition to stimulating interferon-γ production from macrophages, T-cells and NK cells, IL-18 stimulates the production of IL-2 and chemokines. Additionally, it is known to reduce angiogenesis *in vivo*, thus resulting in hypovascularization of the treated tumors [5,6].

As tumor angiogenesis appears to be a key factor controlling tumor growth, various peptide and protein angiogenesis inhibitors have been studied. It has been demonstrated that interferon-α was able to inhibit angiogenesis, but showed some neurotoxicity, including symptoms of somnolence and decreased memory [7]. Other studies have also demonstrated that tumor growth factor TGF-β_2, secreted by human glial tumor cells, may promote tumor growth by enhancement of angiogenesis and depression of T-cell function [8]. TGF-β_2 antagonists may thus prove beneficial in the treatment of human gliomas.

6.2.2. Alzheimer's Disease

Alzheimer's disease is a chronic and progressive neurodegenerative disorder that affects about 10% of individuals over the age of 65. It begins with memory impairment and eventually progresses to dementia, physical impairment and death [9]. The symptoms are well correlated with the degeneration of basal forebrain cholinergic neurons, and consequent cortical and hippocampal cholinergic denervation. The dementia of Alzheimer's disease correlates with the extracellular deposition within the brain of β-amyloid, arising from the release of an insoluble peptide called β-peptide or A4 peptide [10].

The treatment of Alzheimer's disease traditionally involves enhancement of cholinergic transmission through administration of acetycholinesterase inhibitors [11]. Other approaches aim to decrease the accumulation of β-amyloid in order to slow the progression of the disease. Some of these approaches comprise the development of specific γ-secretase. Some peptide-based β-amyloid aggregation inhibitors have been shown to decrease the deposition of β-amyloid *in vivo*, in laboratory settings [12,13].

Nerve growth factor (NGF) appears to be a promising drug for the prevention or reduction of the neuronal degeneration occurring in Alzheimer's disease. *In vivo* studies have shown that administration of NGF reversed the deposition of extracellular amyloid aggregates, and removed the cognitive deficits in mouse models of the disease [14]. Other peptide and protein drugs, such as the vasoactive intestinal peptide (VIP), the pituitary adenylate cyclase activating polypeptide (PACAP) and the peptide histidine-isoleucine (PHI) can protect cells from death induced by β-amyloid, which render them promising agents to treat neurodegeneration [15,16].

6.2.3. Parkinson's Disease

Parkinson's disease is a multicentric neurodegenerative disease affecting approximately 1% of the population older than 60 years [17]. Clinical features of PD include tremor at rest, rigidity, slowness of movement and flexed posture [18]. The main pathological hallmark of Parkinson's disease is the degeneration of dopaminergic neurons in the substantia nigra pars compacta, which causes consequent reduction of dopamine levels in the striatum [19]. Levodopa (L-dopa) still continues to be the most potent drug for improving motor symptoms in this disease [20].

Attempts have been made to regenerate existing dopaminergic neurons. One particularly promising therapeutic approach in this regard involves the use of neurotrophic factors, particularly the glial-derived neurotrophic factor (GDNF), brain-derived neurotrophic factor (BDNF) and insulin-like growth factor (IGF) [21,22]. GDNF in particular has been shown to protect mesencephalic dopaminergic neurons from 6-hydroxydopamine-induced degeneration, after administration via a pump, or repeated injections at the site of action, in animal models of the disease.

6.2.4. Stroke

Around 400 000 people per year suffer from a cerebrovascular accident, or stroke [23]. About 85% of these cases are ischemic strokes, which include thrombotic, embolic and lacunae etiologies [24]. Thrombotic strokes occur when an atheromatous plaque occludes flow through the blood vessel. The atheroma accumulates over time as a result of local platelet aggregation and lipid deposition. Embolic strokes arise from the release of emboli from the heart or a proximal artery atheroma. Lacunar strokes occur in deep regions of the brain such as the thalamus and arise in the vicinity of the penetrating cerebral blood vessels that lack collaterals.

The treatment of the acute ischemic stroke has as its objective the enhancement of cerebral blood flow, thrombolysis and cytoprotection. Given the peptidergic innervation of brain blood vessels, peptide and protein drugs should be particularly promising in regulating cerebral blood flow in stroke. For example, VIP agonists or neuropeptide Y antagonists, leading to vasodilatation, may be effective adjuncts to stroke therapy [25].

Cytoprotection, one of the most important components in the treatment of strokes, is aimed at decreasing the neuronal damage that follows either the excitotoxic discharge of glutamic acid neurons or the brain edema associated with the formation of oxygen-derived free radicals. A major enzymatic protection against these free radicals (such as the superoxide anion radical or the hydroxyl radical) is the enzyme superoxide dismutase, which may represent an effective treatment for ischemic strokes [26].

6.3. TRANSPORT OF PROTEIN AND PEPTIDE DRUGS TO THE BRAIN

6.3.1. Blood–brain Barrier and Blood–cerebrospinal Fluid Barrier

Ideally, drugs used in the treatment of diseases affecting the CNS should reach their site of action after intravenous administration. However, access to the brain is particularly difficult for peptide and protein biopharmaceuticals. Currently, 98% of new drugs discovered for CNS disorder treatment do not enter the brain following systemic administration [27].

Entry of molecules to the brain is regulated by two selective barriers that exist between the CNS and the blood, known as the blood–brain barrier (BBB) and the blood–cerebrospinal fluid (CSF) barrier. The surface area of the human BBB is estimated to be 5000 times greater than that of the blood–CSF barrier; therefore the BBB is considered to be the primary barrier controlling the uptake of drugs into the brain [28]. The dual purpose of the BBB is to provide essential nutrient supply to the brain, but also to ensure a constant internal environment within the CNS. The BBB functions as a gate-keeper to the CNS, strictly limiting transport into the brain through both physical and metabolic barriers [29].

Blood capillaries in the CNS are structurally different from other blood capillaries. The non-brain capillaries have fenestrations between the endothelial cells, through which solutes can move via passive diffusion (Figure 6.1). In brain capillaries, the cerebral endothelial cells do not have fenestrations and are connected by epithelium-like tight junctions which prevent movement via

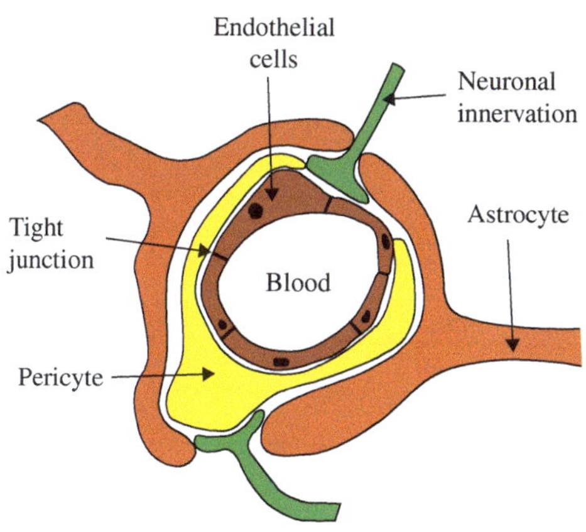

FIGURE 6.1 Structure of the blood–brain barrier. *Reprinted from reference [29] with permission from Elsevier.*

paracellular diffusion pathways. There is also minimal pinocytosis across brain capillary endothelium, which further limits the transport of drugs from the blood to the brain [29,30].

In addition to the tight junctions, the presence of perivascular elements, such as pericytes and astrocytes, contributes to the low permeability of the BBB. Pericytes partially surround the endothelium. Due to their phagocytic properties, they play a major protective role against drug passage when the integrity of the endothelial cells is compromised [31]. They also regulate endothelial cell proliferation [31]. Due to their contractile properties, they mechanically stabilize the endothelium while counteracting the pressure exerted by the erythrocytes [31,32].

In the brain, neurons are supported by a network of glial cells called astrocytes. These cells are characterized by their foot processes closely surrounding the pericytes and capillaries. The endothelial cells and astrocyte foot processes are only separated by a distance of 20 nm. The role of the astrocytes is to regulate endothelial cell differentiation [33] and the ionic balance of the CSF. Finally, microglia cells surround about 85% of the surface of the capillaries and have been described as participating in signaling and transcellular transport processes.

The blood–CSF barrier is found at the circumventricular organs: the choroid plexus, the median eminence of the hypothalamus, the organum vasculosum of the lamina terminalis, the subcommissural organ, the subfornical organ at the roof of the third ventricle and the area postrema at the base of the fourth ventricle. Unlike the BBB, the endothelial cells at the circumventricular organs (with the exception of the subcommissural organ) are fenestrated and allow for the rapid distribution of proteins and small molecules into the immediate interstitial space. The presence of tight junctions linking the choroid plexus epithelial cells retards the further distribution of circulating substances into the CSF [27].

In addition to the permeability barriers described above, highly active enzymes (i.e. carbopeptidases, aminopeptidases, endopeptidases, cholesterases) present in the brain endothelial cells, pericytes and astrocytes, represent a further metabolic component of the BBB and blood–CSF barrier that also restricts the entry of substances to the brain [27].

6.3.2. Transport Mechanisms at the BBB

Due to their polar nature, peptide and protein drugs have been prevented from entering the CNS by passive diffusion. However, many peptides and proteins do cross the BBB mainly due to multiple, inwardly directed, transport systems at the BBB that selectively allow the passage of specific compounds into the CNS. These various transport mechanisms are summarized as follows (Figure 6.2):

Diffusion: In the majority of cases, drug absorption into the CNS occurs by passive diffusion. This is limited to the transcellular route, as the presence of the endothelial tight junctions prevents passive diffusion between the cells.

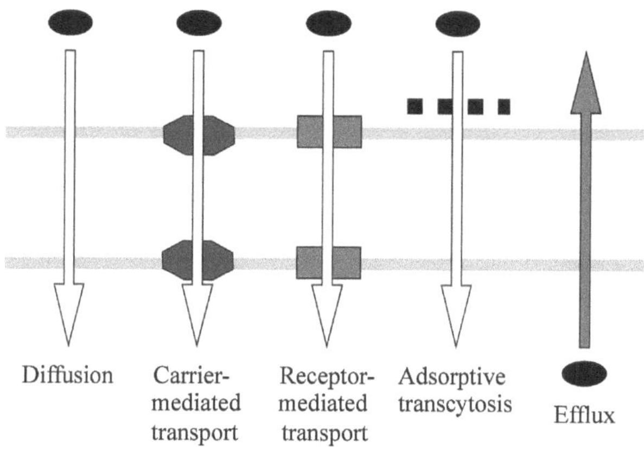

FIGURE 6.2 Schematic of the transport pathways across the blood–brain barrier. *Adapted from reference [34] with permission from Birkhäuser-Verlag, Basel.*

Lipid-soluble drugs move across the lipid-rich plasma membranes of the endothelial cells [35].

Carrier-mediated transport: Several carrier systems have been shown to be present in the brain endothelium, allowing for the selective transport of drugs such as small-molecule peptides, hexoses, monocarboxylic acids and amino acids. This process may involve the formation of transient narrow pores induced by binding of the respective substrate to the carrier, allowing only the passage of the specific molecule [34].

Receptor-mediated transcytosis: Receptor-mediated transcytosis through the BBB involves three sequential steps: receptor-mediated endocytosis at the luminal or blood side of the BBB, movement of the ligand-receptor complex through the endothelial cytoplasm, and receptor-mediated exocytosis of the ligand into the brain interstitial fluid at the brain side of the BBB. Some peptides and proteins such as insulin, insulin-like growth factors, angiotensin II, brain natriuretic peptide and transferrin, can reach the brain via this transport system [36].

Adsorptive transcytosis: Adsorptive-mediated transcytosis is functionally similar to receptor-mediated transcytosis, except that the initial triggering of the endocytosis event at the luminal side of the BBB is accomplished through electrostatic interactions between positively charged peptides and the negatively charged BBB membrane surface, as opposed to interactions with a specific receptor. Cationic peptides and proteins which penetrate the BBB via this transport mechanism include protamine, histone, avidine and cationized albumin [37–39].

Proteins that undergo either receptor-mediated or adsorptive-mediated transcytosis through the BBB may act as BBB transport vectors in the development of chimeric peptides. They are formed by the covalent coupling of a non-transportable peptide to a BBB transport vector (such as transferrin or cationized albumin), which in turn shuttles the non-transportable peptide drug into the CNS.

Efflux transport systems: Efflux transport systems remove substances from the brain or the CSF. P-glycoprotein, present in high density in the luminal membrane of the brain endothelia, is a well-known active efflux transport system at the BBB. It has been involved in preventing the CNS accumulation of a number of peptides and proteins [40]. Similarly, the multidrug resistance protein family (MRP) can also confer multidrug resistance by participating in BBB efflux transport [40].

6.4. STRATEGIES TO ENHANCE THE DELIVERY OF PROTEIN/PEPTIDE TO THE CNS

Most of the peptide and protein drugs currently used for the treatment of CNS disorders have a low molecular weight (comprised between 150 and 500 Daltons), and high lipophilicity. In the most positive case scenario, the drug manages to reach the brain and its target cells, without being cleared by efflux mechanisms or degraded by enzymes. However, the drug may be unable to reach its target because of either being metabolized, cleared, or bound to plasma proteins.

In order to improve the likelihood of a positive delivery outcome to the CNS, peptide drug delivery to the brain may be optimized by following three general approaches:

- Invasive procedures, which include transient osmotic opening of the BBB.
- Physiologically-based strategies, which exploit the various transport mechanisms present at the BBB.
- Pharmacologically-based approaches to increase the passage through the BBB by optimizing the specific biochemical attributes of a compound. This may be accomplished either by chemical modification of the peptide itself, or by the attachment, encapsulation of the peptide in a substance that increases stability, permeability and bioavailability.

6.4.1. Strategies to Increase the Permeability of the BBB

One strategy to improve peptide and protein drug delivery to the CNS consists of combining systemic administration of the drug with transient osmotic opening of the BBB. Osmotic opening of the BBB was one of the earliest techniques to circumvent it for therapeutic purposes [41]. Transient opening or disruption of the BBB may be achieved by the intracarotid infusion of a hyperosmotic solution of mannitol. The hypertonicity causes shrinkage of endothelial cells and opening of the endothelial tight junctions. This method is transitory and the barrier closes within 10–20 min following BBB disruption.

Similarly, solvents such as ethanol at high dose or dimethylsulfoxide, alkylating agents like etoposide, and vasoactive agents such as bradykinin, histamine, have all been used to disrupt the BBB. This technique has been used to increase the uptake of chemotherapeutic agents for the treatment of malignant gliomas. However, this procedure can be highly traumatic and often has serious side-effects, such as seizures, permanent neurological disorders and brain edema. This approach is therefore not recommended as an effective strategy for drug delivery to the CNS [42].

6.4.2. Physiologically-based Strategies

6.4.2.1. Exploitation of Receptor-Mediated and Adsorptive-Mediated Transport Systems using Chimeric Peptide Technology

This strategy involves coupling the peptide or protein drug to a vector, which may also be a protein/peptide, and which normally crosses the BBB either by receptor-mediated transcytosis (transferrin and insulin) or adsorptive endocytosis (cationized albumin). The resulting chimeric peptide/protein is endocytosed at the luminal side of the BBB following the interaction of the transport vector with its corresponding cell surface receptor. It is carried through the membrane and then exocytosed into the brain interstitial fluid. The chemical linker joining the therapeutic agent to the transport vector is cleaved by enzymes, releasing the peptide or protein drug to bind to its appropriate target receptor and initiate a pharmacological action in the brain. Conjugation of peptide and protein drugs to transport vectors can be performed with chemical linkers, polyethylene glycol linkers or avidin-biotin technology.

Receptor-mediated vectors for brain delivery must be specific. Several potential receptors at the BBB may serve this purpose. The plasma protein transferrin is able to bind and undergo endothelial endocytosis in brain capillaries and has proven to be a suitable vector for the transport of numerous peptide and protein drugs across the BBB. The transferrin receptor is constituently expressed at the BBB and is involved in the transport of iron into the brain, thus making it particularly promising in brain-targeted delivery. Alternatively, receptor-specific monoclonal antibodies that undergo receptor-mediated endocytosis at the BBB *in vivo*, such as the murine monoclonal antibody to the transferrin receptor OX26, can also be used as transport vectors in this regard. The chimeric peptide technology has proven to be successful for numerous peptide and protein drugs including VIP analogs, brain-derived neurotrophic factor, adrenocorticotropic hormone analog, doxorubicin, dalargin and cationized albumin [43,44].

6.4.2.2. Exploitation of Carrier-mediated Transport Systems

Numerous nutrient transporters are present at the BBB. The most important ones are:

- peptide carrier systems for small peptides such as enkephalins, thyrotropin-releasing hormone, arginine-vasopressin;

- amino acid carrier systems for glutamate, phenylalanine, leucine and aspartate;
- nucleoside carrier systems for choline and thiamine;
- hexose carrier systems for glucose and mannose [45,46].

The utilization of these transport systems for nutrients and endogenous compounds is a potential strategy for facilitating the delivery of peptide and protein drugs into the CNS. In order to use it, the physicochemical properties of the peptide drugs have to be modified to mimic those of the carried nutrients. As a consequence, the stereochemical requirement for transport by these carriers needs to be known, in order to design peptide and protein drugs as suitable pseudosubstrates, to ultimately enhance their cerebral uptake. For example, the peptide transport system-2 was shown to transport arginine and vasopressin, but not oxytocin [47].

Other nutrient carrier systems could also be useful for delivering peptide and protein drugs to the brain. For example, the large, neutral, amino acid carrier, which accepts a wide variety of pseudosubstrates, has been successfully used to deliver levodopa for the treatment of Parkinson's disease [48]. Levodopa is a lipid-insoluble precursor of dopamine, which crucially contains the carboxyl and α-amino groups needed for transport across the BBB by the large, neutral, amino acid carrier [49]. Similarly, biphalin, a peptide with potent opioid analgesic properties, has been shown to use the same carrier to reach the CNS [50]. On the other hand, some carriers are very selective in their stereochemical requirements and are therefore unsuitable for use for the brain delivery of pseudosubstrates. For example, the BBB glucose transporter GLUT1 puts stringent demands that are difficult to address on the molecules that it will transport.

6.4.3. Pharmacologically-based Strategies

6.4.3.1. Lipidization

The lipophilicity of a drug strongly correlates with its ability to passively cross the BBB.

Lipophilicity may be increased by blocking hydrogen bond-forming functional groups on the drug structure or covalently binding the drug to lipidic moieties, such as long chain fatty acids. The presence of hydroxyl groups on peptides and proteins tends to promote hydrogen bonding with water, leading to a decrease in the partition coefficient and subsequently a decrease in membrane permeability [51].

Methylation has been shown to reduce the overall hydrogen bonding potential of peptides and hence to increase membrane diffusion. For example, the dimethylation of the N-terminal Tyr of the cyclic peptide [2,6-dimethyl-Tyr1, D-pen2, D-Pen5]enkephalin (DPDPE) led to an increase in analgesia in an animal model compared to the unmodified drug [52]. Halogenation of peptides and proteins can also increase lipophilicity and BBB permeability. For example, the halogenation of DPDPE and biphalin has increased BBB permeability, as well as the intensity of the analgesic effect [53,54] compared to the effects obtained when using the unmodified drugs.

Acylation or alkylation of the N-terminal amino acid can also increase the lipophilicity of peptides and proteins. For example, acyl derivatives of [D-ala2, D-Leu5] enkephalin (DADLE), DPDPE and insulin have shown improved transfer through membranes while retaining their pharmacological activity [55–57]. Similarly, glycosylation with various sugar moieties has been shown to increase the concentration of enkephalinic peptides in the CNS after systemic administration [58].

6.4.3.2. Use of Prodrugs

Prodrugs are pharmacologically inactive compounds that result from transient chemical modifications of biologically active species (cf. Chapter 4, page 74; Chapter 5, pages 95–97). They are either conjugated to a molecule carried by a known transporter or to a lipophilicity enhancer, which will be cleaved at the site of action to allow the drug to induce its pharmacological effects. The chemical modification is usually intended to improve membrane permeability or solubility.

Ideally, a prodrug should be enzymatically stable in the blood, but rapidly degraded to the active parent compound once it reaches the target tissue. Esters are therefore particularly suitable for the design of prodrugs for cerebral delivery, due to the abundance of endogenous esterases in the CNS. Esterification and amidation of hydroxyl-, amino- or carboxylic acid-containing peptide and protein drugs are often used in order to enhance the lipophilicity of the drug, and thus its entry into the CNS [59,60]. Hydrolysis of the modifying group would then release the active compound. Recent studies have shown that tertiary butyl esters were stable in plasma while being suitably cleaved in the CNS [61].

Some inconveniences unfortunately limit the use of prodrugs. The increased lipophilicity of the prodrug, needed to cross the BBB, may lead to an undesirable increase of uptake in other tissues than the CNS. This poor selectivity, together with the high risk of forming reactive metabolites, may decrease the therapeutic index of peptide and protein prodrugs.

6.4.3.3. Use of Colloidal Drug Carriers

Colloidal drug carriers used for the delivery of peptide and protein drugs to the CNS include liposomes, nanoparticles and emulsions. They have promising features in common that make them particularly suitable for this application:

- Possibility of targeted delivery;
- Ease of modification of their chemical properties to achieve a specific and selective delivery of the cargo to the intended site of action;

- Masking of the physicochemical properties of the carried peptide and protein drugs;
- Protection of the cargo against enzymatic degradations.

Only liposomes and nanoparticles have been used for brain drug delivery of peptides and proteins so far.

6.4.3.3.1. Liposomes

Liposomes are self-assembling vesicular structures based on one or more lipid bilayers encapsulating an aqueous core (Figure 6.3) (cf. Chapter 2, page 27; Chapter 4, pages 73–74). The major lipidic components of liposomes are usually phospholipids, which are amphiphilic moieties with a hydrophilic head group and two hydrophobic chains. The phospholipid molecules form a closed bilayer sphere in an attempt to shield their hydrophobic groups from the aqueous environment. At the same time, they maintain contact with the aqueous phase via the hydrophilic head group.

On the basis of their size and their number of lipid bilayers, liposomes are generally classified into multilamellar vesicles (MLVs, diameter > 200 nm), large unilamellar vesicles (diameter 100–1000 nm) and small unilamellar vesicles (diameter < 100 nm). Depending on its physicochemical nature, the drug to be carried can be encapsulated in the aqueous phase (hydrophilic drugs), be intercalated into the bilayer (lipophilic drugs) or interact with the surface of the liposome (through electrostatic interactions). Liposomes can therefore be used as carriers for both hydrophilic and lipophilic drugs.

Niosomes, an alternative formulation to liposomes, are unilamellar or multilamellar vesicles prepared by the aqueous dispersion of synthetic, non-ionic amphipathic molecules. They have been developed as more stable

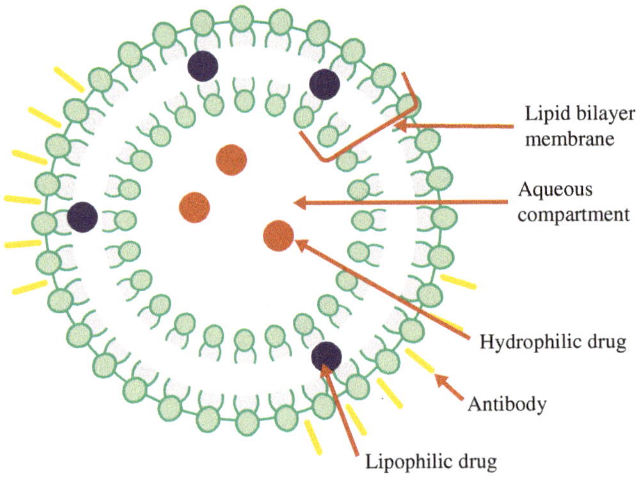

FIGURE 6.3 Generalized structure of a liposome.

alternatives to phospholipid-based liposomes, with which they share the same structure.

One of the key advantages of liposomes and niosomes is the possibility that they could be targeted to the brain by exploiting receptor-mediated transcytosis, by incorporating a targeting moiety on their surface. Various liposomal formulations have been used to transport peptide and protein drugs across the BBB. For example, the intraperitoneal administration of liposomes entrapping GABA led to a decrease of the epileptic activity in rat models of the disease, in contrast to that observed with the free drug [62].

Immunoliposomes bearing an OX26 antibody on their surface have been successfully used to deliver digoxin to the CNS [63]. Glucose-bearing niosomes, encapsulating VIP, have been shown to deliver VIP to the brain after intravenous administration [64], thus demonstrating that the glucose transporter GLUT1 is a useful carrier for efficient drug delivery to the CNS. As a result, liposomes can now be considered as an established peptide and protein drug delivery system, with several formulations now under investigation in clinical trials.

6.4.3.3.2. Nanoparticles

Nanoparticles are colloidal, polymeric colloidal spheres ranging in size from 10 to 1000 μm. They are made of natural or artificial polymers. The term "nanoparticles" encompasses both nanospheres and nanocapsules; the former being made of a continuous matrix, whereas the latter have a core-shell structure (Figure 6.4) [65].

Depending on the method of preparation, the drug can be entrapped within the polymer matrix (in the case of a nanosphere), encapsulated within the core of the nanocapsule by its polymeric shell (in the case of a nanocapsule), or adsorbed or covalently bound to the surface of the nanoparticle. The use of biodegradable materials would allow a sustained release of the peptide and protein drugs at the targeted site over a period of days or weeks after injection, which is particularly suitable for the delivery of drugs to the CNS.

Many formulations of nanoparticles have been described in the literature. For example, polysorbate 80-coated-(poly)butylcyanoacrylate (PBCA) nanoparticles have been used to successfully deliver peptide and protein drugs that otherwise show poor brain diffusion, such as doxorubicin [66,67], loperamide [68], tubocurarine [69] and dalargin [70], into the brain of the mouse and rat where they

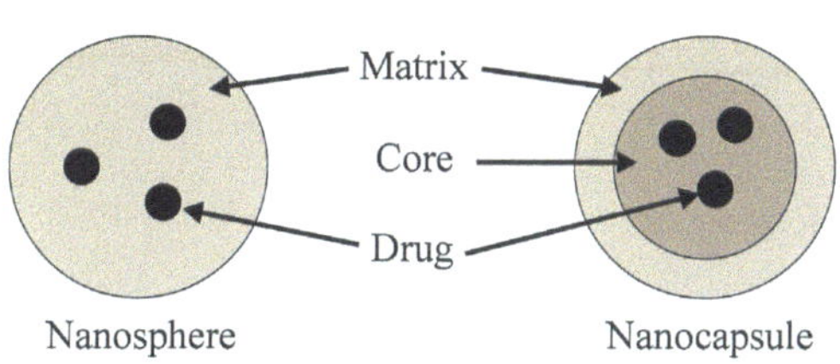

FIGURE 6.4 Generalized structure of a nanosphere and a nanocapsule.

induced pharmacologic effects. A nanoparticulate formulation of doxorubicin enabled a considerable reduction in tumor growth in an experimental rat glioblastoma [71]. The intravenous administration of NGF entrapped within polysorbate 80-coated PBCA nanoparticles led to the drug crossing the BBB, resulting in improved recognition and memory in an amnesia rat model [72].

6.4.3.4. *Inhibition of Proteolytic Enzymes and Efflux Transporters*

Brain uptake of peptide and protein drugs can be improved by reducing their clearance rate from the plasma and the brain. Strategies for achieving this objective would be to either inhibit the proteolytic enzymes or the activity of efflux transporters.

As an example of proteolytic enzyme inhibition, DOPA-carboxylase inhibitors have been shown to improve L-DOPA levels to the CNS. In the CNS, L-DOPA is converted to the active dopamine by DOPA-carboxylase. However, L-DOPA is also converted to dopamine in the peripheral nervous system, as the DOPA-carboxylase is present in abundance there. The resulting hyperdopaminergia is the cause of numerous adverse side-effects, as well as a decrease in the CNS delivery of L-DOPA [73]. In order to overcome this problem, a DOPA-carboxylase inhibitor is co-administered with L-DOPA, resulting in an increase of the plasma half-life of L-DOPA and in improved L-DOPA delivery to the brain [74].

Unfortunately, as many enzyme systems are involved in the degradation of peptides and proteins, a proteolytic enzyme inhibition strategy would not work well for most peptide and protein drugs. Another more applicable strategy would be to mask the enzyme cleavage sites in peptide and protein drugs by the addition of side chains that interfere with enzyme activity. Cyclization, methylation, halogenation and cationization of the peptide and protein drugs have indeed reduced their proteolysis by enzymes.

Expression of drug efflux transporters at the BBB can potentially limit the access of peptide and protein drugs to the brain. The development of inhibitors for the efflux transporters, and the design of peptide and protein drug analogs which cannot be expelled by them should improve the level of drug within the brain. For example, the co-administration of the P-gp blocker valspodar and the drug paclitaxel has resulted in increased brain levels of paclitaxel and consequently decreased tumor volume in mice [75]. The encapsulation of peptide and protein drugs into colloidal delivery systems would also in principle overcome the efflux transporters and thus improve the therapeutic effects of the drug.

6.5. CONCLUSION

This chapter has highlighted the significant advances in brain delivery of peptide and protein drugs made over the past three decades. While many limitations still exist regarding the cerebral delivery of peptide and protein drugs, several strategies show considerable promise for improvement of brain uptake. It is therefore hoped that during the next several years, extensive work in this rapidly expanding field will result in the development of even more potent treatments for CNS diseases.

REFERENCES

[1] M.B. Segal, B.V. Zlokovic, Blood—Brain Barrier: Amino Acids and Peptides, Kluwer Academic publishers, Lancaster and Boston, 1990, pp. 47—123.
[2] W.M. Pardridge, Peptide Drug Delivery to the Brain, Raven Press, New York, 1991.
[3] D.T. Krieger, Brain peptides: What, where, and why? Science 222 (1983) 975—985.
[4] J.P. Benoit, N. Faisant, M.C. Venier-Julienne, P. Menei, Development of microspheres for neurological disorders: From basics to clinical applications, J. Control. Release 65 (2000) 285—296.
[5] S. Lebel-Binay, A. Berger, F. Zinzindohoue, P. Cugnenc, N. Thiounn, W.H. Fridman, et al., Interleukin-18: Biological properties and clinical implications, Eur. Cytokine Netw. 11 (2000) 15—26.
[6] R. Cao, J. Farnebo, M. Kurimoto, Y. Cao, Interleukin-18 acts as an angiogenesis and tumor suppressor, Faseb. J. 13 (1999) 2195—2202.
[7] J. Boëthius, H. Blomgren, V.P. Collins, T. Greitz, H. Strander, The effect of systemic human interferon-alpha administration to patients with glioblastoma multiforme, Acta Neurochrurgica 68 (1983) 239—251.
[8] M.C. Kuppner, M.F. Hamou, S. Bodmer, A. Fontana, N. de Tribolet, The glioblastoma-derived T-cell suppressor factor/transforming growth factor beta 2 inhibits the generation of lymphokine-activated killer (LAK) cells, Int. J. Can. 42 (1988) 562—567.
[9] J.L. Cummings, G. Cole, Alzheimer disease, J. Am. Med. Assoc. 287 (2002) 2335—2338.
[10] B.E. Tomlinson, G. Blessed, M. Roth, Observations of the brains of demented old people, J. Neurol. Sci. 22 (1970) 205—242.
[11] M. Farlow, A clinical overview of cholinesterase inhibitors in Alzheimer's disease, Int. Psychogeriatr. 14 (Suppl. 1) (2002) 93—126.
[12] B. Permanne, C. Adessi, G.P. Saborio, S. Fraga, M.J. Frossard, J. Van Dorpe, et al., Reduction of amyloid load and cerebral damage in a transgenic mouse model of Alzheimer's disease by treatment with a beta-sheet breaker peptide, Faseb. J. 16 (2002) 860—862.
[13] E.M. Sigurdsson, B. Permanne, C. Soto, T. Wisniewski, B. Frangione, *In vivo* reversal of amyloid-beta lesions in rat brain, J. Neuropath. Exp. Neurol. 59 (2000) 11—17.
[14] M.H. Tuszynski, A. Blesch, Nerve growth factor: From animal models of cholinergic neuronal degeneration to gene therapy in Alzheimer's disease, Prog. Brain Res. 146 (2004) 441—449.
[15] D.E. Brenneman, C.Y. Spong, I. Gozes, Protective peptides derived from novel glia proteins, Biochem. Soc. Trans. 28 (2000) 452—455.
[16] S. Onoue, K. Endo, K. Ohshima, T. Yajima, K. Kashimoto, The neuropeptide PACAP attenuates beta-amyloid (1-42)-induced toxicity in PC12 cells, Peptides 23 (2002) 1471—1478.
[17] D. Aarsland, J. Zaccai, C. Brayne, A systematic review of prevalence studies of dementia in Parkinson's disease, Mov. Disord. 20 (2005) 1255—1263.
[18] P. Pal, A. Samii, D. Calne, Cardinal features of early Parkinson's disease, in: S. Factor, W. Weiner (Eds.), Parkinson's Disease: Diagnosis and Clinical Management, Demos Medical Publishing, New York, 2002, pp. 41—56.

[19] A.H. Schapira, Causes of neuronal death in Parkinson's disease, Adv. Neurol. 86 (2001) 155–162.
[20] C.B. Levine, K.R. Fahrbach, A.D. Siderowf, R.P. Estok, V.M. Ludensky, S.D. Ross, Diagnosis and treatment of Parkinson's disease: A systematic review of the literature, Evid. Rep. Technol. Assess. 57 (2003) 1–4.
[21] M. Espejo, B. Cutillas, T.E. Arenas, S. Ambrioso, Increased survival of dopaminergic neurons in striatal grafts of fetal ventral mesencephalic cells exposed to neurotrophin-3 or glial cell line-derived neurotrophic factor, Cell Transplant 9 (2000) 45–53.
[22] J. Guan, R. Krishnamurthi, H.J. Waldvogel, R.L. Faull, R. Clark, P. Gluckman, N-terminal tripeptide of IGF-1 (GPE) prevents the loss of TH positive neurons after 6-OHDA induced nigral lesions in rats, Brain Res. 859 (2000) 286–292.
[23] V. Hachinski, J.W. Norris, The Acute Stroke, F.A. Davis Company, Philadelphia, 1985, pp 1–12.
[24] J.C. Grotta, Current medical and surgical therapy for cerebrovascular disease, N. Engl. J. Med. 317 (1987) 1505–1516.
[25] J.H. Boublik, Central and peripheral significance of neuropeptide Y and its related peptides, Molec. Cell. Endocrinol. 71 (1990) C45–C48.
[26] J.L. Zweier, B.K. Rayburn, J.T. Flaherty, M.L. Weisfeldt, Recombinant superoxide dismutase reduces oxygen free radical concentrations in reperfused myocardium, J. Clin. Invest. 80 (1987) 1728–1734.
[27] W.M. Pardridge, Brain Drug Targeting, Cambridge University Press, Cambridge, 2001.
[28] K.A. Witt, T.P. David, CNS drug delivery: Opioid peptides and the blood–brain barrier, AAPS J. 8 (2006) E76–E88.
[29] W.M. Pardridge, Blood–brain barrier delivery, Drug Discovery Today 12 (2007) 54–61.
[30] M.W.B. Bradbury, Physiology and Pharmacology of the Blood–Brain Barrier, Springer-Verlag, Berlin, 1992.
[31] A. Aigner, S. Wolf, H.G. Gassen, Transport and detoxication: Principles, approaches and perspectives for research on the blood–brain barrier, Angew. Chem. Int. Ed. Engl. 36 (1997) 24–41.
[32] F.N. Miller, D.E. Sims, Contractile elements in the regulation of macromolecular permeability, Fed. Proc. 45 (1986) 84–88.
[33] R.C. Janzer, M.C. Raff, Astrocytes induce blood–brain barrier properties in endothelial cells, Nature 325 (1987) 253–257.
[34] L. Prokai, K. Prokai-Tatrai, Peptide Transport and Delivery into the Central Nervous System, Birkhäuser-Verlag, Basel, 2003.
[35] A. Tsuji, Specific mechanisms for transporting drugs into brain, in: D.J. Begley, M.W. Bradbury, J. Kreuter (Eds.), The Blood–Brain Barrier and Drug Delivery to the CNS, Marcel Dekker, New York, 2000, pp. 121–144.
[36] W.A. Banks, Anorectic effects of circulating cytokines: Role of the vascular blood–brain barrier, Nutrition 17 (2001) 434–437.
[37] W.M. Pardridge, D. Triguero, J. Buciak, Transport of histone through the blood–brain barrier, J. Pharmacol. Exp. Ther. 251 (1989) 821–826.
[38] W.M. Pardridge, R.J. Boado, Enhanced cellular uptake by biotinylated antisense oligonucleotide or peptide mediated by avidin, a cationic protein, FEBS Lett. 288 (1991) 30–32.
[39] A.K. Kumagai, J.B. Eisenberg, W.M. Pardridge, Absorptive-mediated endocytosis of cationized albumin and a beta-endorphin-cationized albumin chimeric peptide by isolated brain capillaries. Model system of blood–brain barrier transport, J. Biol. Chem. 262 (1987) 15214–15219.
[40] W. Löscher, H. Potschka, Role of drug efflux transporters in the brain for drug disposition and treatment of brain diseases, Prog. Neurobiol. 76 (2005) 22–76.

[41] E.A. Neuwelt, K.R. Maravilla, E.P. Frenkel, S.I. Rapaport, S.A. Hill, P.A. Barnett, Osmotic blood–brain barrier disruption. Computerized tomographic monitoring of chemotherapeutic agent delivery, J. Clin. Invest. 64 (1979) 684–688.
[42] E. Garcia-Garcia, K. Andrieux, S. Gil, P. Couvreur, Colloidal carriers and blood–brain barrier (BBB) translocation: A way to deliver drugs to the brain? Int. J. Pharm 298 (2005) 274–292.
[43] W.M. Pardridge, Vector-mediated drug delivery to the brain, Adv. Drug Deliv. Rev. 36 (1999) 299–321.
[44] W.M. Pardridge, Drug and gene targeting to the brain with molecaular Trojan horses, Nat. Rev. Drug Discov. 1 (2002) 131–139.
[45] W.M. Pardridge, Transport of small molecules through the blood–brain barrier: Biology and methodology, Adv. Drug Deliv. Rev. 15 (1995) 5–36.
[46] D.J. Begley, The blood–brain barrier: Principles for targeting peptides and drugs to the central nervous system, J. Pharm. Pharmacol. 48 (1996) 136–146.
[47] W.A. Banks, A.J. Kastin, A. Horvath, E.A. Michals, Carrier-mediated transport of vasopressin across the blood–brain barrier of the mouse, J. Neurosci. Res. 18 (1987) 326–332.
[48] L.A. Wade, R. Katzman, 3-O-methyldopa uptake and inhibition of L-dopa at the blood–brain barrier, Life Sci. 17 (1975) 131–136.
[49] L.A. Wade, R. Katzman, Synthetic amino acids and the nature of L-DOPA transport at the blood–brain barrier, J. Neurochem 25 (1975) 837–842.
[50] P.J. Horan, A. Mattia, E.J. Bilsky, S. Weber, T.P. Davis, H.I. Yamamura, et al., Antinociceptive profile of biphalin, a dimeric enkephalin analog, J. Pharmacol. Exp. Ther. 265 (1993) 1446–1454.
[51] E.G. Chikhale, K.Y. Ng, P.S. Burton, R.T. Borchardt, Hydrogen bonding potential as a determinant of the *in vitro* and in situ blood–brain barrier permeability of peptides, Pharm. Res. 11 (1994) 412–419.
[52] D.W. Hansen Jr., A. Stapelfeld, M.A. Savage, M. Reichman, D.L. Hammond, R.C. Haaseth, et al., Systemic analgesic activity and delta-opioid selectivity in [2,6-dimethyl-Tyr1, d-Pen2, d-Pen5] enkephalin, J. Med. Chem. 35 (1992) 684–687.
[53] S.J. Weber, D.L. Greene, S.D. Sharma, H.I. Yamamura, T.H. Kramer, T.F. Burks, et al., Distribution and analgesia of [3H][d-Pen2, d-Pen5] enkephalin and two halogenated analogs after intravenous administration, J. Pharmacol. Exp. Ther. 259 (1991) 1109–1117.
[54] T.J. Abbruscato, S.A. Williams, A. Misicka, A.W. Lipkowski, V.J. Hruby, T.P. Davis, Blood-to-central nervous system entry and stability of biphalin, a unique double-enkephalin analog, and its halogenated derivatives, J. Pharm. Exp. Ther. 276 (1996) 1049–1057.
[55] A. Bak, O.S. Gudmundsson, G.J. Friis, T.J. Siahaan, R.T. Borchardt, Acyloxyalkoxy-based cyclic prodrugs of opiod peptides: Evaluation of the chemical and enzymatic stability as well as their transport properties across Caco-2 cell monolayers, Pharm. Res. 16 (1999) 24–29.
[56] H. Bundgaard, J. Moss, Prodrugs of peptides. 6. Bioreversible derivatives of thyrotropin-releasing hormone (TRH) with increased lipophilicity and resistance to cleavage by the TRH-specific serum enzyme, Pharm. Res. 7 (1990) 885–892.
[57] H. Asada, T. Douen, M. Waki, S. Adachi, T. Fujita, A. Yamamoto, et al., Absorption characteristics of chemically modified-insulin derivatives with various fatty acids in the small and large intestine, J. Pharm. Sci. 84 (1995) 682–687.
[58] R.D. Egleton, S.A. Mitchell, J.D. Huber, J. Janders, D. Stropova, R. Polt, et al., Improved bioavailability to the brain of glycosylated Met-enkephalin analogs, Brain Res. 881 (2000) 37–46.

[59] M.K. Ghosh, A.K. Mitra, Enhanced delivery of 5-iodo-2′-deoxyuridine to the brain parenchyma, Pharm. Res. 9 (1992) 1173–1176.
[60] D.M. Lambert, M. Geurts, G.K. Scriba, J.H. Poupaert, P. Dumont. Simple derivatives of amino acid neurotransmitters. Anticonvulsant evaluation of derived amides, carbamates and esters of glycine and beta-alanine. J. Pharm. Belg. 50 194–203.
[61] N.H. Greig, P.L. Stahle, H.U. Shetty, S. Genka, V. John, T.T. Soncrant, et al., High-performance liquid chromatographic analysis of chlorambucil tert-butyl ester and its active metabolites chlorambucil and phenylacetic mustard in plasma and tissue, J. Chromatogr. 534 (1990) 279–286.
[62] C. Loeb, E. Benassi, G. Besio, M. Maffini, P. Tanganelli, Liposome-entrapped GABA modified behavioral and electrographic changes of penicillin-induced epileptic activity, Neurology 32 (1982) 1234–1238.
[63] J. Huwyler, D. Wu, W.M. Pardridge, Brain drug delivery of small molecules using immunoliposomes, Proc. Natl. Acad. Sci. USA 93 (1996) 14164–14169.
[64] C. Dufès, F. Gaillard, I.F. Uchegbu, A.G. Schätzlein, J.C. Olivier, J.M. Muller, Central nervous system delivery of vasoactive intestinal peptide (VIP) encapsulated in glucose-bearing niosomes, Int. J. Pharm 285 (2004) 77–85.
[65] J. Kreuter, Nanoparticulate systems for brain delivery of drugs, Adv. Drug. Deliv. Rev. 47 (2001) 65–81.
[66] A.E. Gulyaev, S.E. Gelperina, I.N. Skidan, A.S. Antropov, G.Y. Kivman, J. Kreuter, Significant transport of doxorubicin into the brain with polysorbate 80-coated nanoparticles, Pharm. Res. 16 (1999) 1564–1569.
[67] S.C. Steiniger, J. Kreuter, A.S. Khalansky, A.S. Skidan, A.I. Bobruskin, Z.S. Smirnova, et al., Chemotherapy of glioblastoma in rats using doxorubicin-loaded nanoparticles, Int. J. Cancer 109 (2004) 759–767.
[68] R.N. Alyautdin, V.E. Petrov, K. Langer, A. Berthold, D.A. Kharkevich, J. Kreuter, Delivery of loperamide across the blood–brain barrier with polysorbate 80-coated polybutylcyanoacrylate nanoparticles, Pharm. Res. 14 (1997) 325–328.
[69] R.N. Alyautdin, E.B. Tezikov, P. Ramge, D.A. Kharkevich, Begley, J. Kreuter, Significant entry of tubocurarine into the brain of rats by adsorption to polysorbate 80-coated polybutylcyanoacrylate nanoparticles: An in situ brain perfusion study, J. Microencapsul. 15 (1998) 67–74.
[70] J. Kreuter, R.N. Alyautdin, D.A. Kharkevich, A.A. Ivanov, Passage of peptides through the blood–brain barrier with colloidal polymer particles (nanoparticles), Brain Res. 674 (1995) 171–174.
[71] S.E. Gelperina, A.S. Khalansky, I.N. Skidan, Z.S. Smirnova, A.I. Bobruskin, S.E. Severin, et al., Toxicology studies of doxorubicin-bound to polysorbate 80-coated poly(butyl cyanoacrylate) nanoparticles in healthy rats and rats with intracranial glioblastoma, Toxicol. Lett. 126 (2002) 131–141.
[72] B.A. Abdel Wahab, P.V. Evgenivetch, R.A. Alyautdin, Brain targeting of nerve growth factor using poly(butylcyanoacrylate) nanoparticles, Internet J. Pharmacol. 3 (2005).
[73] W.M. Pardridge, The blood–brain barrier: Bottleneck in brain drug development, NeuroRx 2 (2005) 3–14.
[74] D.L. McLellan, B.C. Dean, Improved control of brittle Parkinsonism by separate administration of levodopa and benserazide, Br. Med. J. 284 (1982) 1001–1002.
[75] S. Fellner, B. Bauer, D.S. Miller, M. Schaffrik, M. Fankhanel, T. Spruss, et al., Transport of paclitaxel (Taxol) across the blood–brain barrier *in vitro* and *in vivo*, J. Clin. Invest 110 (2002) 1309–1318.

Chapter 7

Chemically Modified Polyelectrolytes for Intestinal Peptide and Protein Delivery

Colin Thompson [1] and Woei Ping Cheng [2]

[1] *School of Pharmacy and Life Sciences, The Robert Gordon University, Schoolhill, Aberdeen AB10 1FR, UK,* [2] *School of Pharmacy, University of Hertfordshire, Hatfield AL10 9AB, UK*

Chapter Outline

7.1. INTRODUCTION

Polyelectrolyte complex (PEC) formation represents a simple but very interesting principle with significant importance in protein delivery. It is known that conventional polymer-based nanoparticles that are used for protein delivery are commonly manufactured via the use of solvent evaporation, emulsion, dispersion or polymerization of monomers in the inverse phase microemulsion [1,2]. The process very often requires the use of potentially

Peptide and Protein Delivery. DOI: 10.1016/B978-0-12-384935-9.10007-0

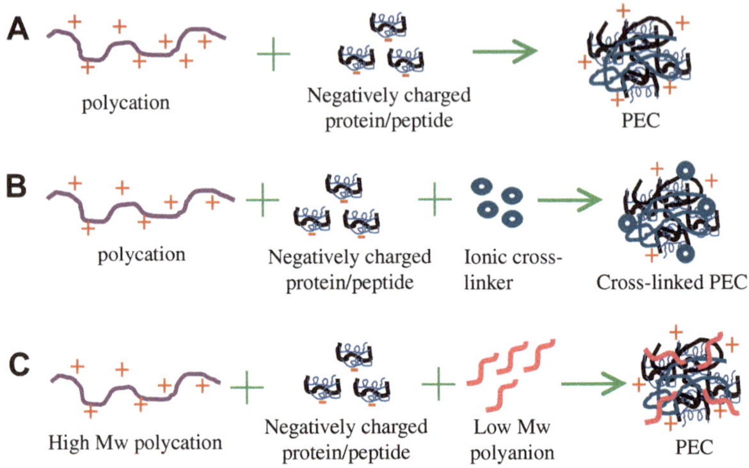

FIGURE 7.1 (A) Spontaneous formation of nano-sized PEC occurs after mixing oppositely charged polymers and proteins in aqueous/buffer media. (B) addition of an ionic cross-linking agent to stabilize the PEC. (C) addition of another oppositely charged low molecular weight polyelectrolyte to form a stable PEC.

toxic organic solvents, heat, vigorous agitation or chemicals that might compromise the stability and biocompatibility of the final products [1–3]. However, the fabrication process of nano-size PEC is easy, and today it represents an attractive alternative to conventional nanoparticulate formulations, which is evident from the recent explosion of literature on the use of PEC in protein delivery [2,4–8]. Generally there are three methods of PEC fabrication as illustrated in Figure 7.1.

Typically, formation of nano-sized PEC occurs spontaneously after mixing oppositely charged polymers and proteins in an aqueous media [2,9–10]. This is the most widely adopted approach in protein delivery [11–18]. However some researchers include an ionic cross-linking agent such as tripolyphosphate, zinc sulphate or calcium chloride to stabilize the PEC [8,19–21], or add another oppositely charged polyelectrolyte to form a stable PEC [3,5,20]. The major driving force of PEC formation is the electrostatic (Coulombic) interaction between the oppositely charged polyelectrolytes. However other intermolecular interactions, such as hydrogen bonding, hydrophobic and van der Waals forces have been reported to play an important role in the formation of these PECs [10,22,23]. Since PEC formation is purely based on non-covalent interaction and primarily driven by Coulombic interaction, there are a number of factors which could affect its formation and stability. The molecular weight of the polyelectrolytes (PE), the stoichiometric ratios and concentration of PE and proteins, the ionic strength and pH of the solution and the temperature will all have an impact on the nature of the complex formation, which will subsequently affect the stability and efficacy of the final formulations [1,9,10].

As mentioned previously, there is a plethora of literature reporting the use of polyelectrolytes for protein delivery, with chitosan being the most commonly reported polyelectrolyte in use. Chitosan in its native state has already been extensively reviewed in the literature [4–6,8] and has been superseded to an extent by its derivatives, namely the quaternary ammonium salts and thiomers. There have also been limited reports of amphiphilic polyelectrolytes being used in this area [17,24–26]. The aim of this chapter is therefore to look at the use of amphiphilic polyelectrolytes and chemically modified chitosan for the gastro-intestinal delivery of protein and peptides, and to discuss how the chemical modifications have impacted on the two important challenges in oral protein/peptide delivery, i.e. protection against intestinal enzymatic degradation and promotion of protein/peptide transport across the gastro-intestinal mucosa.

7.2. CHARACTERIZATION OF POLYELECTROLYTE–PROTEIN COMPLEXES

There are a number of characterization techniques reported for polyelectrolyte–protein complexes ranging from Circular Dichroism (CD), Fourier Transform Infrared Spectroscopy (FT-IR), Differential Scanning Calorimetry (DSC) and static light scattering, among others. An in-depth discussion on the characterization and the physical chemical properties of PECs can be found in the review by Cooper *et al.* [27]. Here, we focus on a few of the main techniques.

7.2.1. Turbidimetric Measurement

The association of polycations and polyanions in aqueous media may result in soluble PEC complexes, precipitation or gelation. It is known that the mixing ratios of proteins and polyelectrolytes influence both the degree of interactions and the properties of PECs so produced [1]. Turbidimetric measurement is a simple and non-invasive technique for determining the optimal mixing ratios of proteins and polyelectrolytes and the onset of precipitation, since in principle the formation of colloidal, high molecular weight complexes should result in an increase of UV light scattering. A number of groups including ourselves have shown that stable, soluble, polyelectrolyte–protein complexes are formed at non-stoichiometric ratios [1,25,28]. Most research shows that at low polymer concentration, large and unstable aggregates are formed due to the excess of protein [1,25,26,28]. An increase of polymer concentration beyond the 1:1 stoichiometric ratio results in water-soluble molecular complexes with smaller particle size. This formation of water soluble PECs is an equilibrium process and is dependent upon:

(i) Electrostatic stabilization – excess high molecular weight polyions (in this case, usually the polyelectrolyte and not the protein) stabilize the complexes by charge repulsion of the nanocomplexes in water [1];
(ii) Steric stabilization – the presence of hydrophilic side chains (e.g. polyethylene oxide) provides colloidal stability between the nano-complexes [2,10].

Proteins and peptides are amphoteric macromolecules, and hence the pH of the aqueous media is important. Above the isoelectric point (pI), the protein will be negatively charged and *vice versa*, hence the manipulation of the pH of the aqueous media will determine the formation of PEC based on electrostatic interaction. Commonly, PECs are prepared in water [3,18,29–31], or occasionally buffer systems including 3-(N-morpholino)propanesulfonic acid buffer (MOPS) [32], phosphate buffer [20] or tris(hydroxymethyl)aminomethane (Tris) buffer [25,26,28,33].

7.2.2. Dynamic Light Scattering Techniques

As is the case for the conventional nanoparticles that are used for protein delivery, the most commonly used technique to characterize the PEC is dynamic light scattering [10,32]. This technique measures the hydrodynamic size of the PEC formation, hence providing invaluable information on the stability of the complexes. Depending on the type of polyelectrolyte and the mixing ratios of protein and polyelectrolyte, typically a stable, soluble PEC has a hydrodynamic size of between 100–400 nm in aqueous/buffer media [1,18,25,26,28,32,33]. This measurement is generally accompanied by zeta potential measurement – another important parameter, since the major driving force of PEC formation is electrostatic interaction. Changes in zeta potential often indicate the onset of complexation which could either increase [28] or decrease [25] depending on the nature of the charge on both the polymer and protein.

7.2.3. Microscopic Techniques

Scanning electron microscopy (SEM), transmission electron microscopy (TEM) and atomic force microscopy (AFM) are the three most commonly used microscopic techniques to visualize these complexes. They show that, generally, most complexes are discrete, spherical nanoparticles with smooth surfaces [1,19,31]. The absence of agglomeration in the liquid environment shown by AFM shows these PECs are possibly stabilized by charge repulsion [1,33]. Interestingly, other morphologies such as vesicles [25,32], nanoparticles with a dense core and nanoparticles with a "fluffy" appearance have also been reported [25]. Unlike most of the complexes formed between polyelectrolyte and proteins, these PECs with different morphologies were formed by amphiphilic polyelectrolytes. It is thought the presence of hydrophobic moieties on these polyelectrolytes have a significant impact on the morphologies of the complexes so produced [25,26].

7.3. TYPES OF CHEMICALLY MODIFIED POLYELECTROLYTES FOR PROTEIN AND PEPTIDE DELIVERY

7.3.1. Amphiphilic Polyelectrolytes

Amphiphilic polyelectrolytes (AP) have both hydrophilic and hydrophobic components [34]. Like conventional amphiphiles, i.e. small molecular weight surfactants, they form nano-sized self-assemblies above their critical aggregation concentration (CAC) in aqueous media. Unlike small molecular weight surfactants – which typically form micelles – these AP can self-assemble into polymeric micelles, vesicles, disc-like structure or nanoparticles in aqueous media due to the diverse polymer architecture [35–38]. The hydrophobic component of the AP forms the core of the particles and is shielded from the aqueous environment by the hydrophilic component [37]. These polymers have been used widely in the solubilization of hydrophobic drugs for the past two decades [37,39–41] but only a handful of studies have reported on the use of these self-assembled polyelectrolytes for protein and peptide delivery.

As mentioned in the introduction, the principle mechanism of PEC formation is electrostatic interaction between oppositely charged polyelectrolytes and proteins. Most peptides and proteins, however, typically exhibit amphiphilic character due to the presence of multiple hydrophobic and hydrophilic amino acid residues in their primary sequence. Amphiphilic polymers therefore may have unique potential in peptide and protein delivery due to their ability to interact with such peptides or proteins via both electrostatic and hydrophobic association. Our group recently looked at the use of amphiphilic polyallyalmine (PAA) for insulin and salmon calcitonin (sCT) delivery [25,26,42,43] while others, such as Thomas Kissel's group, have used poly(vinyl alcohol) graft copolymers for insulin delivery [16,17,28,44–48], and Sakuma and colleagues have reported on polystyrene grafted copolymers for sCT delivery [11,24,29,49–52].

7.3.1.1. Polyallylamine with Hydrophobic and Hydrophilic Modification

Our group has synthesized a range of novel amphiphilic polyelectrolytes based on PAA, with a view to complexing them with insulin in order to aid its oral delivery [34]. PAA was modified by the addition of hydrophobic moieties such as a C_{16} alkyl chain – cetyl (Ce), a C_{16} acyl chain – palmitoyl (Pa) or a cholesteryl (Ch) graft at 2.5 or 5% mole ratios [34]. Further alteration of the PAA backbone was achieved by quaternization of its primary amines (Q)

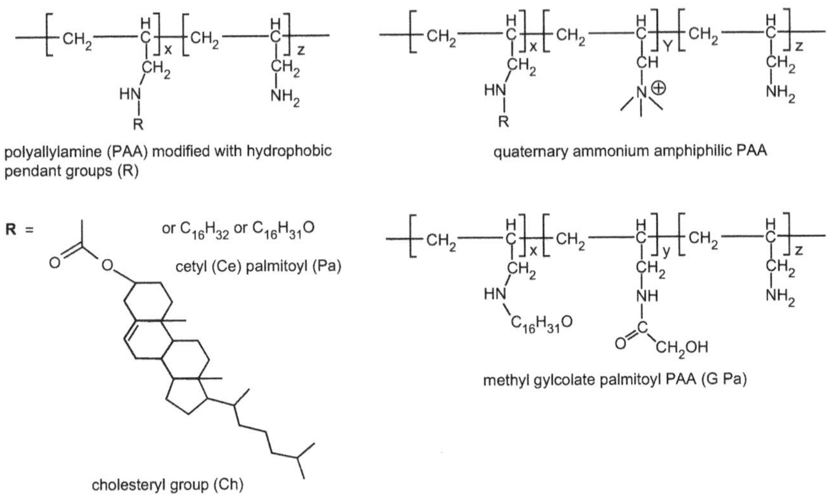

FIGURE 7.2 Structure of poly(allylamine) graft polymers.

(Figure 7.2). This was done in order to increase the cationic charge of the AP, as well as to increase its aqueous solubility [34]. These positively charged APs spontaneously form PECs with negatively charged insulin, resulting in particle sizes of around 100–200 nm in Tris buffer. We have shown that the presence of hydrophobic and quaternary ammonium moieties had a marked impact on the complexation efficiency (%CE), the morphology of the PEC, its ability to protect insulin against enzymatic degradation, and its interaction with Caco-2 cell monolayers.

The complexation efficiency (%CE) of each AP with insulin was determined using HPLC [25,26]. It was found that the presence of a Pa chain resulted in the highest %CE (78–92%), while PAA modified with Ce and Ch had less than 40% CE at a 2:1 polymer:insulin mass ratio [25,26]. TEM showed that most of the complexes existed as discrete nanoparticles, whereas QPa-insulin complexes were "fluffy" aggregates [25,26]. The observed differences can be attributed to the presence or absence of insulin on the surface of the PEC. The zeta potential of QPa decreased by more than half in the presence of insulin, indicating the partial neutralization of its positive charge due to electrostatic interactions on complexation. Although Pa APs demonstrated the highest %CE, their presence had varying effects on the enzymatic degradation of insulin *in vitro*. Pa and QPa were shown to offer protection against both trypsin and pepsin. However, rather than protecting insulin against α-chymotrypsin they actually enhanced its degradation [25].

Interestingly, it was discovered that PAA modified with Ce and Ch hydrophobic moieties offered protection against both trypsin and α-chymotrypsin, but not against pepsin [26]. Trypsin and α-chymotrypsin attack specific target sites on the insulin molecule [53], while pepsin is a non-specific proteolytic enzyme that cleaves most of the peptide bonds in the protein [54]. It is thought that the target sites for α-chymotrypsin were exposed, due to the unfolding of the insulin molecules upon the formation of a PEC with Pa AP, whilst the exposure of the target sites on insulin was reduced with Ce and Ch AP. However, the actual interaction between these APs and insulin, as well as with a range of enzymes, would need to be further investigated using either CD, FT-IR or other techniques. It is clear however, that to be effective via the oral route, either a combination of AP with different hydrophobic moieties or different grafts on the same backbone may be needed, if the technique is to offer insulin greater enzymatic protection.

To explore the potential of these APs for oral delivery, we investigated their interaction with human colorectal carcinoma cells (Caco-2), which are known to mimic the gut epithelial cells when differentiated [55]. As a whole, the presence of hydrophobic moieties and quaternary ammonium moieties reduced the cytotoxicity of parent PAA, although there are differences in the IC_{50} depending on the type of hydrophobic grafts that are used [34]. A surprising trend was observed when comparing QCe and QPa AP; it was found that QPa is three times less cytotoxic than QCe, despite the fact that both have C_{16} hydrocarbon chains [34]; the only difference in structure between the two APs was the additional carbonyl group in the Pa grafts. This would indicate very specific differences in cellular interaction with APs of different polymer architecture. Indeed, examination of the cellular interaction between rhodamine-labeled AP (as AP-insulin PEC) and Caco-2 cells demonstrated a difference between AP with QCe or QPa grafts (Figure 7.3). QPa AP appeared to be taken up by a non-calcium dependent process, which was only partially diminished by inhibition of active transport with sodium azide (a recognized active transport inhibitor) [1]. However, QCe AP uptake appeared to be completely inhibited by sodium azide.

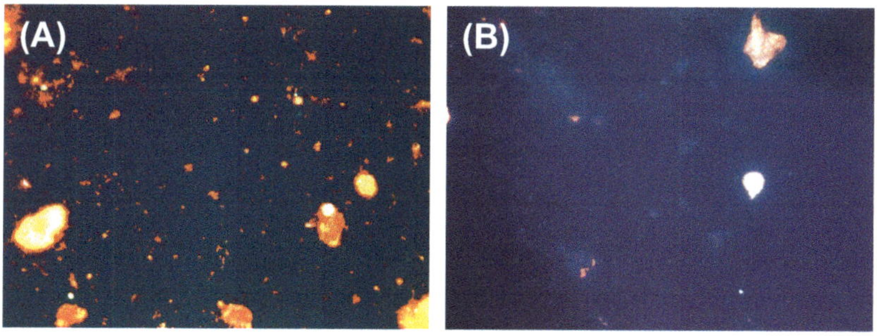

FIGURE 7.3 Fluorescent microscopy images of Caco-2 cells after incubation with polymer-insulin PEC for 2 h at 37°C. Cells treated with QPa-insulin (A) or QCe-insulin (B) PEC for 2 h after pre-treatment with sodium azide for 0.5 h (Mag. ×400 (to eyepiece)).

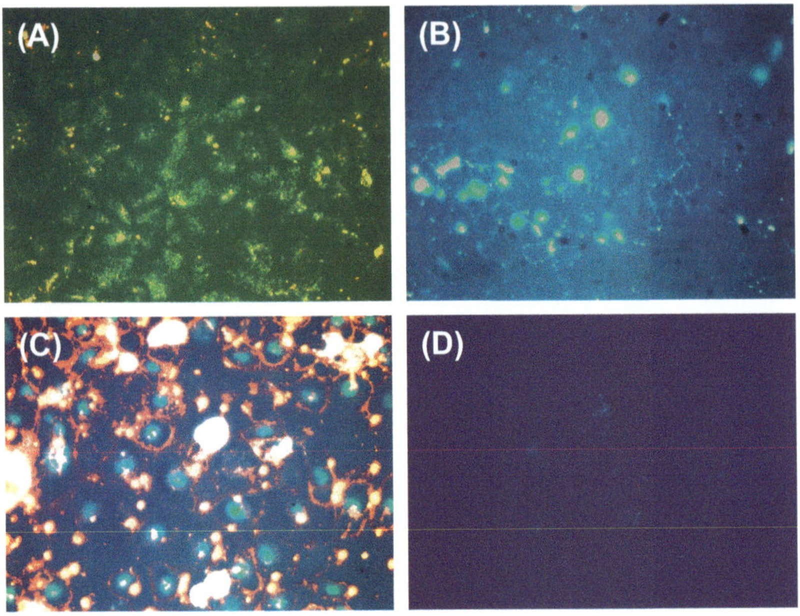

FIGURE 7.4 Fluorescent microscopy images of Caco-2 cells after incubation with polymer-insulin PEC for 2 h at 37°C. Cells treated with QPa-insulin (A) or Pa-insulin (B) PEC for 2 h; cells pre-treated with calcium-free media for 1 h then exposed to Pa-insulin (C) or QPa-insulin; (D) PEC in calcium-free media (Mag. ×400 (to eyepiece)).

Quaternary ammonium moieties have an impact on the extent of cellular uptake. Absence of calcium in the cell culture media prevented any uptake by non-quaternized Pa AP, but did not affect QPa uptake. In addition, the absence of quaternized groups resulted in the localization of Pa AP to the cell membrane, while QPa AP was found in the cytoplasm (Figure 7.4). Our work also showed that, unlike free insulin, where cellular uptake is dependent on insulin receptor interaction, the uptake of these polymer-insulin complexes occurred in the absence of insulin receptors [43]. Importantly the application of trypan blue to AP treated cells demonstrated minimal cell death, and quenching of extracellular fluorescence indicated that AP uptake was not due to cell damage [1,43].

Transepithelial electrical resistance (TEER) data on QPa indicated that it was able to reversibly open tight junctions with a 30–40% drop within the first 30 minutes. This reduction was maintained for 2 h and the effect was reversible – TEER values returned to around 90% of their initial values after removal of polymer solution (Figure 7.5). This is important in terms of safety, as the reversible effect indicates the polymer did not permanently destroy the tight junctions. Transport of insulin through the Caco-2 monolayers followed the same pattern, with the majority of insulin being transported within the first 30 minutes [55].

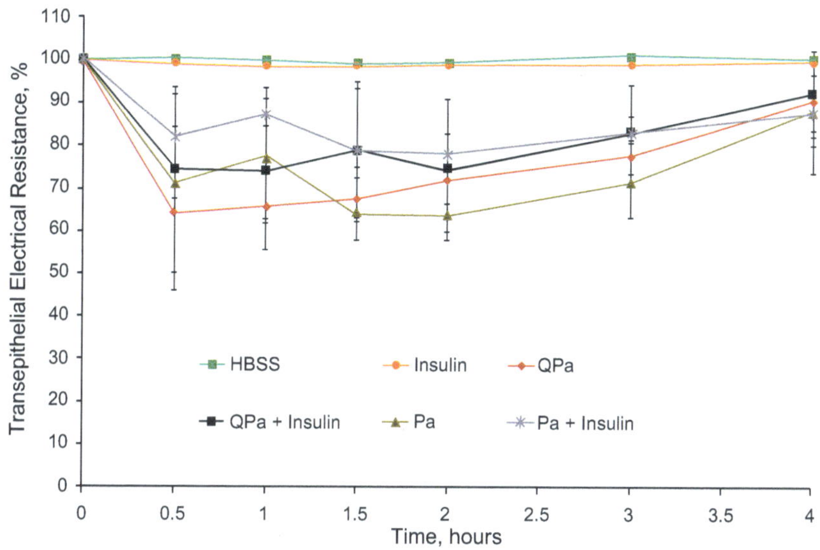

FIGURE 7.5 Transepithelial electrical resistance of Caco-2 cells exposed to: HBSS buffer (negative control); Insulin (3 μg mL^{-1}); Pa (6 μgmL^{-1}); QPa (48 μg mL^{-1}); Pa + Insulin complexes (6:3 μg mL^{-1}) or QPa + Insulin complexes (48:3 μg mL^{-1}) for 2 h. Samples were removed, and cells washed (×3), at 2 h and replaced with fresh media (n = 3; ± S.D.).

To date, most research has focused on complexation between polyelectrolytes and peptides of opposite charge. Our aim is to find out whether a cationic amphiphilic polyelectrolyte (QPa) could be complexed with a cationic peptide (salmon calcitonin (sCT)), and still retain the physicochemical properties that are compatible with maintenance of the sCT bioactivity. There are reports which suggest that polyelectrolyte–peptide complexation can indeed be achieved when both components have the same charge, due to the presence of localized patches of opposite charge on the peptide surface [56], or due to hydrophobic interactions or hydrogen bonding between polymer and peptide [29].

sCT has a pI of 10.2 and is positively charged at physiological or acidic pH. Although sCT is positively charged, complexation was still achieved with QPa in the current study, resulting in complexes with a narrow particle size distribution of hydrodynamic diameter of approximately 200 nm [42]. Overall, QPa complexation had a positive effect on the physical stability of sCT at both pH 5 and 7.4. The sCT *in vitro* bioactivities were also evaluated on the formulations after 7 days, which were kept in the dark at room temperature. Free sCT was completely degraded by day 7, while the complexes maintained markedly higher bioactivity (40%) than free sCT at both pH 5.0 and pH 7.4 [42]. This is an additional advantage to the use of PEC, which could confer better physical stability to the protein formulation. To investigate the interaction between QPa and cationic sCT, we used a fluorescent hydrophobic probe which detects pockets of hydrophobic domains on either sCT or QPa [57,58]. Both compounds on their own exhibited a reduction in the maximum emission wavelength suggesting the presence of hydrophobic domains.

However, further reduction in the maximum emission wavelength was observed upon complexation. This suggests that the changes in conformation on formation of the complex exposed additional hydrophobic surfaces for dye binding. From TEM images, it would appear that the polymer and sCT interact and alter the conformation of both. QPa normally forms discrete, dense nanoparticles alone or "fluffy" nanoparticles when complexed with insulin [25,34]. When complexed with sCT however, there was an unusual alteration in the conformation, resulting in a bilayered vesicular structure. It is possible that sCT undergoes unfolding in the presence of QPa, resulting in the exposure of hydrophobic domains. Assuming that these domains intercalate between the Pa chains of QPa, such interactions may lead to increased backbone flexibility and the subsequent production of liposome-like bilayer vesicles [35,36].

Complexes were also shown to protect sCT from trypsin, α-chymotrypsin and elastase degradation *in vitro*. This is interesting as it would indicate that QPa is able to offer protection against α-chymotrypsin, unlike its interaction with insulin. This may be due to the differences in interaction between QPa and sCT, and between QPa and insulin, and the differences in α-chymotrypsin target sites between the two proteins. All complexes showed an ability to reduce calcium levels *in vivo* in rats when given intravenously as well as via intrajejunal instillation. The ability to reduce calcium blood levels was comparable to that of sCT alone. Our studies suggest that when proteins and AP of similar charge are complexed, hydrophobic association or hydrogen bonding would perhaps be the predominant interaction. The beneficial effect of using amphiphilic polyelectrolytes in comparison to polyelectrolytes might be the additional hydrophobic association, which could stabilize the PEC when it is in the gastro-intestinal tract (GIT).

It has been shown that the dissociation of the PEC, which relies solely on electrostatic interaction, might occur when it is being exposed to acidic pH values below protein/peptide pI values, and the dilution and counterion effects of gastric fluid [19]. Although *in vitro* studies show potential in terms of facilitating protein transport across intestinal epithelial cells and protection against enzymatic degradation, the complex physiology and anatomical barriers of the GIT, and the presence of food and bile salts, still need to be taken into consideration, and hence future work will look at the ability of these complexes to deliver proteins orally *in vivo*.

7.3.1.2. Amine Modified Polyesters

Poly(lactic-co-glycolic acid) (PLGA) has been widely used to prepare microparticles for protein/peptide delivery. However, release from these hydrophobic matrices is normally triphasic with a considerable burst due to large pore sizes in their matrix and repulsion due to the differences in their polarity with most proteins/peptides [59]. Additionally, the acidic microenvironment produced by their degradation, together with their hydrophobic character can cause protein denaturation [59]. Therefore, grafting (g) poly(vinyl alcohol) (PVA) polymers with poly(lactic acid) (PLA) or PLGA has been attempted in order to increase the overall hydrophilicity of the polymer. It is intended that this should overcome the inherent problems associated with protein/peptide delivery from PLA/PLGA matrices [44,45,47,60].

Utilizing the benefit of the above PVA-g-PLA polymer, Simon and colleagues have further modified the polymer by attaching amine groups, diethylaminopropylamine (DEAPA) to the PVA backbone. This has enabled the design of an amphiphilic polyelectrolyte with an overall positive charge and further increased aqueous solubility (Figure 7.6) [28]. The increase in aqueous solubility and amphiphilic nature allow them to self-assemble under mild conditions (as with the PAA graft polymers discussed in section 7.3.1.1) thereby removing the need to use the complex manufacturing procedures and organic solvents that can have a detrimental effect on proteins/peptides [17,28,48].

A library of polymers with varying degrees of DEAPA and PLA attachment were produced in order to find the optimal AP for complexation with insulin [28]. It was shown that the degree of hydrophilic modification (DEAPA) affects the stability of PEC formation (< 500 nm diameter), while the hydrophobic modification (PLA) did not have an impact. This is because the increase in charge density caused an increased in Coulombic interactions with insulin at lower polymer concentrations. The lack of effect of alteration of PLA attachment would suggest that hydrophobic interactions played little part in the formation of a PEC with insulin. Interestingly, hydrophobic interactions are important in the level of insulin loading in PEC – increasing the degree of PLA attachment resulted in an increase in loading from 20 to 98%.

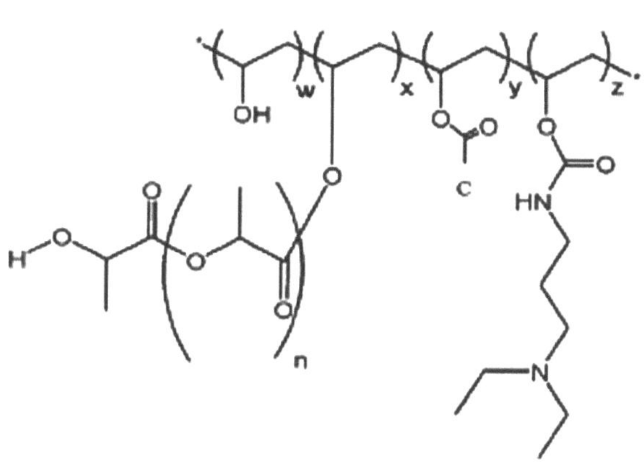

FIGURE 7.6 Structure of amine modified PVA-g-PLGA. *Reprinted from reference [28] with permission from the American Chemical Society.*

Similarly, an increase in DEAPA grafting also lead to an increase in insulin loading due to increased electrostatic interactions. This increase with increasing hydrophobic character of the AP could have been due to more interaction with the hydrophobic β-chain of the insulin molecules. It is thought that insulin unfolds in the presence of AP, and as such the β-chain, which is normally folded inside the molecule, is exposed and so able to hydrophobically interact with the AP. The change in insulin loading with changes in DEAPA/PLA grafting was further clarified by isothermal titration calorimetry. There was a linear increase in binding coefficient between the AP and insulin with increases in DEAPA and PLA grafting [28].

This AP was subsequently investigated for oral delivery of insulin [17]. Enzymatic degradation studies against trypsin were carried out, and the APs were found to limit insulin degradation. Insulin alone was degraded by around 23% within 1 h, while a PEC composed of AP with high degree of PLA grafting (48 lactide chains/PVA backbone) limited degradation to only 5% (Figure 7.7). Again, this was probably due to a greater level of hydrophobic interactions between AP and insulin due to the presence of PLA grafts. In this case the interactions appear to have shielded trypsin's target sites in the β-chain of the insulin molecules. As with amphiphilic PAA, this is an unusual and promising form of protection from enzymatic degradation, since traditionally enzyme inhibitors/chelating agents have had to be used in oral peptide delivery systems

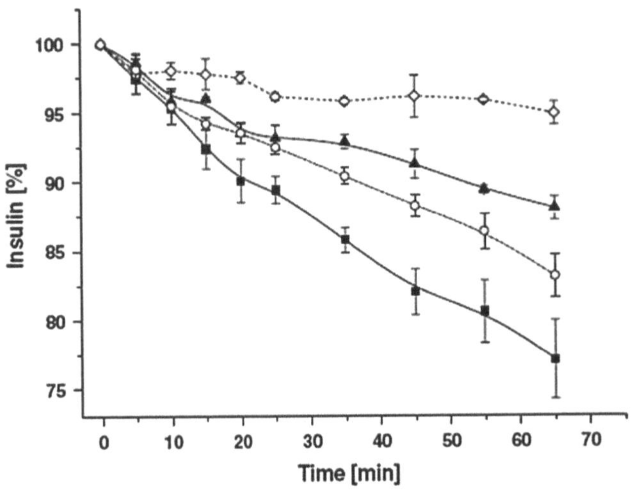

FIGURE 7.7 Enzymatic degradation of insulin (1.0 mg/ml) by trypsin. The nanocomplexes of insulin and P(26)- (- - ○- -), P(26)-1_{LL}- (—▲—), respective P(26)-2_{LL} – polymer (- - ◇ - -) exhibited a significant protection against enzymatic attack in comparison to the unprotected insulin control solution (—■—). P(26)-1_{LL} has 26 amino and 32 lactide groups/PVA molecule; P(26)-2_{LL} has 26 amino and 48 lactide groups/PVA molecule. *Reprinted from reference [17] with permission from Elsevier.*

in order to minimize protein degradation [13,61]. However further work is required to determine if these APs would be just as effective against α-chymotrypsin, pepsin and elastase, since these enzymes have a larger number of target sites on the insulin molecule compared to trypsin. It is difficult to predict the performance of these PECs *in vivo* without knowing how they interact with all major GIT proteases.

The APs with a high degree of PLA grafting (48 lactide chains/PVA backbone) were also shown to reduce the TEER of a Caco-2 cell monolayer by a greater amount than those with lower PLA grafting (32 lactide chains/PVA backbone) or those without PLA attachment. As previously discussed, it is thought that tight junction opening ooccurs due to electrostatic interaction between tight junction proteins and the charged AP. However in this case, the addition of hydrophobic grafts markedly increased the reduction in TEER, which suggests that hydrophobic interactions are equally important. It may also be that charge displacement by PLA grafts to the surface of the PEC [28] resulted in a greater cationic charge being available for tight junction/protein interaction. In addition it is possible that hydrophobic interactions between PLA and cell membranes disrupt tight junctions since it has been shown that APs with a higher degree of hydrophobic grafting appeared to have greater cytotoxicity.

Lactate dehydrogenase (LDH) release from Caco-2 cells was monitored after application of each polymer, and only APs with a high degree of PLA grafts were shown to promote LDH release compared to the control of buffer alone (Figure 7.8). The exact mechanism of interaction is unclear, but it may be that high levels of PLA chains could anchor into the cell membrane and cause pore formation and damage the cell membrane. However, it seems that this effect is reversible. Once AP is removed from cells and is replaced with fresh cell culture media, TEER values returned to 100% of their normal values within 23.5 h of polymer removal and cell washing (Figure 7.9).

Insulin transport across and uptake into a Caco-2 monolayer was also greater for an AP with a high degree of PLA grafting (48 lactide chains/PVA backbone); around 1% of the total insulin administered to the apical side of the monolayer reached the basal compartment of the transwell plate after 2.5 h. However, only 0.2% reached the basal chamber in the same time for AP without hydrophobic grafting. Cell internalization of PECs with a high degree of PLA grafting accounted for 25% of total administered insulin, whereas PVA with no PLA grafts only resulted in 1% of total insulin concentration being internalized. In other words the greater the hydrophobic character of the AP the greater the ability to facilitate both paracellular transport and cell internalization. It would be interesting if the authors could conduct *in vivo* studies to determine the ability to reduce glucose levels via oral administration and elucidate the exact mechanism of how these APs promote the oral absorption of protein.

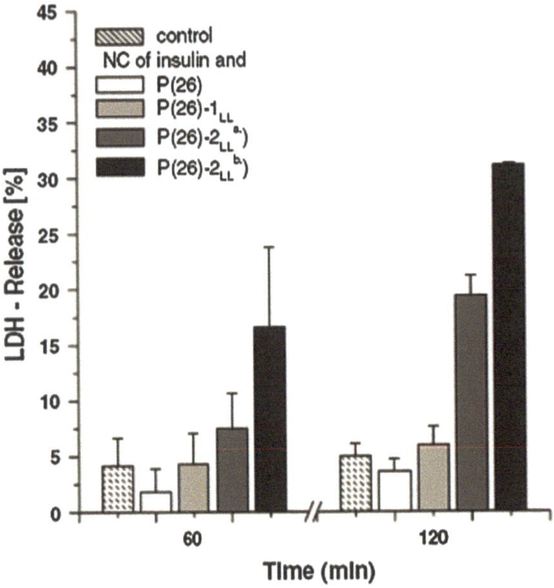

FIGURE 7.8 LDH release after 60 and 120 min of buffer control and PEC composed of insulin and the respective polymers. PECs of P(26)-2_{LL} with (a) 2.125 mg/ml and (b) 6.25 mg/ml exhibited a concentration dependent increase of LDH release. P(26)-1_{LL} has 26 amino and 32 lactide groups/PVA molecule; P(26)-2_{LL} has 26 amino and 48 lactide groups/PVA molecule. *Reprinted from reference [17] with permission from Elsevier.*

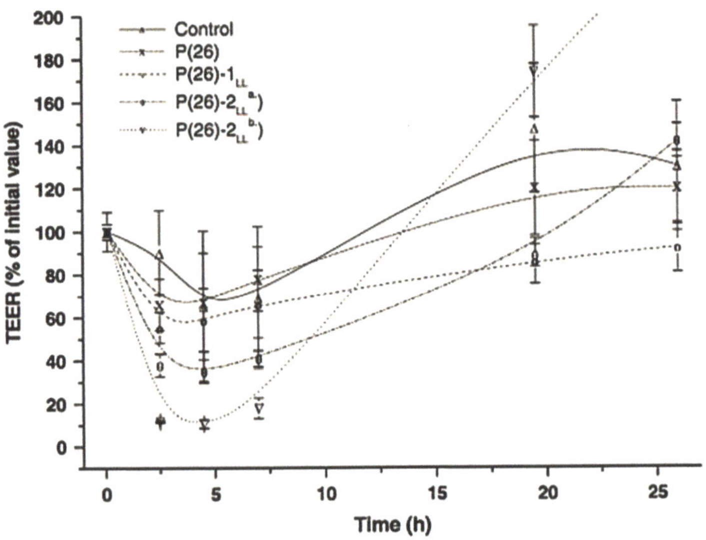

FIGURE 7.9 Reversibility of TEER decrease over 26 h. Layers were incubated with nanocomplexes for 150 min followed by removal of PEC, washing three times and further incubation with culture medium. Applied were a control solution (—Δ—) and PEC composed of insulin and/P(26) polymer (- -X- -), /P(26)-1_{LL} polymer (—O—), /P(26)-2_{LL} polymer (a) with 2.125 mg/ml (—θ—), (b) with 6.25 mg/ml (- - -∇- - -). P(26)-1_{LL} has 26 amino and 32 lactide groups/PVA molecule; P(26)-2_{LL} has 26 amino and 48 lactide groups/PVA molecule. *Reprinted from reference [17] with permission from Elsevier.*

7.3.1.3. Polystyrene with Hydrophilic Branches

In section 7.3.1.1 and 7.3.1.2, we discussed the use of APs which consisted of a hydrophilic backbone modified with hydrophobic moieties. Sakuma and colleagues have produced a novel series of graft copolymers which are unusual in that they consist of a hydrophobic backbone (polystyrene) with hydrophilic grafts [24]. A library of these APs has been produced by altering the hydrophilic grafts and was used to systemically investigate the effect of different charges on the ability of these APs to complex with sCT for intestinal delivery (Figure 7.10).

sCT has been used as a model peptide to determine the ability of these APs to facilitate oral delivery [29]. PECs produced from each AP were shown to have a different level of sCT loading – in the order PMAA (100%) > PNVA (84%) > PVAm (60%) > PNIPAAm (19%). The degree of loading was thought to be affected by electrostatic, hydrophobic interaction and hydrogen bonding between sCT and the AP. PMAA, a polyanion, had the highest loading due to its ability to form electrostatic bonds as well as hydrophobic interactions and hydrogen bonds with cationic sCT. Cationic PVAm and non-ionic PNVA and PNIPAAm were only able to complex with sCT via hydrophobic interactions and hydrogen bonding, which resulted in a lower loading [29]. Contrary to the popular belief that PECs are generally formed between polyelectrolytes and proteins of opposite charges, this work correlates with our work on amphiphilic PAA demonstrating that complexation can take place for an AP and a protein with similar charges, or with a non-ionic AP.

Oral administration of these sCT to rats was able to lower blood calcium to either a greater extent or for a longer period in comparison to an oral solution of free sCT (except PNVA which had no effect on blood calcium levels). sCT alone was able to lower calcium levels by around 0.1 mM after 1 h, however calcium levels returned to pretreatment levels after 2 h. Interestingly the effect of the type of hydrophilic grafting has an impact on the pharmacokinetic profile

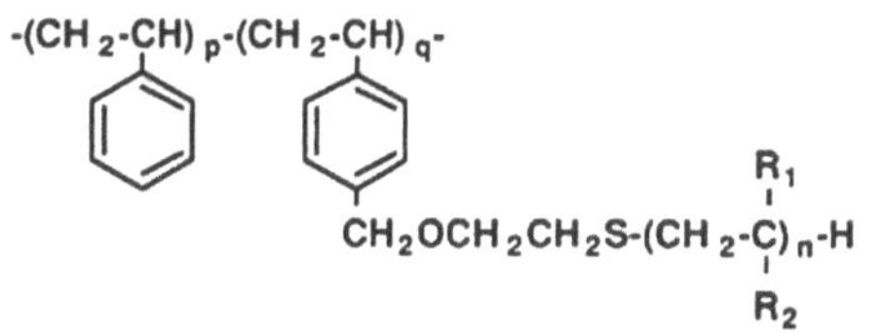

	R_1	R_2
PNIPAAm nanoparticles	H	$CONHCH(CH_3)_2$
PNVA nanoparticles	H	$NHCOCH_3$
PVAm nanoparticles	H	NH_2
PMAA nanoparticles	CH_3	COOH

FIGURE 7.10 Structure of polystyrene based amphiphilic polymers. *(From [11], reproduced with the permission of the copyright owner.)* PNIPAAm, poly(N-isopropylacrylamide); PNVA, poly (N-vinylacetamide); PVAm, poly(vinylamine); PMAA, poly(methylacrylic acid).

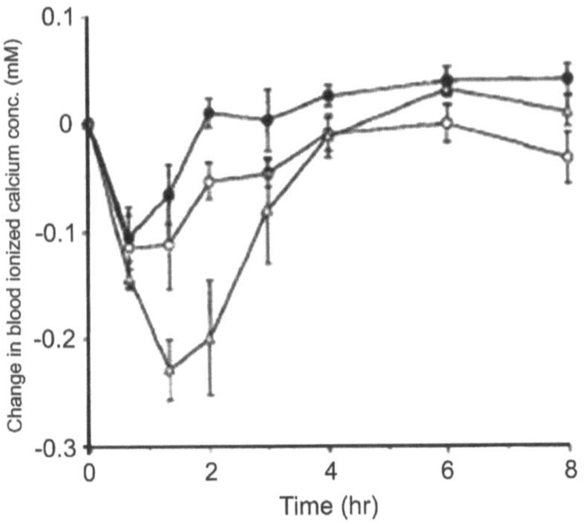

FIGURE 7.11 Concentration-time profiles of ionized calcium in blood after oral administration of sCT aqueous solution (●), a mixture of sCT and PMAA nanoparticles (○) and a mixture of sCT and PNIPAAm nanoparticles (△) in rats (0.25 mg sCT/2.5 ml dosing solution/kg rat). The nanoparticle concentration in the dosing solution was 10 mg/ml. The PNIPAAm nanoparticle size and molecular weight of macromonomers were 750 nm and 3500, respectively. Each value represents the mean ± S.E. *Reprinted from reference [29] with permission from Elsevier.*

of sCT. Non-ionic PNIPAAm PEC was able to cause a greater reduction in calcium levels compared to both PVam and PMMA, but the effect only lasted for 4 h (Figure 7.11). Although this was promising, however, the hypocalcaemic effects facilitated by AP PEC were not taken to be pharmacologically significant.

In order to achieve greater reductions in calcium levels, two approaches were attempted: either altering the dosing regimen, or the structure of PNIPAAm PEC. Splitting the dose of PNIPAAm PEC in half, and then administering each half dose 40 minutes apart had the cumulative effect of a 3-fold increase in hypocalcaemic effect (Figure 7.12) [49]. Alternatively, PVAm were attached to PNIPAAm to confer a cationic charge to the PEC. This again increased hypocalcaemia 3-fold depending on the ratio of VAm and NIPAAm to styrene: the greater the ratio of VAm and NIPAAm to styrene the greater the reduction in calcium (Figure 7.13) [51]. The introduction of a cationic charge to PNIPAAm PEC would suggest that, as with other polyelectrolytes discussed above, the presence of a charged group facilitates paracellular transport. However no work has been carried out to show whether these APs open tight junctions. Indeed the authors state that these APs act by increasing sCT absorption due to their ability to form mucoadhesive bonds with GI mucosa, and limiting enzymatic degradation in the GIT [11,29,49,50,52].

These PECs can maintain sCT proximity to GI epithelium for prolonged periods, which increases sCT uptake by the transcellular route [11]. PNIPAAm, PVAm and PMAA PEC were shown to have slower GI transit times than PNVA PEC (which exerted no hypocalcaemic effect) with transit rates through the stomach of 0.62, 0.35 and 0.87 compared to 1.42% of dose/minute,

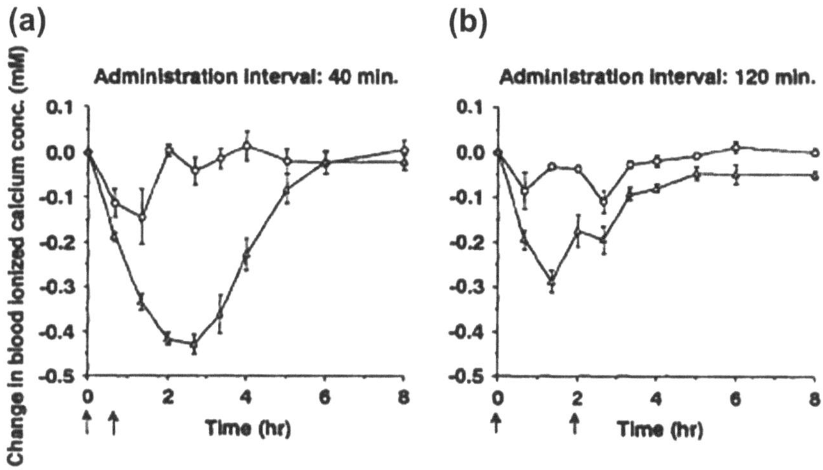

FIGURE 7.12 Concentration-time profiles of ionized calcium in blood after oral administration of sCT solution (○) and a mixture of sCT and PNIPAAm nanoparticles (△) in rats (0.25 mg sCT with 25 mg nanoparticles in 2.5 ml dosing solution/kg of rat body weight). A dose of each dosing solution was halved and the halves were given orally 40 and 120 min apart, as denoted by the upward arrows. The PNIPAAm nanoparticle size and the molecular weight of the macromonomers were 530 nm and 3500, respectively. Water was used as the solvent. Each value represents the mean ± S.E. *Reprinted from reference [49] with permission from Elsevier.*

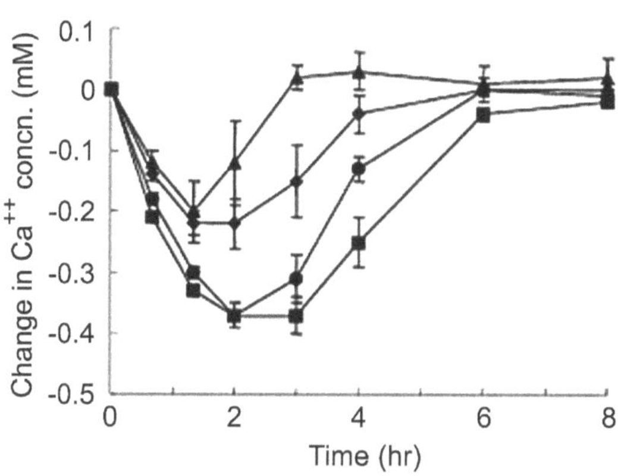

FIGURE 7.13 Concentration–time profiles of ionized calcium in blood after oral administration of a mixture of sCT and PNIPAAm–PVAm nanoparticles (NIPAAm macromonomer: VAm macromonomer:styrene=1.5:0.5:80) (●), a mixture of sCT and PNIPAAm–PVAm nanoparticles (1.5:0.5:10) (■), a mixture of sCT and PNIPAAm–PVAm nanoparticles (1:1:80) (▲) and a mixture of sCT and PNIPAAm–PVAm nanoparticles (0.5:1.5:80) (◆) in rats (0.25 mg of sCT with 25 mg of nanoparticles/2.5 ml/kg). Each value represents the mean ± S.E. of five experiments. *Reprinted from reference [51] with permission from Elsevier.*

respectively [11]. The mucoadhesive strength of the non-ionic PNIPAAm was shown to be greater than that of cationic PVAm and anionic PMAA, indicating that hydrogen bonding and hydrophobic interactions with the mucus layer played a greater part in mucoadhesion than did electrostatic interactions (Figure 7.14) [11]. This ability to form stronger mucoadhesive bonds could explain the greater hypocalcaemic effect found with PNIPAAm PEC, since it allows those PEC to remain *in situ* on the GI mucosa for longer. The addition of PVAm to its structure may have further increased its mucoadhesive capabilities by conferring a cationic charge [51].

Enzymatic protection was offered by these APs due to their varying abilities to reduce the activity of endogenous proteases (trypsin and pepsin) [50]. PNIPAAm and PVAm PEC offered complete protection against pepsin degradation, while PMAA PEC protected sCT against trypsin degradation. Only PVAm PEC offered protection against both enzymes [50]. Again, unlike other polyelectrolytes, such as the amine modified polyesters, these AP appear to display a novel ability to reduce enzyme activity rather than shield proteins/peptides target sites [17,25,26] and/or chelate metal ions from enzymes [62]. It is thought that they prevented the enzymes from interfacing with target sites on sCT by entrapping them [50]. The difference in protection against the two enzymes was determined by the polymer architecture which is able to entrap each enzyme, and in the case of trypsin how much sCT was loaded into the

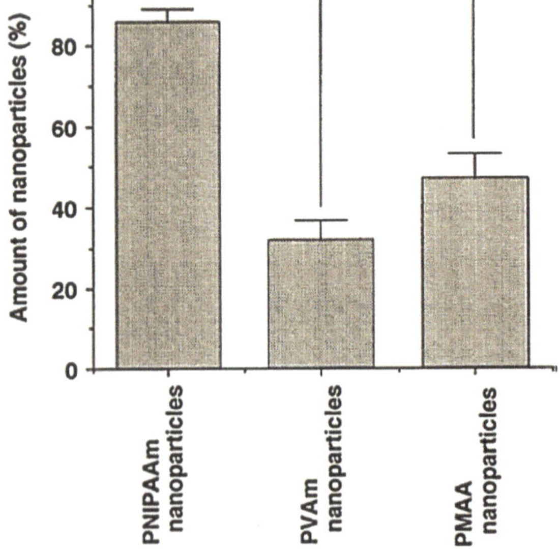

FIGURE 7.14 Amounts of nanoparticles remaining in the intestinal lumen. Rats were anaesthetized with urethane, and nanoparticles dispersed in a phosphate buffered solution of pH 6.8 were injected into the 10-cm intestinal loop (10 mg of nanoparticles in 1 ml dosing solution:kg rat). The radioactivity of the dosing solution was adjusted to 0.3–0.5 kBq:ml. At 5 min after injection, the phosphate buffered solution was perfused for 10 min at 1 ml/min. The radioactivities of the dissected loop and the perfusate were measured by gamma scintigraphy. Each value represents the mean ± S.E. (n = 3). *Reprinted from reference [11] with permission from Elsevier.*

PEC. PMAA PEC prevented degradation by trypsin as they had 100% sCT loading, and so were able to reduce trypsin activity from around 200 BAEE IU/mL to less than 50 BAEE IU/mL (Figure 7.15). However, even though PVAm

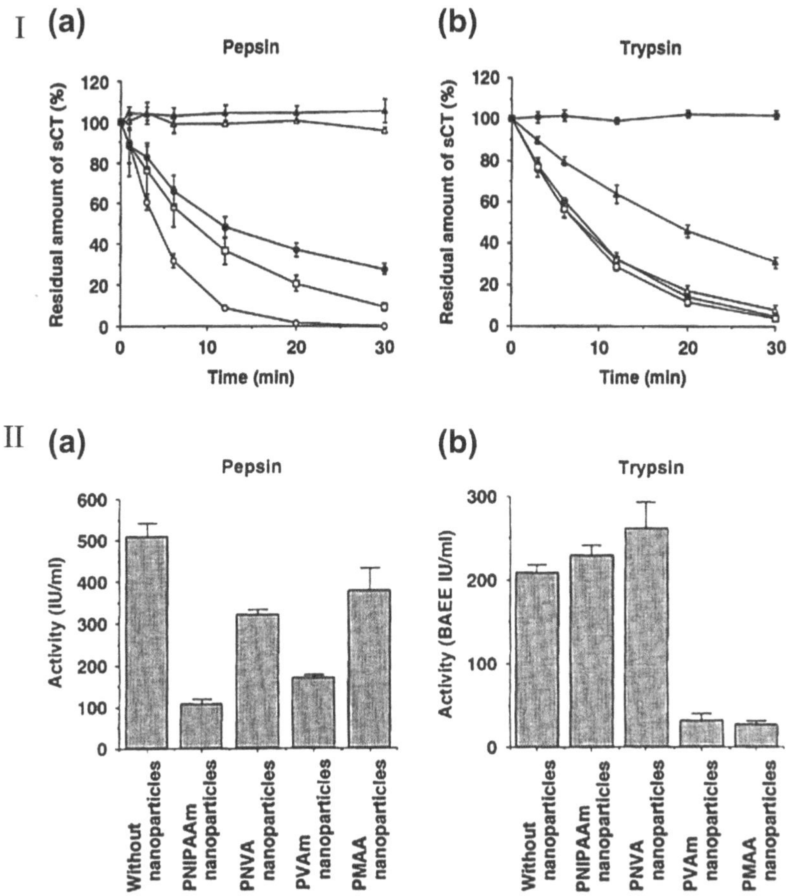

FIGURE 7.15 (I) Residual amount of sCT after incubation of sCT alone (○), a mixture of sCT and PNIPAAm nanoparticles (△), a mixture of sCT and PNVA nanoparticles (□), a mixture of sCT and PVAm nanoparticles (▲) and a mixture of sCT and PMAA nanoparticles (●) with pepsin (a) and with trypsin (b) in a hydrochloric acid-sodium chloride buffered solution of pH 1.2 and a phosphate buffered solution of pH 6.8, respectively, at 37°C. The initial concentrations of sCT, nanoparticles, pepsin and trypsin were 0.1 mg/ml, 10 mg/ml, 1000 IU/ml and 3 BAEE IU/ml, respectively. Each value represents the mean ± S.D. of three experiments. (II) Enzymatic activity after pepsin (a) and trypsin (b) were incubated in a hydrochloric acid-sodium chloride buffered solution of pH 1.2 and a phosphate buffered solution of pH 6.8, respectively, in the presence or absence of nanoparticles at 37°C for 20 min. The initial concentrations of sCT, nanoparticles, pepsin and trypsin were 0.1 mg/ml, 10 mg/ml, 1000 IU/ml and 300 BAEE IU/ml, respectively. Each value represents the mean ± S.D. of three experiments. *Reprinted from reference [50] with permission from Elsevier.*

PEC reduced trypsin activity to a similar level as with PMAA, the lower level of sCT loading meant that the unloaded sCT was still exposed to the remaining active trypsin molecules. This resulted in around 60% of sCT being degraded after 30 minutes.

The ability to protect against pepsin activity appeared to be independent of sCT loading, as PNIPAAm offered almost the same level of protection against pepsin as PVAm even though sCT loading was three times greater in PVAm PEC [50] (Figure 7.15). The level of protection against pepsin degradation is surprising given that it is a non-selective protease, whereas trypsin has selected target sites. It might have been expected that sCT loading would play a bigger role in determining degradation by pepsin rather than trypsin. The fact that the opposite was true would indicate that the mechanism of entrapment of the two enzymes by PEC differed to such an extent that even non-loaded sCT could be protected from pepsin. The form of interaction with PNIPAAm and PVAm would appear to have altered the conformation of pepsin such that it was unable to interact with sCT at all. In the case of trypsin it would appear that any changes in its conformation in the presence of PVAm had a very limited effect on its ability to degrade non-loaded sCT.

The inability of PNIPAAm PEC to protect against trypsin degradation may explain why its effect in lowering blood calcium was more transitory than PVAm and PMAA PEC. sCT may have been degraded more rapidly in the intestine when administered with PNIPAAm, and so even though it formed stronger mucoadhesive bonds, it was unable to facilitate sCT absorption quickly enough to prevent notable degradation in the intestine. However, the addition of PVAm grafts to PNIPAAm would appear to have overcome this problem. The presence of PVAm could increase protection against trypsin and sCT loading while maintaining mucoadhesive properties [51]. To date, polystyrene with hydrophilic grafts were the only APs which have been administered orally *in vivo* and shown considerable potential for the oral delivery of proteins/peptides. No further published work is available on these polymers, but it would be interesting to see if they were as effective in protecting against enzymatic breakdown and promoting oral absorption when used in conjunction with other peptides.

7.3.2. Chemically Modified Polycations (Chitosan Derivatives)

Chitosan is a naturally occurring polysaccharide derived from chitin which is extracted from the shells of crustaceans [61] (cf. Chapter 3, pages 52–54). It has good biodegradability, biocompatibility and bioadhesive properties, making it ideal for pharmaceutical applications [21,61,63]. Chitosan has come to prominence in the field of peptide/protein delivery as it has been shown to promote the penetration of proteins/peptides across both cultured Caco-2 cell monolayers [64,65], nasal mucosa *in vivo* [66–68] and intestinal mucosa (in pH 5.6 media) [65,69]. It facilitates the paracellular transport of hydrophilic macromolecules across cell layers via the opening of tight junctions. This occurs via electrostatic interactions between its protonated NH_2 groups and the negatively charged groups found in mucus, tight junction proteins and glycoproteins on cell surfaces [70–73]. The ability of chitosan to open tight junctions was the result of the reduction of ZO-1 proteins and a change in F-actin morphology from filaments to globules in tight junctions [63,73]. These changes have been shown to be less toxic to cells than the traditional permeation enhancers, i.e. carbomers and surfactants [74].

However, chitosan is a weak base with a pK_a of 5.5, and as such is only soluble under acidic conditions [21,75]. Therefore its use as a permeation enhancer in protein/peptide delivery is limited to acidic conditions where it carries a cationic charge and exists in its uncoiled state [72,75,76]. The effect of charge on cellular interaction was confirmed when the charge on chitosan solutions was partially reduced by the presence of heparin, which resulted in a reduced ability to open tight junctions between Caco-2 cells [63].

Acid salt forms of chitosan can be formed by recrystallizing chitosan base in a variety of acids, e.g. chitosan HCl and chitosan glutamate [63]. Acidic solutions of these salts have been shown to increase the nasal absorption of insulin *in vivo,* whereby blood glucose levels in rats fell by up to 74% of their initial value within 120 minutes of administration [67]. Gels of chitosan HCl with the peptide buserelin have also been shown to increase its bioavailability (to 5.1%) in rats after intraduodenal administration [69]. This salt form of chitosan was only effective in promoting absorption from the intestine as it was preformed as a gel in pH 6.7 media.

When Thanou *et al.* attempted to promote buserelin absorption in rats by intraduodenal administration of a chitosan HCl dispersion (pH 7.2) there was negligible uptake of the peptide [77]. The chitosan HCl precipitated at neutral pH and hence had very limited ability to interact with intestinal epithelial cells, as these salt forms are insoluble at neutral and basic pHs [63,74,78]. They only appear to be suited as absorption enhancers when pre-formed as gels. When present in this gel form they appear to be able to maintain their structure and therefore their positive charge at intestinal pH – which they cannot do as simple solutions/dispersions. Derivatives of chitosan have therefore been produced in order to maximize its potential as a peptide/protein absorption enhancer, including the production of quaternary ammonium salts and thiolated forms [7,61,79].

7.3.2.1. Chitosan with Quaternary Ammonium Moieties

A number of different forms of quaternized chitosan have been produced, including: *N*-diethylmethyl chitosan (DEMC) [21], triethyl chitosan (TEC) [21,75,80], dimethylethyl chitosan (DMEC) [75,80] and, the most popular to date in protein/peptide delivery, *N*-trimethyl chitosan chloride (TMC) (Figure 7.16) [19,21,61,63,64,72–75,77–79,81–83]. They all contain quaternary ammonium moieties which confer a largely pH-independent,

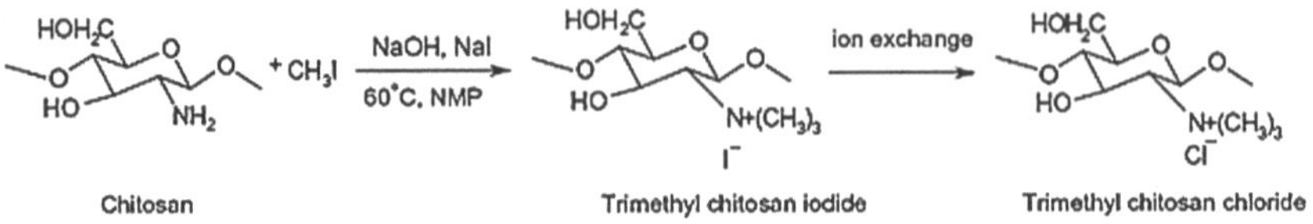

FIGURE 7.16 Structure of chitosan and synthesis of TMC. *Reprinted from reference [61] with permission from Elsevier.*

permanent, positive charge; quaternization of the primary amines at C-2 on chitosan chains can be achieved by using methyl iodide to methylate the primary amines [84]. This means that these derivatives are soluble under both neutral and alkaline conditions, and hence are potentially more suited as intestinal absorption promoters. These derivatives are able to maintain electrostatic interactions with tight junctions between cells at neutral and alkaline pH and so, unlike chitosan, may be able to facilitate paracellular transport of macromolecules in the intestine either as a macromolecular dispersion ("free" solution) [63,64,75] or as part of a PEC delivery system [19,75].

The ability of quaternized chitosan salts to interact with cell membranes and junctions is determined to a large extent by the percentage of quaternization (%Q) where the higher the %Q, the higher the observed level of transport of buserelin across a Caco-2 monolayer. This correlates well with the bioavailability of buserelin when administered intraduodenally [73,74,77]. Thanou *et al.* found that both transport across a Caco-2 monolayer and absolute bioavailability of buserelin increased with TMC %Q from 40% to 60% (Figure 7.17) [77]. This compares with Kotzé *et al.* who achieved a high total dose of buserelin passing across a Caco-2 monolayer using TMC with only 12

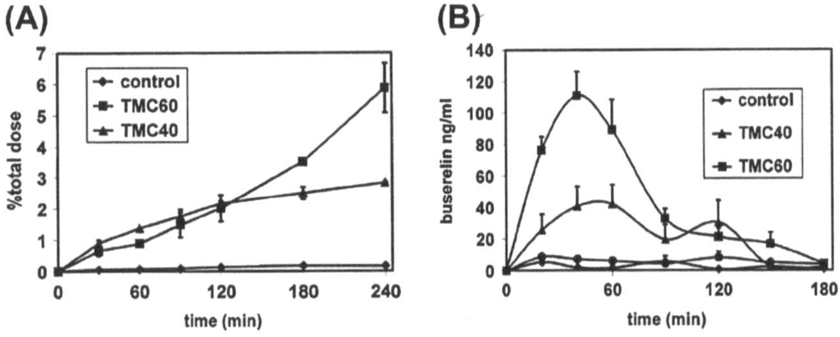

FIGURE 7.17 (A) Transport of buserelin acetate across Caco-2 cell monolayers, without co-application of the polymers (control) and with co-application of 1.0% TMC40 and TMC60 (pH = 7.2; n = 3; mean ± S.E.). (B) Buserelin serum levels after intraduodenal administration of buserelin acetate (500 mg/rat) alone (control) and after co-administration of buserelin with 1.0% TMC40, 1.0% TMC60 or 1.0% chitosan HCl. (pH = 7.2; n = 6; mean ± S.E.). *Reprinted from reference [77] with permission from Springer.*

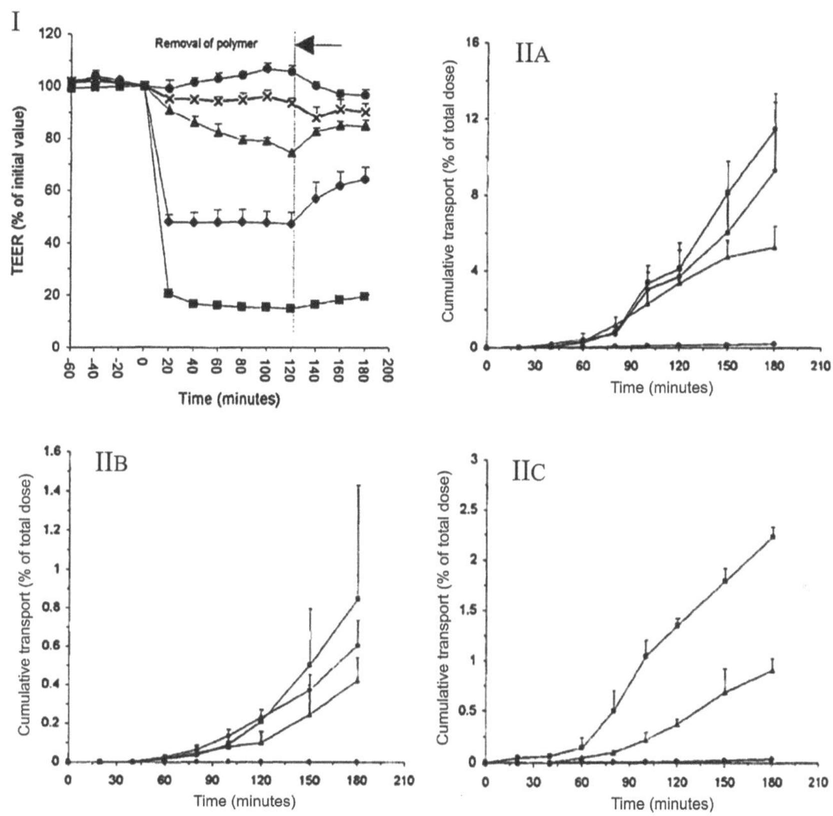

FIGURE 7.18 (I) Effect of TMC concentration on TEER of Caco-2 cell monolayers. Each point represents the mean ± S.D. of three experiments. Control (X), TMC 1%(●), TMC 1.5%(▲), TMC 2% (◆), TMC 2.5% (■). (II) Each point represents the mean ± S.D. of three experiments. (A) Effect of TMC on the cumulative transport of [^{14}C]-mannitol in Caco-2 cell monolayers. (B) Effect of TMC on the cumulative transport of FD-4 in Caco-2 cell monolayers. (C) Effect of TMC on the cumulative transport of buserelin in Caco-2 cell monolayers. Control (●), TMC 1.5%(▲), TMC 2% (◆), TMC 2.5% (■). *Reprinted from reference [64] with permission from Springer.*

%Q (Figure 7.18) [64]. This high degree of buserelin transport with a low level of %Q could be explained by the difference in TMC concentrations used in the two studies: Thanou *et al.* used 1% solutions, while Kotzé *et al.* used 2.5% solutions of TMC [64,77]. Therefore the concentration of TMC also plays a role in the extent to which it can facilitate paracellular transport. The higher the concentration used the greater the charge density present, and so the greater the chance of forming electrostatic bonds with cellular proteins and facilitating junction opening [64].

It has been stated that the optimal %Q is around 50% to facilitate the paracellular transport of hydrophilic molecules [85,86]; the lower the %Q the lower the charge density on the TMC molecules and therefore the weaker their interactions with tight junction and surface proteins [73,85,86]. The use of high %Q values (above 50 %) could lead to a reduced interaction with the mucus layer found on epithelial cells, which could inhibit mucoadhesion and also subsequently reduce the cellular junction protein interaction [85,86]. Synman *et al.* found that bioadhesion to a dried mucin solution decreased as %Q increased from 22% to 48% [87]. It was found that the greater the %Q, the greater the charge on chitosan molecules and so the greater the interchain repulsion. This can lead to changes in chain conformation and a decrease in chain flexibility, which will limit interpenetration of chains into the mucus layer. The increased steric hindrance of the methyl groups now present on the amine group also limits the electrostatic interaction between positive charges on TMC molecules and the negative charges in the mucus. This lack of chain flexibility and reduced ability to form mucoadhesive bonds at higher %Q values mean that a balance must be struck between the need for high enough %Q values to maintain solubility over a wide pH range, and the need to minimize reductions in chain flexibility, which could impinge upon intimate cellular contact and interaction.

However, when using TMC, the transport of proteins across a mucosa is still limited by the molecular weight of the protein. Kotzé *et al.* found that the ability of TMC solutions (2.5% (w/v)) to facilitate hydrophilic molecule transport across a Caco-2 monolayer was limited by the molecular weight of the transported molecule [64]. The transport of [^{14}C]-mannitol was greater than that of buserelin. This occurred even though the TMC solution reduced TEER by around 80% of its initial value, and maintained that reduction for as long as it was in contact with the cells (more than 120 minutes) (Figure 7.18). The higher the molecular weight of a hydrophilic molecule, the lower is its ability to pass across cell layers via the paracellular route, apparently regardless of how low TEER becomes or for how long the reduction is maintained. Therefore the ability of polymer solutions to increase peptide/protein transport is limited by their ability to only facilitate paracellular transport [64,80].

In the above case, TMC and the hydrophilic molecules were administered as separate solutions to the monolayers. However when polymer and peptide/protein solutions are pre-mixed prior to administration they form PECs, which may be able to facilitate both paracellular and transcellular transport [75,80]. When formulated as PEC they can still interact electrostatically with cell proteins while also being taken up by non-specific endocytosis/transcytosis as long as they have a positive charge and are less than 200 nm in size [80]. However there is some disagreement in the literature over this. Sadeghi *et al.* found that chitosan and its derivatives in solution enhanced insulin transport across Caco-2 monolayers to a greater extent than PEC formulations produced at the same polymer concentrations [75]. They found that polymer solutions co-applied to Caco-2 cells with insulin produced greater falls in TEER and corresponding higher insulin levels were found in the transwell basal chambers (Figure 7.19).

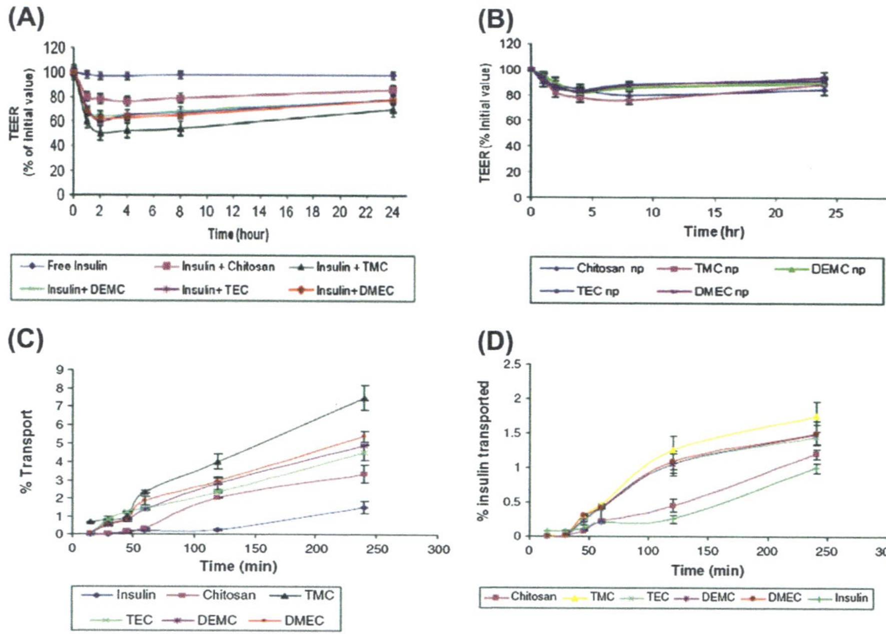

FIGURE 7.19 (A) Effects of free-soluble polymers on TEER of Caco-2 cell monolayer. After 240 min, the monolayers were rinsed with HBSS–HEPES and culture medium was applied on the monolayers. Data are expressed as mean ± S.D. of 3 experiments the difference ($p < 0.05$) is considered significant. (B) Effects of nanoparticles on TEER of Caco-2 cell monolayer. After 240 minutes, the monolayers were rinsed with HBSS–HEPES and culture medium was applied on the monolayers. Data are expressed as means ± S.D. of three experiments the difference ($p < 0.05$) is considered significant. (C) Cumulative transport of insulin across Caco-2 cell monolayers using free-soluble polymers in HBSS–HEPES medium. Data are expressed as means ± S.D. of three experiments the difference $p < 0.05$ is considered significant. (D) Cumulative transport of insulin across Caco-2 cell monolayers using nanoparticles in HBSS–HEPES medium. Data are expressed as means ± S.D. of three experiments the difference $p < 0.05$ is considered significant. *Reprinted from reference [75] with permission from Elsevier.*

These differences can in part be explained by the reduction in polymer zeta potential upon PEC formation. The formation of the PEC is driven by Coulombic interactions between the positively charged groups on the polymer and the negatively charged groups on insulin [75]. Therefore a degree of charge reduction on the polymer chains will take place on PEC formation (TMC zeta potential fell from 43 mV to 22 mV on PEC formation). The smaller reduction in TEER when cells were treated with PEC, compared to polymer only solutions, could account for the fall in insulin transport with PEC. There would be limited opportunities for paracellular transport with PEC due to the higher TEER values. Transcellular transport may also have occurred with the PEC, but this could have been limited due to the fact that some PEC formulations were greater than 200 nm in size and had polydispersity index values of greater than 0.3. This could have meant that there was a limit to how many particles were of the required size and charge to facilitate transcellular transport. The PEC could have also become trapped inside the Caco-2 cells in their Golgi apparatus or endoplasmic reticulum [88]. This appeared to be confirmed by the amount of insulin neither in the apical or basal chambers of the transwell plate after 4 h incubation with the PEC formulations (27% with TMC-insulin PEC) (Table 7.1).

However, given that PECs can facilitate both transcellular and paracellular transport, they may be more suited to peptide/protein delivery than co-administration with polymer solutions. This may be particularly important with high molecular weight proteins, e.g. human growth hormone (22124 g/mole), as the transport of even dextran (4400 g/mole) is severely limited by paracellular transport alone (< 1% of total dose after 3 h incubation with 2.5% TMC solution in basal compartment of transwell plate) [64].

The reported differences in insulin transport between Bayat *et al.* and Sadeghi *et al.* may also be due to the conditions used during the transport experiments [75,80]. While Sadeghi *et al.* carried out *in vitro* work using a Caco-2 cell line, Bayat *et al.* carried out transport experiments on *ex vivo* rat colon mucosa and *in vivo* in the rat colon. While Caco-2 cells are known to mimic the gut wall when differentiated, the use of an *ex vivo* mucosa should give a better indication of *in vivo* performance. Indeed Bayat *et al.* found that PECs were able to both increase insulin transport across the *ex vivo* mucosa and reduce rat blood glucose to a greater extent than co-administration of insulin with polymer solutions (Table 7.2) [80].

It may have been that the sustained intimate contact with the intestinal mucosa provided by PEC for insulin could have helped to set up a concentration gradient for insulin across the GIT wall. The insulin and polymer should be co-localized as long as there are electrostatic forces between them. This means that the insulin was in close proximity to the intestinal wall, and in particular to the parts of the wall where the quaternized chitosan was exerting its permeability-enhancing effects. This may not have been the case in the co-administration of separate polymer and insulin solutions, as the polymer solutions and insulin

TABLE 7.1 Insulin concentration in the apical and basolateral compartments of transwell plates cultured with Caco-2 cells (incubation time for each sample before washing and replacing with fresh media was 240 min, n = 3). Reprinted from reference [75] with permission from Elsevier

Polymer	μg Insulin apical part (t_0)	μg Insulin basolateral part	μg Insulin apical part (t = 240 min)	μg Insulin apical + basolateral	% Insulin missing
Insulin	2000	30 ± 0.3	1960 ± 20	1990 ± 20.3	0.5
Chitosan	2000	67 ± 0.5	1874 ± 53	1941 ± 53.5	2.9
TMC	2000	150 ± 0.7	1690 ± 62	1840 ± 62.7	8.0
DMEC	2000	108 ± 0.4	1788 ± 75	1896 ± 75.4	5.2
DEMC	2000	90 ± 0.3	1657 ± 75	1747 ± 76.3	12
TEC	2000	98 ± 0.4	1758 ± 76	1856 ± 74.4	7.2
Chitosan (np)	2000	24 ± 0.1	1591 ± 66	1615 ± 66.1	19
TMC (np)	2000	35 ± 0.1	1380 ± 46	1415 ± 46.1	29
DMEC (np)	2000	30 ± 0.2	1456 ± 55	1476 ± 55.2	26
DEMC (np)	2000	29 ± 0.1	1490 ± 65	1519 ± 65.1	24
TEC (np)	2000	29 ± 0.2	1440 ± 74	1469 ± 57.2	27

solutions were not a single formulation as with PEC. Therefore there was no guarantee *in vivo* that the polymer solutions would have increased the intestinal residence time for insulin, or kept it in close proximity to the epithelial wall. The polymer solutions may have adhered to parts of the intestinal wall and increased their permeability. However the majority of the insulin may have been degraded or washed out of the duodenum without the opportunity to pass across the membrane as the insulin may not have been co-localized at the areas the polymer solutions adhered to the wall.

Caco-2 cells are widely accepted as a good model for screening the ability of polymers to facilitate intestinal absorption [89,90]. However, these cells do not produce a mucus layer on differentiation [91]. When chitosan solutions were tested using HT29-H goblet cells, which do produce a mucus layer, and rat ileum, they had a reduced effect on drug perfusion than when applied to Caco-2 monolayers [91]. It was demonstrated that if the mucus layer of HT29-H cells was removed prior to addition of the chitosan solutions then drug perfusion increased. Therefore the presence of a mucus layer limits the ability of chitosan

TABLE 7.2 Mean blood glucose concentration after ascending colon administration of the nanoparticles and free insulin to diabetic rats (mg/dL, n = 4). Reprinted from reference [80] with permission from Elsevier

Test sample	0 min	After 30 min	After 60 min
Insulin	395.7 ± 9.87	388.3 ± 25.3	399.2 ± 12.5
Insulin + DMEC 2%	405.8 ± 31.2	379.8 ± 25.8	365.5 ± 18.8
Insulin + TEC 2%	412.1 ± 26.5	396.5 ± 30.1	381.6 ± 35.2
Insulin + CS 2%	409.5 ± 27.1	401.2 ± 21.1	395.6 ± 18.4
DMEC-INS nano	401.2 ± 26.5	310.7 ± 35.5	264.5 ± 17.6
TEC-INS nano	405.6 ± 32.2	321.2 ± 28.1	285.4 ± 22.3
CS-INS nano	403.3 ± 29.1	365.6 ± 44.1	326.3 ± 29.4
DMEC-INS nano + DMEC 2%	389.3 ± 19.3	299.1 ± 22.6	232.3 ± 18.9
TEC-INS nano + TEC 2%	401.7 ± 31.1	310.2 ± 29.8	249.6 ± 33.5
CS-INS nano + CS 2%	414.5 ± 27.9	379.2 ± 31.2	311.6 ± 21.3

solutions to interact with tight junction proteins. Chitosan, as a macromolecular dispersion, cannot diffuse effectively through the mucus layer. It appeared to be trapped within the mucus layer due to the formation of electrostatic interactions with mucus. The use of Caco-2 cells does not take into account the effect of a mucus layer on the ability of a polymer to increase intestinal epithelial wall permeability. As stated above, Bayat *et al.* found that PECs were much more effective *in vivo* in lowering blood glucose than simple polymer solutions. This could be because the PECs may have had an easier passage through the mucus layer, and an increased level of intimate contact with cell surfaces than macromolecular polymer solutions. Their reduced zeta potentials in comparison to free solutions may limit interaction and entrapment within mucus layers.

In order to facilitate oral protein/peptide delivery, the formulation used must not only increase permeation across the GIT, but also protect proteins/peptides from enzymatic degradation [19,65,72]. Traditional enzyme inhibitors are polyanions, e.g. carbomers, and they function by chelating Ca^{2+} ions [92]. These ions are essential to maintain the structure and function of endogenous enzymes, and their temporary chelation can inhibit local enzyme activity [72].

Carbomers also facilitate paracellular transport by chelating some ions, which can lead to irreversible conformational changes in cell morphology and therefore permanent damage to the GIT epithelium [77]. However, quaternized chitosan salts do not inhibit enzyme activity in this way as they are polycations [72].

It has been shown that chitosan and its salts have no direct effect on enzyme activity [65,72]. It is thought that the protection against enzymatic degradation of proteins and peptides is due to a combination of electrostatic and steric hindrance that prevents the enzyme getting access to its target sites, as well as potential changes in protein/peptide folding which again limit the availability of enzyme target sites [15,19,80,93]. TMC solutions have been shown to have a very limited ability to inhibit α-chymotrypsin, one of the proteases found in the intestine, even up to concentrations of 2.5% (w/v) [72]. However, TMC PECs have been shown to protect insulin from trypsin degradation *in vitro* for up to an hour, with around 60% of initial insulin concentration remaining after this time [19]. Indeed this was greater than the protection offered by cross-linked TMC hydrogel nanoparticles, which showed only 40% of the insulin remaining after 60 minutes incubation with trypsin. This is surprising given that the hydrogel nanoparticles should have offered a physical barrier to insulin degradation, while the PECs were a colloidal dispersion. It would appear that the interactions of the hydrogel nanoparticles with insulin were weaker than those of the PEC, due to the competitive interaction between insulin and tripolyphosphate (the anionic cross-linker) for the cationic charges on TMC [19].

The stability of chitosan PEC in the GIT has not been fully investigated. Some limited work has been carried out to determine this stability in simulated gastric and intestinal fluid [19]. This showed that TMC-insulin PECs are completely unstable in simulated gastric fluid, due to the change in charge on insulin below its pI and the ionic strength of the media (high counterion effect). In neutral and alkaline conditions insulin is anionic. However, below its pI insulin is in a cationic state, which means that the Coulombic interactions between quaternized chitosan salts and insulin are lost, causing the structure of the PEC to collapse [19]. However, the same instability was found with the same PEC added with a cross-linker, tripolyphosphate. Indeed these cross-linked PECs were less stable in simulated intestinal fluid (SIF) than the original PEC. PECs were able to maintain their structure at volume ratios of < 1:2 (PEC:SIF), whereas these cross-linked PEC dissociated at volume ratios of 1:1. This was perhaps due to the weaker interaction between insulin and TMC in the presence of tripolyphosphate, an anionic molecule.

Therefore, as mentioned in section 7.3.1.1, it would appear that any delivery system that is solely reliant on electrostatic interactions between its component parts to maintain its structure may be largely ineffective as an oral protein/peptide delivery system. APs may be more suitable as they can form hydrophobic as well as electrostatic interactions with peptides/proteins which may in part negate any counterion effects experienced in the GIT. Alternatively, chemically modified polyelectrolytes may need to be part of a solid delivery system, i.e. tablets/capsules, which are resistant to gastric pH values/counterion effects. This would suggest that the *in vivo* work carried out using these chitosan derivatives does not give a true picture of their performance as oral protein/peptide absorption enhancers. The *in vivo* work carried out using these systems has to date involved direct administration to the colon/duodenum/jejunum [63,74,80,81]. This means that the systems were not exposed to acidic pH values below protein/peptide pI values, and to the diluting and counterion effects of gastric fluid. There may be still be some dissociation of PEC on intraintestinal administration due to the ionic strength and volume of intestinal fluid, but it may be that enough PEC remained stable to facilitate protein/peptide uptake and hence a positive result was observed in these studies.

Some work has been carried out to investigate the ability of TMC to act as an absorption enhancer in tablet/granule form [82]. The use of a solid dosage form is desirable in peptide/protein transport as peptides and proteins are inherently unstable in aqueous conditions. Van der Merwe *et al.* produced minitablets and granules containing TMC and the peptide desmopressin, which could be filled into gelatine capsules [82]. These multiple unit dosage forms (MUDF) are thought to be more suitable to protein/peptide delivery than single unit dosage forms (SUDF). In general, MUDF are more effective than SUDF at prolonging GIT residence time and achieving a more reproducible therapeutic effect, while minimizing the risk of dose dumping [82,94,95].

It was found that the release of TMC and desmopressin could be controlled by the use of tetraglycerol pentastearate. The use of this surfactant resulted in a formulation which released TMC quickly and thereafter desmopressin in pH 7.2 phosphate buffer. This was thought to be important because TMC must be released quickly in order that it could spread over and adhere to the GIT epithelium, and subsequently open tight junctions before the release of desmopressin. This is to ensure that the maximum amount of desmopressin could be transported across the GIT before it became unstable or was degraded in the intestinal fluid. However, at most, only around 40% of the desmopressin within the formulations was released within 2 h. Additionally, no release studies were carried out using SIF or SGF to determine the effect that ionic strength and pH would have on the release. Also, the gelatine capsules that were used were not coated in a gastric-pH insoluble coating (e.g. Eudragit®), which would probably be required in order to promote intestine-only release and absorption [96]. These formulations would probably act in a similar way to the co-administration of peptide and polymer solutions, and so would not be able to promote transcellular transport.

7.3.2.2. Thiolated Chitosan

Thiolated forms of chitosan, or thiomers, consist of compounds containing thiol groups conjugated to chitosan via amide (e.g. cysteine HCl [31]) or amidine (2-iminothiolane [97]) bonds. Formation of amide bonds needs to be carried out at pH values below 5 and/or in an inert atmosphere to avoid oxidation of the thiol groups. However, even if carried out under these conditions only a third of the thiol groups are free in the recovered product. The remainder have cross-linked to form disulfides [31]. Therefore amidine bonds are sometimes favored since the thiol groups are protected against cross-linking during the coupling reaction [98].

Thiolated chitosans are able to chelate metal ions of endogenous enzymes – which native chitosan or its quaternized salts cannot. Their increased ability to protect against enzymatic degradation is necessary if they are to be effective in delivering proteins and peptides orally [61,62]. Thiolated forms of chitosan have been found to form stronger mucoadhesive bonds than unmodified chitosan [97]. They are able to do this at pH values above 5 due to the increased charge on the thiol groups at elevated pH values; at pH values above 5 the thiol groups are oxidized from SH to S^-. This confers an additional ability to form disulphide bonds with cysteine-rich regions in the mucal glycoproteins [61,99–101]. Thiomers can form intra- and intermolecular bonds between thiol groups at pH values above 5. This means that they can form gels *in situ* in the intestine, thereby increasing the residence time in the GIT and releasing the proteins and peptides in a controlled manner [13,97,102,103].

This increased ability to form mucoadhesive bonds can also enhance mucosal wall permeability to proteins and peptides; the increased duration of mucoadhesive attachment and subsequently prolonged interaction with tight junction proteins can maintain increased mucosal wall permeability [91,98]. The increased contact time with mucosal membranes may also set up a peptide/protein concentration gradient across mucosal walls, thereby further increasing the absorption of the peptide/protein [98]. Thiolated chitosans may also be able to inhibit the ATPase activity of the P-glycoprotein pump in the intestine, which could limit the efflux of therapeutic proteins and peptides and further increase their absorption [79].

However, the ability of all mucoadhesive delivery systems is limited by the turnover of mucus within the body, which occurs every 12–24 h in humans [13]. This means that there is a limit to the length of time a delivery system can remain attached to mucosal surfaces, and hence a limit to the length of time they can effect protein and peptide delivery into the body.

Polymers with thiol groups can also have unfavorable reactions with the thiol/disulphide bonds within other protein/peptide structures [98], potentially destabilizing them, which will alter their activity. This can be avoided if the potential interaction between the thiomer and protein/peptide is minimized by reducing the charge on the polymer thiol groups. This is accomplished by either limiting water content of the delivery system, maintaining the pH of the delivery system below pH 5 or situating the thiol groups of the polymer near non-ionic or cationic groups [98].

Thiolated chitosan has predominantly been used to produce minitablets for the oral delivery of either sCT [13,14] or insulin [103]. These formulations have proved effective in promoting absorption of insulin or sCT across the GIT. In the case of sCT, delivery to both the stomach and intestine has been shown to reduce plasma calcium levels to around 90% of their initial levels for up to 12 h after administration (Figure 7.20) [14]. This is surprising given that the available surface area for absorption in the intestines is so much greater than that in the stomach. Traditionally, the intestines are considered the ideal site for absorption of therapeutic compounds, due to the presence of villi and microvilli in their walls which increases the surface area available for absorption.

The difference in release profile of calcitonin from thiolated chitosan matrices in HCl and a pH 6.8 buffer may explain why the reduction in calcium levels appeared to be the same whether targeted at the stomach or intestines. When the tablets were in HCl, around 90% of calcitonin was released within 4 h [14], in pH 6.8 around 50% was released in 8 h [13]. The slower release at pH 6.8 may be due to the increased inter- and intramolecular bonding between thiomer molecules, brought about by an increased oxidation of thiol groups, and forming a more cohesive, gelled matrix. Therefore this may have limited the ability of calcitonin to pass across the intestinal wall before the detachment of the tablet due to mucin turnover thus limiting their ability to take advantage of the larger surface area for absorption.

Guggi *et al.* suggested that the stomach is in fact a better site for peptide and protein absorption due to the lack of enzymatic exposure compared to the

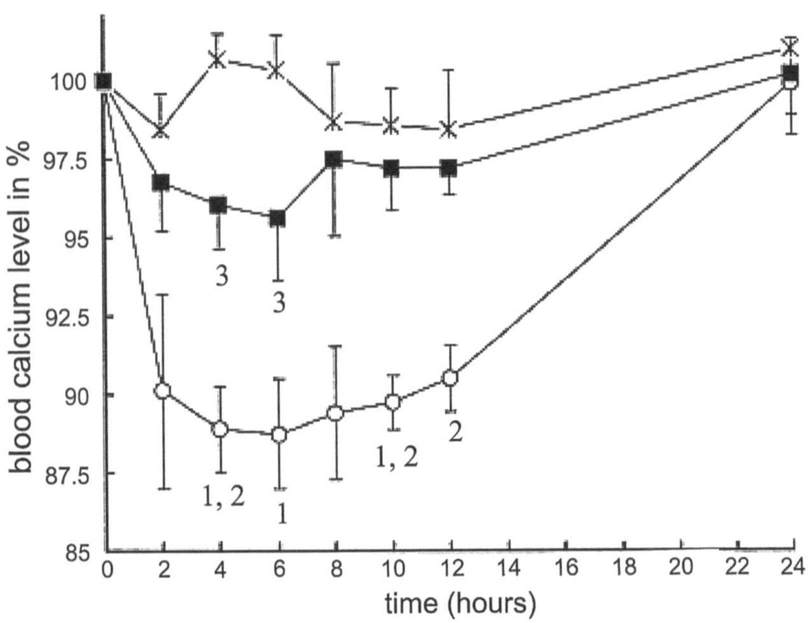

FIGURE 7.20 Decrease in plasma calcium level as a biological response for the salmon calcitonin bioavailability in fasted rats after oral administration of dosage form A (○), of dosage form B (■) and of dosage form C (X) all containing 50 μg of the peptide drug. Indicated values are the mean results from five rats ± S.D.; 1, different from dosage form C, $P < 2 \times 10^{-6}$; 2, different from dosage form B, $P < 0.0001$; 3, different from dosage form C, $P < 0.005$. *Reprinted from reference [14] with permission from Elsevier.*

intestines (pepsin in the stomach [14]; elastase, trypsin and α-chymotrypsin in the intestines), and the fact that gastric transit time would not affect onset and duration of absorption. It is known that thiolated chitosan was able to form mucoadhesive bonds at low pH in the stomach, as 4 h after administration thiolated chitosan tablets were still found in rats' stomachs. This proved that the effect on plasma calcium levels was due to absorption through the stomach, and that peptides like calcitonin can be absorbed from the stomach. However, it was necessary to include pepstatin-A conjugated to chitosan in the minitablets to reduce calcitonin degradation in the stomach. Additionally, when formulated for delivery in the intestines (by using a Eudragit® coating) it was necessary to include inhibitors for elastase, trypsin and α-chymotrypsin conjugated to chitosan in the tablets [13].

Reduced glutathione was also included in the tablets which could further increase GI wall permeability. Reduced glutathione inhibits protein tyrosine phosphatase which maintains tight junctions. It is possible to use it in conjunction with thiomers because they minimize oxidation which causes its deactivation [98]. However, these formulations do not appear to be as effective as I/V administration of sCT which achieved a reduction to around 80% of initial calcium levels for 8 h.

In the case of insulin delivery to the intestines, the use of enzyme inhibitors and reduced glutathione in combination with thiolated chitosan in minitablet form resulted in reduced glucose levels for up to 24 h after oral administration to rats even after feeding them at 12 h. The reduction in glucose levels was slower and lower than with I/V or S/c insulin (reduced to 50% within 1 h after oral administration compared to a reduction to 20% 2 h after I/V or S/c administration). However, after injection, glucose levels increased to 80% of initial levels after 4 h and then returned to 100% or greater thereafter.

Thiolated chitosan has also been produced with quaternary ammonium moieties (TMC SH) in order to deliver insulin orally in the form of PECs (Figure 7.21) [31]. Quaternization was carried out in order for the thiolated chitosan to remain soluble in media above pH 5. It was found that insulin release and mucoadhesion could be controlled by changing molecular weight and the degree of quaternization. Release rates could be reduced by increasing these factors: TMC SH 30 kDa 15 %Q released 100% of insulin in 1 h, 30 %Q released 100% of insulin in 2 h; TMC SH 500 kDa 30 %Q released 80% of insulin in 4 h. Mucoadhesion increased with molecular weight and %Q as well. TMC SH PECs were shown to have a much greater mucoadhesion than TMC PECs *in vivo*: TMC PEC had a mucoadhesive capacity (relative to their initial mass) of around 20% while TMC SH PEC had a mucoadhesive capacity of between 70 and 80% at equivalent molecular weight and %Q in rat GITs. Permeation enhancement of insulin *in vitro* (Caco-2 cells), *ex vivo* (rat ileum) and *in vivo* (rat) was found to be optimal at 200 kDa and 30 %Q. At higher molecular weights, steric hindrance was thought to limit cellular interactions and therefore reduce the thiolated chitosans' ability to increase the permeability of cell layers to insulin.

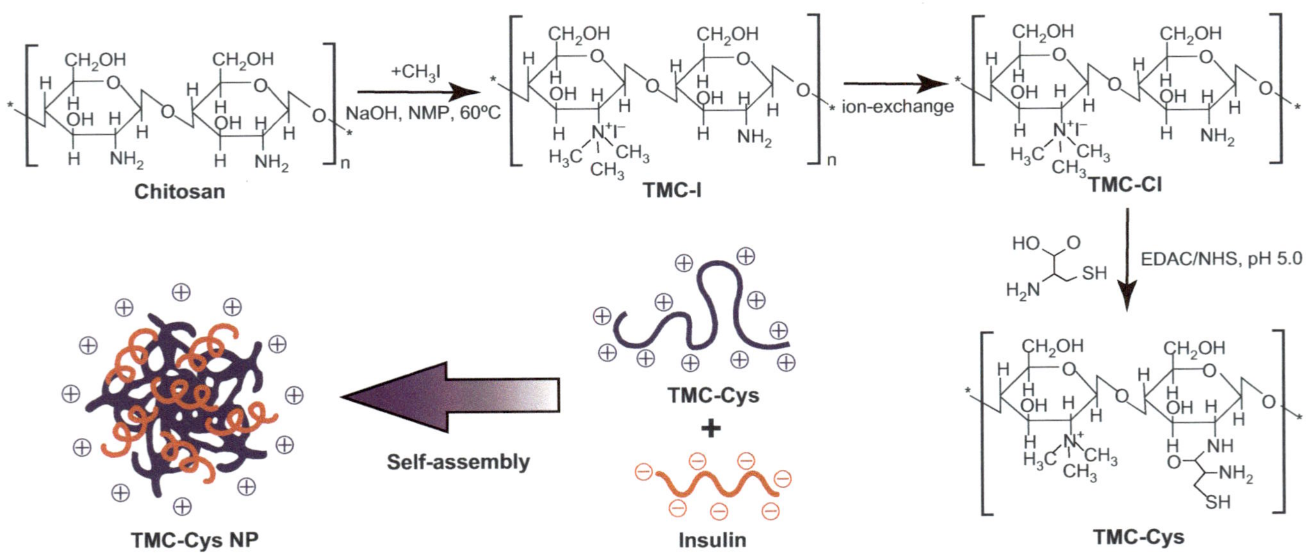

FIGURE 7.21 Structure and synthesis of cysteine grafted chitosan with quaternary ammonium moieties (TMC-Cys) and structure of TMC-Cys complexes (TMC-Cys NP). *Reprinted from reference [31] with permission from Elsevier.*

The use of PEC was also found to increase the permeation of insulin between 2.5–3.5 times more than equivalent polymer solutions; as was stated above this is probably due to the ability of the PEC to facilitate both transcellular and paracellular transport. TMC SH PECs were shown to increase permeation of insulin across the ileum and Caco-2 cells compared to TMC PEC. This again was due to the increased interactions between TMC SH and the mucus, tight junctions and membranes compared to TMC. *In vivo*, TMC SH PECs were shown to reduce glucose levels for up to 12 h, while TMC PECs reduced glucose levels only up to 8 h and to a lesser extent (Figure 7.22). The authors also demonstrated uptake via Peyer's patches which suggests that these PECs could be taken up by the lymphatic system as well as via paracellular and transcellular routes. The uptake again was greater with TMC SH than TMC due to the greater mucoadhesive interaction of TMC SH with Peyer's patches.

It would appear that the combination of thiol groups and quaternary ammonium moieties on the same molecule may provide the inherent advantages of both functional groups and overcome the limitations of chitosan as a delivery system. TMC SH has high mucoadhesive interactions with mucus layers and electrostatic interactions with cell membranes and junctions, as well as with therapeutic proteins. They have the ability to chelate metal ions, which can deactivate some peptidases, and they retain a degree of solubility across the pH values that are found in the GIT. When in PEC form they may be able to facilitate absorption across the GIT via paracellular and transcellular routes and via the lymphatic system.

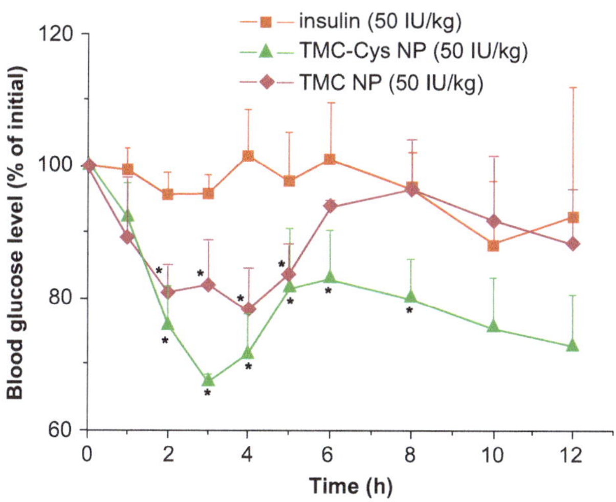

FIGURE 7.22 Serum glucose levels in normal rats following oral administration of TMC NP and TMC-Cys NP at an insulin dose of 50 IU/kg. Indicated values were means ± S.D. of four experiments. Significant difference from insulin solution: $p < 0.05$. *Reprinted from reference [31] with permission from Elsevier.*

7.4. CONCLUSION

Chemically modified polyelectrolytes have been shown to form nanocomplexes spontaneously with a range of peptides under mild conditions. Most studies showed that they have the ability to facilitate both paracellular and transcellular transport of peptides *in vitro* by either opening tight junctions via electrostatic or hydrophobic interactions with cells, or internalization via endocytosis. The additional mucoadhesive properties of chitosan that has been modified with quaternary ammonium moieties and/or thiolated groups increases the permeation of the proteins/peptides across Caco-2 cell monolayers *in vitro*, which have translated to promising oral peptide delivery systems. However, enzyme inhibitors are still needed to obtain a better pharmacokinetic profile.

With regard to the protection against enzymatic degradation, most of the polyelectrolytes have only a limited ability to protect against enzymatic breakdown. They can, however, offer some form of electrostatic or hydrophilic shielding of some target sites. The exceptions to this are polystyrene based amphiphilic polymers which were able to entrap and reduce the activity of a range of enzymes, and thiolated chitosan, where the presence of negative charge allows the chelation of some metal ions which are essential for enzymatic activity.

The amphiphilic polyelectrolytes have the advantage of being able to interact with a protein/peptide via both hydrophobic and electrostatic associations, whereas chitosan modified with quaternary ammonium moieties or thiolated groups can only form electrostatic interactions. The former is beneficial since it allows complexation with proteins of opposite as well as similar charges. In addition, the hydrophobic association can potentially prevent the destabilization of PEC in the GI tract environment, where high concentrations of ions and variations in pH might affect electrostatic interaction between protein and polyelectrolytes. However, to date there are only limited *in vivo* data on the oral administration of amphiphilic polyelectrolyte–protein complexes due to their relatively recent application in this area. Although chemically modified polyelectrolytes have shown considerable potential in the intestinal delivery of proteins and peptides, a thorough understanding of the exact mechanism of the interactions is needed to design complexes of superior functionalities for oral delivery. Further understanding is needed regarding the specific groups on the protein that associate with the polyelectrolytes, the number of protein molecules being complexed to one polyelectrolyte, the presence of interpolymer network and insulin, etc.

REFERENCES

[1] S. Mao, O. Germershaus, D. Fischer, T. Linn, R. Schnepf, T. Kissel, Uptake and transport of PEG-graft-trimethyl-chitosan copolymer-insulin nanocomplexes by epithelial cells, Pharm. Res. 22 (2005) 2058–2068.

[2] S.M. Hartig, R.R. Greene, M.M. Dikov, A. Prokop, J.M. Davidson, Multifunctional nanoparticulate polyelectrolyte complexes, Pharm. Res. 24 (2007) 2353–2369.
[3] C. Weber, A. Drogoz, L. David, A. Domard, M.-H. Charles, B. Verrier, et al., Polysaccharide-based vaccine delivery systems: Macromolecular assembly, interactions with antigen presenting cells, and *in vivo* immunomonitoring, J. Biomed. Mater. Res. A 93 (2010) 1322–1334.
[4] K. Bowman, K.W. Leong, Chitosan nanoparticles for oral drug and gene delivery, Int. J. Nanomed. 1 (2006) 117–128.
[5] M. George, Abraham, T. Emilia, Polyionic hydrocolloids for the intestinal delivery of protein drugs: Alginate and chitosan – a review, J. Control. Release 114 (2006) 1–14.
[6] M. de la Fuente, N. Csaba, M. Garcia-Fuentes, M.J. Alonso, Nanoparticles as protein and gene carriers to mucosal surfaces, Nanomedicine 3 (2008) 845–857.
[7] M. Werle, H. Takeuchi, A. Bernkop-Schnurch, Modified chitosans for oral drug delivery, J. Pharm. Sci. 98 (2009) 1643–1656.
[8] M. Amidi, E. Mastrobattista, W. Jiskoot, W.E. Hennink, Chitosan based delivery systems for protein therapeutics and antigens, Adv. Drug Deliv. Rev. 62 (2010) 59–82.
[9] H. Dautzenberg, Polyelectrolyte complex formation in highly aggregating systems 1. Effect of salt: Polyelectrolyte complex formation in the presence of NaCl, Macromolecules 30 (1997) 7810–7815.
[10] A. Shovsky, I. Varga, R. Makuska, P.M. Claesson, Formation and stability of water soluble, molecular polyelectrolyte complexes: Effects of charge density, mixing ratio, and polyelectrolyte concentration, Langmuir 25 (2009) 6113–6121.
[11] S. Sakuma, R. Sudo, N. Suzuki, H. Kikuchi, M. Akashi, M. Hayashi, Mucoadhesion of polystyrene nanoparticles having surface hydrophilic polymeric chains in the gastrointestinal tract, Int. J. Pharm. 177 (1999) 161–172.
[12] M. Thanou, J.C. Verhoef, P. Marbach, H.E. Junginger, Intestinal absorption of octreotide: *N*-trimethyl chitosan chloride (TMC) ameliorates the permeability and absorption properties of the somatostatin analogue *in vitro* and *in vivo*, J. Pharm. Sci. 89 (2000) 951–957.
[13] D. Guggi, C.E. Kast, A. Bernkop-Schnurch, *In vivo* evaluation of an oral salmon calcitonin-delivery system based on a thiolated chitosan carrier matrix, Pharm. Res. 20 (2003) 1989–1994.
[14] D. Guggi, A.H. Krauland, A. Bernkop-Schnurch, Systemic peptide delivery via the stomach: *In vivo* evaluation of an oral dosage form for salmon calcitonin, J. Control. Release 92 (2003) 125–135.
[15] A.H. Krauland, M.J. Alonso, Chitosan/cyclodextrin nanoparticles as macromolecular drug delivery system, Int. J. Pharm. 340 (2007) 134–142.
[16] C.B. Packhaeuser, T. Kissel, On the design of an in situ forming biodegradable parenteral depot systems based on insulin loaded dialkylaminoalkyl-amine-poly(vinyl alcohol)-g-poly (lactide-co-glycolide) nanoparticles, J. Control. Release 123 (2007) 131–140.
[17] M. Simon, I. Behrens, L.A. Dailey, M. Wittmar, T. Kissel, Nanosized insulin-complexes based on biodegradable amine-modified graft polyesters poly[(vinyl-3-(diethylamino)-propylcarbamate-co-(vinyl acetate)-co-(vinyl alcohol)]-graft-poly(l-lactic acid): Protection against enzymatic degradation, interaction with Caco-2 cell monolayers, peptide transport and cytotoxicity, Eur. J. Pharm. Biopharm. 66 (2007) 165–172.
[18] G. Perera, M. Greindl, T. Palmberger, A. Bernkop-Schnurch, Insulin-loaded poly(acrylic acid)–cysteine nanoparticles: Stability studies towards digestive enzymes in the intestine, Drug Deliv. 16 (2009) 254–260.

[19] A. Jintapattanakit, V.B. Junyaprasert, S. Mao, J. Sitterberg, U. Bakowsky, T. Kissel, Peroral delivery of insulin using chitosan derivatives: A comparative study of polyelectrolyte nanocomplexes and nanoparticles, Int. J. Pharm. 342 (2007) 240–249.
[20] M. Huang, S.N. Vitharana, L.J. Peek, T. Coop, C. Berkland, Polyelectrolyte complexes stabilize and controllably release vascular endothelial growth factor, Biomacromolecules 8 (2007) 1607–1614.
[21] A.M.M. Sadeghi, F.A. Dorkoosh, M.R. Avadi, P. Saadat, M. Rafiee-Tehrani, H.E. Junginger, Preparation, characterization and antibacterial activities of chitosan, *N*-trimethyl chitosan (TMC) and *N*-diethylmethyl chitosan (DEMC) nanoparticles loaded with insulin using both ionotropic gelation and polyelectrolyte complexation methods, Int. J. Pharm. 355 (2008) 299–306.
[22] S. Dragan, M. Cristea, C. Luca, B.C. Simionescu, Polyelectroylte complexes. 1. Synthesis and characterisation of some insoluble polyanion–polycation complexes, J. Polym. Sci. Pol. Chem. 34 (1996) 3485–3494.
[23] C. Tribet, Complexation between amphiphilic polyelectrolytes and proteins: From necklaces to gels, in: R. Tsetska (Ed.), Physical Chemistry of Polyelectrolytes, Surfactant Science Series, Vol. 99, Marcel Dekker Inc., New York, 1999, pp. 687–741.
[24] S. Sakuma, M. Hayashi, M. Akashi, Design of nanoparticles composed of graft copolymers for oral peptide delivery, Adv. Drug Deliv. Rev. 47 (2001) 21–37.
[25] C.J. Thompson, L. Tetley, I.F. Uchegbu, W.P. Cheng, The complexation between novel comb shaped amphiphilic polyallylamine and insulin – Towards oral insulin delivery, Int. J. Pharm. 376 (2009) 46–55.
[26] C.J. Thompson, L. Tetley, W.P. Cheng, The influence of polymer architecture on the protective effect of novel comb shaped amphiphilic poly(allyl amine) against *in vitro* enzymatic degradation of insulin – Towards oral insulin delivery, Int. J. Pharm. 383 (2010) 216–227.
[27] C.L. Cooper, P.L. Dubin, A.B. Kayitmazer, S. Turksen, Polyelectrolyte–protein complexes, Curr. Opin. Colloid Interface Sci. 10 (2005) 52–78.
[28] M. Simon, M. Wittmar, U. Bakowsky, T. Kissel, Self-assembling nanocomplexes from insulin and water-soluble branched polyesters, poly[(vinyl-3-(diethylamino)-propylcarbamate-co-(vinyl acetate)-co-(vinyl alcohol)]-graft-poly(l-lactic acid): A novel carrier for transmucosal delivery of peptides, Bioconjugate Chem. 15 (2004) 841–849.
[29] S. Sakuma, N. Suzuki, H. Kikuchi, K. Hiwatari, K. Arikawa, A. Kishida, et al., Oral peptide delivery using nanoparticles composed of novel graft copolymers having hydrophobic backbone and hydrophilic branches, Int. J. Pharm. 149 (1997) 93–106.
[30] K. Gupta, M. Ganguli, S. Pasha, S. Maiti, Nanoparticle formation from poly(acrylic acid) and oppositely charged peptides, Biophys. Chem. 119 (2006) 303–306.
[31] L. Yin, J. Ding, C. He, L. Cui, C. Tang, C. Yin, Drug permeability and mucoadhesion properties of thiolated trimethyl chitosan nanoparticles in oral insulin delivery, Biomaterials 30 (2009) 5691–5700.
[32] A. Wittemann, T. Azzam, A. Eisenberg, Biocompatible polymer vesicles from biamphiphilic triblock copolymers and their interaction with bovine serum albumin, Langmuir 23 (2007) 2224–2230.
[33] K.-M. Jin, Y.-H. Kim, Injectable, thermo-reversible and complex coacervate combination gels for protein drug delivery, J. Control. Release 127 (2008) 249–256.
[34] C.J. Thompson, C. Ding, X. Qu, Z. Yang, I.F. Uchegbu, L. Tetley, et al., The effect of polymer architecture on the nano self-assemblies based on novel comb-shaped amphiphilic poly(allylamine), Colloid Polym. Sci. 286 (2008) 1511–1526.

[35] W. Wang, L. Tetley, I.F. Uchegbu, The level of hydrophobic substitution and the molecular weight of amphiphilic poly-l-lysine-based polymers strongly affects their assembly into polymeric bilayer vesicles, J. Colloid Interface Sci. 237 (2001) 200–207.
[36] W. Wang, X.Z. Qu, A.I. Gray, L. Tetley, I.F. Uchegbu, Self-assembly of cetyl linear polyethylenimine to give micelles, vesicles and dense nanoparticles, Macromolecules 37 (2004) 9114–9122.
[37] W.P. Cheng, A.I. Gray, L. Tetley, T.L.B. Hang, A.G. Schatzlein, I.F. Uchegbu, Polyelectrolyte nanoparticles with high drug loading enhance the oral uptake of hydrophobic compounds, Biomacromolecules 7 (2006) 1509–1520.
[38] X.Z. Qu, L. Omar, T.B. Hang Le, L. Tetley, K. Bolton, K.W. Chooi, et al., Polymeric amphiphile branching leads to rare nanodisc shaped planar self-assemblies, Langmuir 24 (2008) 9997–10004.
[39] G. Kwon., T. Okano, Polymeric micelles as new drug carriers, Adv. Drug Deliv. Rev. 21 (1996) 107–116.
[40] M. Talelli, C.J.F. Rijcken, C.F. van Nostrum, G. Storm, W.E. Hennink, Micelles based on HPMA copolymers, Adv. Drug Deliv. Rev. 62 (2010) 231–239.
[41] M. Yokoyama, Polymeric micelles as a new drug carrier system and their required considerations for clinical trials, Expert Opin. Drug Deliv. 7 (2010) 145–158.
[42] W.P. Cheng, C.J. Thompson, S.M. Ryan, T. Aguirre, L. Tetley, D.J. Brayden, *In vitro* and *in vivo* characterisation of a novel peptide delivery system: Amphiphilic polyelectrolyte-salmon calcitonin nanocomplexes, J. Control. Release 147 (2010) 289–297.
[43] C.J. Thompson, R. Knott, W.P. Cheng, The use of quaternised amphiphilic polymer to modify insulin uptake in Caco-2 cells towards oral insulin, J. Pharm. Pharmacol. 61 (Suppl 1) (2009) A40–A41.
[44] A. Breitenbach, T. Kissel, Biodegradable comb polyesters: Part I. Synthesis, characterization and structural analysis of poly (lactide) and poly (lactide-co-glycolide) grafted onto water-soluble poly (vinyl alcohol) as backbone, Polymer 39 (1998) 3261–3271.
[45] A. Breitenbach, K.F. Pistel, T. Kissel, Biodegradable comb polyesters: Part II. Erosion and release properties of poly (vinyl alcohol)-g-poly (lactide-co-glycolide), Polymer 41 (2000) 4781–4792.
[46] T. Jung, A. Breitenbach, T. Kissel, Sulfobutylated poly(vinyl alcohol)-graft-poly(lactide-co-glycolide)s facilitate the preparation of small negatively charged biodegradable nanospheres, J. Control. Release 67 (2000) 157–169.
[47] L.A. Dailey, M. Wittmar, T. Kissel, The role of branched polyesters and their modifications in the development of modern drug delivery vehicles, J. Control. Release 101 (2005) 137–149.
[48] M. Simon, M. Wittmar, T. Kissel, T. Linn, Insulin containing nanocomplexes formed by self-assembly from biodegradable amine-modified poly(vinyl alcohol)-graft-poly(l-lactide): Bioavailability and nasal tolerability in rats, Pharm. Res. 22 (2005) 1879–1886.
[49] S. Sakuma, N. Suzuki, H. Kikuchi, K. Hiwatari, K. Arikawa, A. Kishida, et al., Absorption enhancement of orally administered salmon calcitonin by polystyrene nanoparticles having poly(N-isopropylacrylamide) branches on their surfaces, Int. J. Pharm. 158 (1997) 69–78.
[50] S. Sakuma, N. Suzuki, H. Kikuchi, K. Hiwatari, K. Arikawa, A. Kishida, et al., Stabilization of salmon calcitonin by polystyrene nanoparticles having surface hydrophilic polymeric chains, against enzymatic degradation, Int. J. Pharm. 159 (1997) 181–189.
[51] S. Sakuma, N. Suzuki, R. Sudo, K. Hiwatari, A. Kishida, M. Akashi, Optimized chemical structure of nanoparticles as carriers for oral delivery of salmon calcitonin, Int. J. Pharm. 239 (2002) 185–195.

[52] S. Sakuma, R. Sudo, N. Suzuki, H. Kikuchi, M. Akashi, Y. Ishida, et al., Behavior of mucoadhesive nanoparticles having hydrophilic polymeric chains in the intestine, J. Control. Release 81 (2002) 281–290.

[53] R.J. Schilling, A.K. Mitra, Degradation of insulin by trypsin and alpha-chymotrypsin, Pharm. Res. 8 (1991) 721–727.

[54] P. Calceti, S. Salmaso, G. Walker, A. Bernkop-Schnurch, Development and *in vivo* evaluation of an oral insulin-PEG delivery system, Eur. J. Pharm. Sci. 22 (2004) 315–323.

[55] C.J. Thompson, W.P. Cheng, P. Gadad, K. Skene, M. Smith, G. Smith, et al., Uptake and transport of novel amphiphilic polyelectrolyte-insulin nanocomplexes by Caco-2 cells – towards oral insulin. In preparation.

[56] J. Xia, P.L. Dubin, Y. Kim, B.B. Muhoberac, V.J. Klimkowski, Electrophoretic and quasi-elastic light scattering of soluble protein–polyelectrolyte complexes, J. Phys. Chem. 97 (1993) 4528–4534.

[57] S. Gautier, M. Boustta, M. Vert, Alkylated poly(l-lysine citramide) as models to investigate the ability of amphiphilic macromolecular drug carriers to physically entrap lipophilic compounds in aqueous media, J. Control. Release 60 (1999) 235–427.

[58] M.A.H. Capelle, R. Gurny, T. Arvinte, A high throughput protein formulation platform: Case study of salmon calcitonin, Pharm. Res. 26 (2009) 18–128.

[59] E.S. Lee, K.-H. Park, D. Kang, I.S. Park, H.Y. Min, D.H. Lee, et al., Protein complexed with chondroitin sulfate in poly(lactide-co-glycolide) microspheres, Biomaterials 28 (2007) 2754–2762.

[60] K.F. Pistel, A. Breitenbach, R. Zange-Volland, T. Kissel, Brush-like branched biodegradable comb polyesters: Part III. Protein release from microspheres of poly (vinyl alcohol)-graft-poly (d, l-lactic-co-glycolic acid), J. Control. Release 73 (2001) 7–20.

[61] V.K. Mourya, N.N. Inamdar, Chitosan-modifications and applications: Opportunities galore, React. Funct. Polym. 68 (2008) 1013–1051.

[62] A. Bernkop-Schnürch, G. Walker, Multifunctional matrices for oral peptide delivery, Crit. Rev. Ther. Drug Carr. Syst. 18 (2001) 459–501.

[63] M. Thanou, J.C. Verhoef, H.E. Junginger, Oral drug absorption enhancement by chitosan and its derivatives, Adv. Drug Deliv. Rev. 52 (2001) 117–126.

[64] A.F. Kotzé, H.L. Leuβen, B.J. de Leeuw, B.G. de Boer, J.C. Verhoef, H.E. Junginger, *N*–Trimethyl chitosan chloride as a potential absorption enhancer across mucosal surfaces: *In vitro* evaluation in intestinal epithelial cells (Caco-2), Pharm. Res. 14 (1997) 1197–1202.

[65] H.L. Leuβen, C.O. Rentel, A.F. Kotzé, C.-M. Lehr, A.G. de Boer, J.C. Verhoef, et al., Mucoadhesive polymers in peroral peptide drug delivery. IV: Polycarbophil and chitosan are potent enhancers of peptide transport across intestinal mucosae *in vitro*, J. Control. Release 45 (1997) 14–23.

[66] L. Illum, N.F. Farraj, S.S. Davis, Chitosan as a novel nasal delivery system for peptide drugs, Pharm. Res. 11 (1994) 1186–1189.

[67] T.J. Aspden, L. Illum, Ø. Skaugrud, Chitosan as a nasal delivery system: Evaluation of insulin absorption enhancement and effect on nasal membrane integrity using rat models, Eur. J. Pharm. Sci. 4 (1996) 23–31.

[68] A.M. Dryer, M. Hinchcliffe, P. Watts, J. Castile, I. Jabbal-Gill, R. Nankervis, et al., Nasal delivery of insulin using novel chitosan formulations: A comparative study in two animal models between simple chitosan formulations and chitosan nanoparticles, Pharm. Res. 19 (2002) 998–1008.

[69] H.L. Leuβen, B.J. de Leeuw, M.W.E. Langemeyer, A.G. de Boer, J.C. Verhoef, H.E. Junginger, Mucoadhesive polymers in peroral drug delivery VI. Carbomer and chitosan

improve the intestinal absorption of the peptide drug buserelin *in vivo*, Pharm. Res. 13 (1996) 1666–1670.

[70] C.-M. Lehr, J.A. Bouwstra, E.H. Schacht, H.E. Junginger, *In vitro* evaluation of mucoadhesive properties of chitosan and some other natural polymers, Int. J. Pharm. 78 (1992) 43–48.

[71] C.-M. Lehr, J.A. Bouwstra, H.E. Junginger, A surface energy analysis of mucoadhesive performance by spreading coefficients, Eur. J. Pharm. Sci. 1 (1993) 19–30.

[72] A.F. Kotzé, B.J. de Leeuw, H.L. Leuβen, A.G. de Boer, J.C. Verhoef, H.E. Junginger, Chitosans for enhanced delivery of therapeutic peptides across intestinal epithelia: *In vitro* evaluation in Caco-2 cell monolayers, Int. J. Pharm. 159 (1997) 243–253.

[73] I.M. van der Lubben, J.C. Verhoef, G. Borchard, H.E. Junginger, Chitosan and its derivatives in mucosal drug and vaccine delivery, Eur. J. Pharm. Sci. 14 (2001) 201–207.

[74] M.M. Thanou, A.F. Kotzé, T. Scharringhausen, H.L. Leuβen, A.G. de Boer, J.C. Verhoef, et al., Effect of degree of quaternization of *N*-trimethyl chitosan chloride for enhanced transport of hydrophilic compounds across intestinal Caco-2 cell monolayers, J. Control. Release 64 (2000) 15–25.

[75] A.M.M. Sadeghi, F.A. Dorkoosh, M.R. Avadi, M. Weinhold, A. Bayat, F. Delie, et al., Permeation enhancer effect of chitosan and chitosan derivatives: Comparison of formulations as soluble polymers and nanoparticulate systems on insulin absorption in Caco-2 cells, Eur. J. Pharm. Biopharm. 70 (2008) 270–278.

[76] A.F. Kotzé, H.L. Leuβen, A.G. de Boer, J.C. Verhoef, H.E. Junginger, Chitosan for enhanced intestinal permeability prospects for derivatives soluble in neutral and basic environments, Eur. J. Pharm. Sci. 7 (1998) 145–151.

[77] M. Thanou, B.I. Florea, M.W.E. Langemeyer, J.C. Verhoef, H.E. Junginger, N-Trimethylated chitosan chloride (TMC) improves the intestinal permeation of the peptide drug Buserelin *in vitro* (Caco-2 cells) and *in vivo* (rats), Pharm. Res. 17 (2000) 27–31.

[78] A.F. Kotzé, H.L. Leuβen, B.J. de Leeuw, B.G. de Boer, J.C. Verhoef, H.E. Junginger, Comparison of the effect of different chitosan salts and *N*-trimethyl chitosan chloride on the permeability of intestinal epithelial cells (Caco-2), J. Control. Release 51 (1998) 35–46.

[79] M. Werle, M. Hoffer, Glutathione and thiolated chitosan inhibit multidrug resistance P-glycoprotein activity in excised small intestine, J. Control. Release 111 (2006) 41–46.

[80] A. Bayat, F.A. Dorkoosh, A.R. Dehpour, L. Moezi, B. Larijani, H.E. Junginger, et al., Nanoparticles of quaternized chitosan derivatives as a carrier for colon delivery of insulin: *Ex vivo* and *in vivo* studies, Int. J. Pharm. 356 (2008) 259–266.

[81] M. Thanou, J.C. Verhoef, J.H.M. Verheijden, H.E. Junginger, Intestinal absorption of octreotide using trimethyl chitosan chloride: Studies in pigs, Pharm. Res. 18 (2001) 823–828.

[82] S.M. van der Merwe, J.C. Verhoef, A.F. Kotzé, H.E. Junginger, *N*-Trimethyl chitosan chloride as absorption enhancer in oral peptide drug delivery. Development and characterization of minitablet and granule formulations, Eur. J. Pharm. Biopharm. 57 (2004) 85–91.

[83] S.M. van der Merwe, J.C. Verhoef, J.H.M. Verheijden, A.F. Kotzé, H.E. Junginger, Trimethylated chitosan as polymeric absorption enhancer for improved peroral delivery of peptide drugs, Eur. J. Pharm. Biopharm. 58 (2004) 225–235.

[84] A. Domard, M. Rinaudo, C. Terrassin, New method for the quaternization of chitosan, Int. J. Biol. Macromol. 8 (1986) 105–107.

[85] C. Jonker, J.H. Hamman, A.F. Kotzé, Intestinal paracellular permeation enhancement with quaternized chitosan: *In situ* and *in vitro* evaluation, Int. J. Pharm. 238 (2002) 205–213.

[86] J.H. Hamman, C.M. Schultz, A.F. Kotzé, *N*-trimethyl chitosan chloride: Optimum degree of quaternization for drug absorption enhancement across epithelial cells, Drug Dev. Ind. Pharm. 29 (2003) 161–172.

[87] D. Synman, J.H. Hamman, A.F. Kotzé, Evaluation of the mucoadhesive properties of N-trimethyl chitosan chloride, Drug Dev. Ind. Pharm. 29 (2003) 59–67.

[88] P. Artursson, T. Lindmark, S.S. Davis, L. Illum, Effect of chitosan on the permeability of monolayers of intestinal epithelial cells (Caco-2), Pharm. Res. 11 (1994) 1358–1361.

[89] H. Lennernas, K. Palm, U. Fagerholm, P. Artursson, Comparison between active and passive drug transport in human intestinal epithelial (Caco-2) cells *in vitro* and human jejunum *in vivo*, Int. J. Pharm. 127 (1996) 103–107.

[90] T. Lindmark, N.G. Schipper, L. Lazorova, A.G. de Boer, P. Artursson, Absorption enhancement in intestinal epithelial Caco-2 monolayers by sodium caprate: Assessment of molecular weight dependence and demonstration of transport routes, J. Drug Target. 5 (1998) 215–223.

[91] N.G.M. Schipper, K.M. Varum, P. Stenberg, G. Ocklind, H. Lennernas, P. Artursson, Chitosans as absorption enhancers of poorly soluble drugs: 3: Influence of mucus on absorption enhancement, Eur. J. Pharm. Sci. 8 (1999) 335–343.

[92] H.L. Lueßen, B.J. de Leeuw, D. Pérard, C.-M. Lehr, A.G. de boer, J.C. Verhoef, et al., Mucoadhesive polymers in peroral peptide drug delivery. I. Influence of mucoadhesive excipients on the proteolytic activity of intestinal enzymes, Eur. J. Pharm. Sci. 4 (1996) 117–128.

[93] A. Vila, A. Sanchez, M. Tobio, P. Calvo, M.J. Alonso, Design of biodegradable particles for protein delivery, J. Control. Release 78 (2002) 15–24.

[94] N. Rouge, E.T. Cole, E. Doelker, P. Buri, Screening of potentially floating excipients for minitablets, STP Pharma Sci. 7 (1997) 386–392.

[95] N. Rouge, J.-C. Leroux, E.T. Cole, E. Doelker, P. Buri, Prevention of the sticking tendency of floating minitablets filled into hard gelatin capsules, Eur. J. Pharm. Biopharm. 43 (1997) 165–171.

[96] W.A. Chan, C.D. Boswell, Z. Zhang, Comparison of the release profiles of a water soluble drug carried by Eudragit-coated capsules in different *in vitro* dissolution liquids, Powder Technol. 119 (2001) 26–32.

[97] A. Bernkop-Schnürch, M. Hornof, T. Zoidl, Thiolated polymers—thiomers: Synthesis and *in vitro* evaluation of chitosan-2-iminothiolane conjugates, Int. J. Pharm. 260 (2003) 229–237.

[98] A. Bernkop-Schnurch, A.H. Krauland, V.M. Leitner, T. Palmberger, Thiomers: Potential excipients for non-invasive peptide delivery systems, Eur. J. Pharm. Biopharm. 58 (2004) 253–263.

[99] V.M. Leitner, G.F. Walker, A. Bernkop-Schnurch, Thiolated polymers: Evidence for the formation of disulphide bonds with mucus glycoproteins, Eur. J. Pharm. Biopharm. 56 (2003) 207–214.

[100] M. Roldo, M. Hornof, P. Caliceti, A. Bernkop-Schnurch, Mucoadhesive thiolated chitosans as platforms for oral controlled drug delivery: Synthesis and *in vitro* evaluation, Eur. J. Pharm. Biopharm. 57 (2004) 115–121.

[101] K. Kafedjiiski, A.H. Krauland, M.H. Hoffer, A. Bernkop-Schnurch, Synthesis and *in vitro* evaluation of a novel thiolated chitosan, Biomaterials 26 (2005) 819–826.

[102] A. Bernkop-Schnurch, G. Walker, H. Zarti, Thiolation of polycarbophil enhances its inhibition of intestinal brush border membrane bound aminopeptidase N, J. Pharm. Sci. 90 (2001) 1907–1914.

[103] A.H. Krauland, D. Guggi, A. Bernkop-Schnurch, Oral insulin delivery: The potential of thiolated chitosan-insulin tablets on non-diabetic rats, J. Control. Release 95 (2004) 547–555.

Chapter 8

Nanoparticle Mediated Oral Delivery of Peptides and Proteins: Challenges and Perspectives

M.R. Rekha and Chandra P. Sharma
Biosurface Technology Division, Biomedical Technology Wing, Sree Chitra Tirunal Institute for Medical Sciences & Technology, Thiruvananthapuram, Kerala, India

Chapter Outline

Peptide and Protein Delivery. DOI: 10.1016/B978-0-12-384935-9.10008-2

8.1. INTRODUCTION

Drug carrier systems are now as important as the drug itself. Controlled release provides prolonged delivery of a drug while maintaining its blood concentration within therapeutic limits. Drug delivery systems can thus influence the pharmacological activity by modulating its release from the carrier. Other advantages include increased patient compliance (given a reduction in the frequency of administration), non-invasive routes of administration, minimized local and systemic side-effects, and thus a reduced toxicity profile. Nanosized, controlled, drug delivery systems can deliver the drugs to the site of action in a predesigned manner thereby minimizing side-effects as well as enhancing the bioavailability of that drug [1–4].

Increasing the bioavailability of orally administered therapeutic peptides and proteins is still a challenging and unachieved goal. By virtue of their small size and a high surface area, nanoparticle-mediated oral peptide delivery is believed to enhance the bioavailability of these proteinaceous drugs. The purpose of this chapter is to review the different nanoparticulate matrices, the stability of nanoparticles and the encapsulated peptides/proteins, the issues related to the bioavailability, biocompatibility and toxicity of the nanoparticles, and other related topics such as techniques used to characterize the delivery systems and clinical applications.

8.2. BIOLOGICAL BARRIERS IN ORAL PEPTIDE/PROTEIN DELIVERY

The development of a successful oral peptide/protein delivery system has been actively investigated for past few decades and is still a challenging area of research. Recent advances in biochemistry and biotechnology have led to the discovery and mass production of therapeutic peptides and proteins. However, their bioavailablity via oral delivery is still almost nil because of the many barriers in the gastro-intestinal (GI) tract, such as proteolytic degradation and the inability of these macromolecules to penetrate the intestinal cell wall [5,6].

8.2.1. Gastro-intestinal Barriers

The major proteolytic enzyme in the stomach is pepsin and those in the small intestine include trypsin, chymotrypsin, carboxypeptidase and elastase. In the gastro-intestinal tract, enzymatic degradation can occur at the lumen, brush border, mucosal mesh, etc. Metabolism of proteinaceous drugs inevitably reduces their bioavailability. The small intestine is generally the major site for absorption of drugs, but enzymatic activity of proteases is also higher in this region than any other part of the GI tract. The efficiency and mechanism of drug absorption changes between the different segments of the small intestine. Towards the distal sections, the villi become smaller and fewer, thereby

TABLE 8.1 Summary of the gastro-intestinal barriers to peptide/protein absorption

	Intestine	
Stomach	**Enzymatic (enzymes)**	**Absorption**
Acidic environment Nonspecific proteolytic enzymes like pepsin	Trypsin Chymotrypsin Elastase Pepsin Carboxypeptidase A Carboxypeptidase B Aminopeptidase	A limiting step is the very poor absorption of macromolecules either through the enterocytes or tight junctions

reducing the surface area of absorption. For instance, it is reported that in the upper jejunum the mucosal area per cm of serosal length is 98 cm^2, whereas it is only 20 cm^2 at the ileal side [7]. A summary of these GI barriers is given in Table 8.1.

8.2.2. Mucosal Barriers

Mucus plays an important role in determining the absorption and bioavailability of orally administered drugs, the major component of which is the secreted glycoprotein mucin. In contrast to proteases, the mucin lining presents a physical barrier rather than a chemical one. It is reported that the mucin layer is thickest in the stomach and colon, whereas in the small intestine the thickness varies depending on the extent of digestive activity [8]. The digestive enzyme activity of trypsin and chymotrypsin of the duodenum decreases to almost one-half in the jejunum and to one-third in the ileum [8].

Aoki *et al.* investigated the contribution of the mucous/glycocalyx layers in rat small intestine as a diffusional or enzymatic barrier to the absorption of insulin by *in vitro* studies [9]. Mucosal layers were removed from duodenum, jejunum and ileum using hyaluronidase, without affecting the integrity of the epithelium or cell membrane. The transport of insulin across the hyaluronidase-treated and untreated intestinal segments from these regions was investigated. The authors reported that the apparent permeability coefficient (P_{app}) of insulin for the hyaluronidase-treated segment was significantly higher than that of the control group (treated with phosphate buffered saline, PBS) in all small intestinal regions. The P_{app} of insulin in both groups increased in the order duodenum $<$ jejunum $<$ ileum. Their studies also suggested the possibility of mucous/glycocalyx layers acting as an enzymatic but not a diffusional barrier, irrespective of the intestinal region. Similar observations were reported by Morishita *et al.* using an *in situ* absorption study with different intestinal segment loops [10]. They reported an increase in insulin absorption from ileum, the distal part of the small intestine.

8.3. APPROACHES FOR ORAL PEPTIDE/PROTEIN DELIVERY

Many approaches aimed at enhancing the absorption of orally administered peptides and proteins have been examined, namely: enzyme inhibitors, absorption enhancers and chemical modification of the peptide/protein [11−14]. Over the past few decades, microparticles and (more recently) nanoparticles have been employed to overcome the intestinal barriers and improve the bioavailability of orally administered peptides and proteins.

8.3.1. Enzyme Inhibitors

Conventional enzyme inhibitors have been investigated as a means for preventing the proteolytic degradation of the co-administered peptide/protein drug, thereby enhancing their bioavailability. In the case of co-administration, the inhibitors are not restricted to a particular intestinal site, thus, the disadvantage associated with the co-administration of enzyme inhibitors in long-term therapy is their deleterious side-effects. This mainly involves the poor digestion of food proteins, leading to pancreatic hypertrophy and hyperplasia. In part, this is because the digestive activity and capacity of the small intestine is very high, and large quantities of protease inhibitors will be required to have the desired protective effect. When the intestinal enzyme activity is inhibited and digestion reduced, one consequence is hypersecretion of enzymes from the pancreas, since the release of digestive enzymes is based on a feedback mechanism [15]. As a result, current research is focused on polymer-based localized enzyme inhibition, which will be discussed in detail in the following section.

8.3.2. Absorption Enhancers

Intestinal drug absorption can be improved by altering paracellular and transcellular transport pathways using permeation enhancers (Figure 8.1) [16−19]. Permeation enhancers have different mechanisms of action, including the induction of a loss of integrity of the cell−cell tight junctions, changes in mucus viscosity, modulation of membrane structure and fluidity − associated with changes in the permeation of proteins through the cell membrane or through receptor mediated endocytosis.

Most commonly investigated permeation enhancers include bile acids, fatty acids and dicarboxylic acids. In addition, the peptide/protein may be

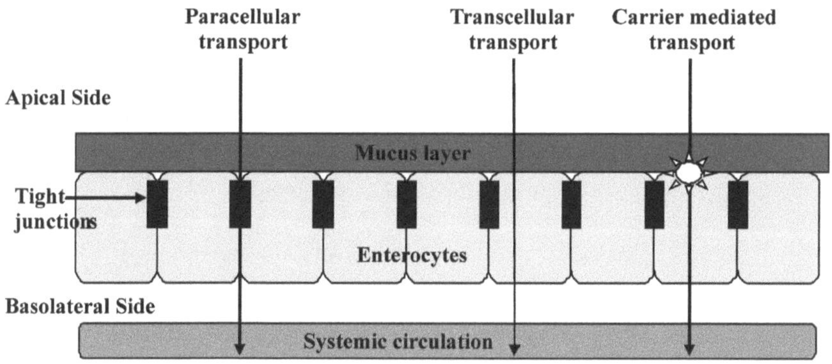

FIGURE 8.1 Schematic summary of the transport pathways across intestinal cell epithelium.

conjugated with ligands such as vitamin B12, transferrin and lectin(s) in order to bind to specific cell epithelial receptors and harness carrier mediated transcytosis. The major permeation enhancers used for oral drug delivery are summarized in Table 8.2.

The main barrier to the paracellular diffusion of macromolecules like proteins and peptides across the epithelial cell layer is the tight junction complex. Tight junctions are composed of transmembrane proteins, namely: occludin, claudins and junctional adhesion molecules (JAM) which intercalate with corresponding proteins from adjacent cells to form the intercellular barrier. These proteins associate with peripheral membrane proteins including the membrane proteins zonula occludens (ZO-1 to 3), which join the transmembrane proteins to the actin cytoskeleton. ZO-1 and occludin are associated with a stimulus-induced tight junction disassembly and concomitant increase in paracellular permeability. It is reported that even though the degree of permeability at the tight junctions varies significantly within different epithelia, the tight junctions are generally impermeable to molecules with radii larger than 11–15 Å [20]. This would include most polypeptides and certainly proteins.

However, various groups have attempted to study the effect and mechanism of chemical permeation enhancers and their correlation with induced toxicity. Most of these studies have highlighted the toxic effects of these enhancers [21–23]. Hence, in long-term therapy, for example in case of diabetes, daily administration of permeation enhancers could cause serious side-effects. Other than toxicity issues, another major problem associated with tight junction loosening (even transiently) is that while transport is enhanced for peptide and protein drugs, this is also true for the pathogenic viruses, toxic peptides, etc., that are naturally present in the GI tract [21]. The tight junctions and their possible modulation for peptide and protein delivery are discussed in further detail in Chapter 9.

TABLE 8.2 The major classes of intestinal permeation enhancers

Class of permeation enhancer		Chemical name
Surfactants	Anionic	Sodium lauryl sulfate Sodium decyl sulfate Sodium octyl sulfate
	Cationic	Cetyltrimethyl ammonium bromide Decyltrimethyl ammonium bromide Benzyldimethyl dodecyl ammonium chloride Myristyltrimethyl ammonium chloride
	Zwitterionic	Cocamidopropyl betaine Decyldimethyl ammonio propane sulfonate Myristyldimethyl ammonio propane sulfonate Palmityldimethyl ammonio propane sulfonate
	Nonionic	Triton-X-100
Fatty acids		oleic acid caprylic acid capric acid lauric acid hexanoic acid
Fatty acid derivatives		Acylcarnitines Acylcholines Isopropyl myristate Isopropyl palmitate Methyl palmitate Diethyl sebaccate Lauryl amine
Bile salts		sodium deoxycholate sodium glycocholate sodium taurocholate sodium taurodihydrofusidate sodium glycodihydrofusidate
Nitrogen containing rings		Caprolactam Methyl pyrrolidone Octyl pyrrolidone Methyl piperazine Phenyl piperazine
Chelators		EDTA Citric acid Salicylates Dicarboxylates

8.3.3. Chemical Modification of Peptides and Proteins

Peptides and proteins, via the application of advanced techniques in recombinant technologies and biochemistry, can be conjugated with other polymeric moieties such as polyethylene glycol (PEG); this is generally called PEGylation. From a therapeutic perspective, there are three main reasons for doing this [24]:

(i) PEGylation alters the interaction of the therapeutic molecule with the blood proteins, especially with respect to its surface charge and size. This may also affect its receptor-binding capacities and filtration through the glomeruli. Generally, PEGylation decreases the rate of clearance of the therapeutic peptide from the systemic circulation.
(ii) Therapeutic proteins, even of recombinant origin (e.g. humanization of monoclonal immunoglobulins), elicit immune responses on repeated exposure *in vivo*, particularly with regard to generating "neutralizing" antibodies. PEG is an aqueously soluble polymer with a large hydrated volume and raised chemical potential. This property of PEG sterically masks the protein's immunogenic epitopes.
(iii) PEG is reported to protect the proteins from enzymatic attack; again, this is thought to be a consequence of its steric effect ("shielding" the protein). PEGylation is discussed further in Chapter 11.

Conjugation may also involve ligands which are targets for receptor-mediated endocytosis. The transferrin receptor is one such example; transferrin is a protein that occurs naturally in the human body and is employed in iron transport. Transferrin receptor-mediated uptake of various drugs, drug delivery matrices and gene delivery vectors has been widely investigated by various groups. With respect to the topic of this book, Shah and Shen conjugated insulin with transferrin via a disulfide linkage, and investigated the carrier-mediated transport of insulin across Caco-2 cell monolayers [25]. They observed that transport of the conjugated insulin was mediated via the transferrin receptor and not through the insulin receptor. The authors found that insulin-transferrin (In-Tf) transport across the Caco-2 cell monolayers increased by 5- to 15-fold compared to free insulin. From the same group, Xia *et al.* investigated the efficiency of In-Tf conjugates in lowering the blood glucose level of diabetic rats [26]. They observed that the In-Tf conjugate, if administered either subcutaneously or orally, was capable of lowering blood glucose levels by 70% of the initial value. The In-Tf complex also demonstrated a delayed onset with prolonged activity in the lowering of blood glucose levels. Kavimandan *et al.* took this study further, and investigated the stability of In-Tf in the presence of the proteolytic enzymes trypsin and chymotrypsin, demonstrating increased stability of insulin to proteolysis [27]. The authors stated that the exact mechanism was not clearly understood, but it is suggested that the stability may be due to steric hindrance (a shielding effect) from the transferrin moiety.

8.4. NANOPARTICLE BASED APPROACHES

In its broadest definition, "nano" may be defined, with respect to nanoparticles, as sizes ranging from 10 to 1000 nm. Other definitions will describe nanoparticles as being around 100 nm or less, with the term "sub-micron" used for particles of several hundred nanometers in diameter. Nano-based protein or peptide delivery systems cover various categories, such as polymeric nanoparticles, liposomes and solid lipid nanoparticles, inorganic particles, dendrimers and polymeric micelles (cf. Chapter 5, pages 97–98). Nanoparticles possess very high surface area to volume ratios, thus allowing for intimate interaction between the surface of the particle and the gastro-intestinal epithelium (and mucus). In accordance with the Stokes-Einstein relation, the diffusion coefficient for nano-sized carriers (compared to microspheres) should in principle facilitate diffusion further into the mucus layer, enabling nanoparticles to reach the cells of the epithelial layer. The particle size and surface properties, namely, their relative hydrophobicity/hydrophilicity, charge and polymer functional groups, are the main factors which affect the particles' ability to prolong their transit time in the GI tract and protect the encapsulated peptide/protein from degradation. Huge efforts are being made worldwide in developing non-invasive, oral peptide/protein delivery systems. Many exhaustive reviews on the various types of oral peptide delivery exist in literature [28–31]. This section deals with nanoparticle-mediated approaches.

8.4.1. Polymeric Nanoparticles

Numerous methods have been reported for the fabrication of nanoparticles and a wide variety of polymer choices exist. The most highly investigated matrices for oral peptide delivery are those based on biodegradable polymers, whether synthetic or natural in origin. Depending upon the manufacturing technique used, the polymeric nanoparticles obtained can be classified into nanospheres and nanocapsules. Nanospheres are matrix-type particles whose entire mass is solid. These particulate systems are characterized by a size that ranges from several tens to a few hundred nanometers.

Nanocapsules have a reservoir form in which a solid material shell surrounds a core, which is liquid or semisolid at room temperature (15–25°C) [32]. Generally, the polymer shell that surrounds the liquid core is formed by polymerization, which takes place at the interface between the dispersed and continuous phase of the emulsion, or by precipitation of a preformed polymer at the surface of emulsion droplets. Nanocapsules are vesicular systems in which the drug is confined to a cavity consisting of an inner liquid core surrounded by a polymeric membrane. Drugs may be absorbed at the sphere surface or encapsulated within the particle where the drug is dissolved, entrapped, encapsulated, adsorbed or conjugated [33–35].

The selection of polymers depends on the chosen method of administration, the active biological to be loaded, the desired release profile, the intention to target

TABLE 8.3 Some examples of polymers used in the fabrication of nanoparticles for oral peptide and protein delivery

Polymer type	Reference(s)
Polyalkylcyanoacrylates	[36–38]
Polymethacrylic acid/acrylates	[39–42]
Polysebacicanhydride-co-polyfumaric anhydride	[43–44]
PLGA	[45–47]
Chitosan and derivatives	[48–53]
Alginate	[54]

specific tissues, the desired rate of particle degradation and the biocompatibility. For drug delivery purposes the polymeric material needs to meet physicochemical and biological requirements optimized for these specific applications. Of these requirements, biocompatibility, safety and biodegradability into non-toxic metabolites is of crucial importance. Table 8.3 outlines some of the polymers which are currently used in the fabrication of nanoparticles.

With advances in polymer chemistry, it is now possible to derivatize the polymer backbone or side chains with functional groups, such as carboxyl, amino or thiol groups. Alternatively, functional groups may be grafted to the particle surface. In either case, these functional groups facilitate conjugation to moieties which in turn may be used to target the nanoparticles to a specific tissue, or promote tight junction opening, or mucoadhesion, for example.

8.4.1.1. Enzyme Inhibitory Properties

The nanoparticulate drug delivery system can be predesigned so that it will release the peptide/protein only when it is in an appropriate biological environment. For example, hydrogels swell by imbibing water from the surroundings which will correspondingly open or widen the pore size of the hydrogel mesh, thereby allowing the release of the encapsulated drug from the matrix. Hydrogels possess the ability to swell due to the hydrophilic nature of the polymers within them, and the presence of either physical or chemical cross-links [55]. The hydrogels will basically remain intact without dissolving, but gradually release the loaded drug into the surrounding environment. If stimuli-sensitive polymers are used for developing the drug delivery matrix, the release of the drug will be based on the trigger from the biological environment. If a pH sensitive polymer is used, then depending on the type of charged group present, the release of drug may be promoted in either acidic or alkaline milieu. The introduction of ionizable carboxylic acids or amino groups will cause the matrices to be triggered by an environmental stimulus such as pH. By utilizing the pH shift that occurs as an oral drug delivery system is transported from the acidic environment of the stomach to the near-neutral duodenum and/or mildly alkaline small intestine, both drug release and adhesion can be triggered [55].

The integration of these biomaterials into a drug delivery system that utilizes the polymer as a carrier is also facilitated by the synthesis of microparticles or nanoparticles. Depending upon the polymer used in the matrix, the system can either shrink or swell upon a change in any of the biological environmental factors such as pH, ionic strength or temperature [55]. If anionic groups are present in the delivery system, then in the stomach the particle matrix will shrink due to the acidic environment (pH 1−2) and reduce diffusion; in contrast, drug will be released in the small intestine where the pH is above 6. Scientists have now developed smart scaffolds which respond to external stimuli such as temperature, pH, light, electric field, magnetic field, chemicals and ionic strength [55].

Trypsin plays an important role in initiating the degradation of orally administered protein/peptide drugs and in activating the proenzymes of proteases. Since many proteolytic enzymes such as trypsin and chymotrypsin have a requirement for a Ca^{2+} or Zn^{2+} ion within their active site (as a cofactor for the enzymatic activity), chelation of these cations using anionic polymers is a current approach under considerable investigation. The inhibitory properties of poly(acrylates) on intestinal proteases were first reported by Hutton *et al.* [56]. They observed a reduction in albumin degradation by a mixture of proteases in the presence of carbomer 934P. Lueßen *et al.* reported that mucoadhesive polymers such as polycarbophil and carbomer display an inhibitory effect on the proteolytic activity of trypsin, α-chymotrypsin and carboxypeptidase A, because of their chelation with divalent metal cations [57].

Polymethacrylic acid based drug delivery systems have gained significant attention due to their mucoadhesive properties, proteolytic enzyme inhibition properties, tight junction opening, drug absorption enhancement, etc. Methacrylic acid free radical polymerization in the presence of chitosan and PEG resulted in the formation of particles of size < 1 μm. These pH-sensitive, polymethacrylic acid-chitosan-polyethylene glycol (PCP) nanoparticles were prepared under mild aqueous conditions via polyelectrolyte complexation, by using a water-soluble initiator without the need for organic solvents or surfactants/steric stabilizers [39]. These particles exhibited good encapsulation efficiency as well as a pH dependent release profile and trypsin inhibition (though this was less than the reference polymer, carbopol). A similar observation was reported by Qian *et al.* who developed methylmethacrylate copolymer nanoparticles with hydrophilic polymeric chains by the free radical polymerization of methylmethacrylate with *N*-isopropylacrylamide (NIPAAm), *N*-methacrylic acid (MAA), *N*-trimethylaminoethylmethacrylate chloride (TMAEMC) or *N*-dimethylaminoethylmethacrylate hydrochloride

(DMAEMC) [58]. These nanoparticles were incubated with pepsin and trypsin at 37°C for 20 min and their enzyme inhibition was then determined. The authors observed that the activity of pepsin decreased to 27% in the presence of MMA–NIPAAm nanoparticles, and the activity of trypsin decreased to 39% in the presence of MMA–MAA nanoparticles. Similar observations have been reported by others. For example, particles made of polymers with a high content of acrylic acid (hydrophilic) provided better loading, stability, and release of human calcitonin, which retained *in vivo* biological activity [57].

Bernkop-Schnürch *et al.* have extensively investigated the possibility of using thiomeric polymers as oral peptide delivery systems [59,60]. It is now established that the thiomers are promising candidates within the group of enzyme inhibiting polymers. From their work on thiolation of polycarbophil it was concluded that by the covalent attachment of cysteine to PCP, the inhibitory effect of the polymer towards carboxypeptidase A, carboxypeptidase B and chymotrypsin could be significantly improved [59,60]. PCP-Cys also had a significantly greater inhibitory effect than unmodified PCP on the activity of aminopeptidase N (both in solution and present on intact intestinal mucosa). The authors explain this observation by postulating that cysteine residues on the polymer bind to the Zn^{2+} ion cofactor of the peptidase, thereby bringing about enzyme inhibition. The authors conclude that, on comparison with the delivery systems based on the co-administration of enzyme inhibitors, the thiomers offer the advantage that the inhibitory effect can be concentrated and localized to the delivery system. Hence, the advantage is that the side-effects caused by the inhibition of digestion (of food) as well as feedback regulation, which in turn leads to an increase in enzymatic secretion, can be avoided.

8.4.1.2. Permeation Enhancement

The bioavailability of the drug depends upon its permeability across the intestinal cell wall. To gain understanding of the nature of gastro-intestinal permeability for a particular drug delivery system, the methods that are routinely used include *in vivo* or *ex vivo* intestinal perfusion in a suitable animal model, and *in vitro* permeability methods using excised intestinal tissues, or monolayers of suitable epithelial cells, e.g. Caco-2 cells which are utilized to monitor tight junction modulation for example [61,62].

Both anionic and cationic polymer-based nanoparticles are reported to have tight junction modulating capacities. Tight junction integrity and drug permeation across the Caco-2 cell monolayers are correlated, which gives an indication of how these particles may perform *in vivo*. The mechanism by which anionic polymers open tight junctions is by chelating Ca^{2+} and the cationic polymers by binding of the amino groups to Zonula Occludens, thus causing disruption in the actin cytoskeleton [63,64].

Sadeghi *et al.* developed quaternized derivatives of chitosan: trimethyl chitosan (TMC), dimethylethyl chitosan (DMEC), diethylmethyl chitosan (DEMC) and triethyl chitosan (TEC) [65]. The effect of these chitosan derivatives on the permeability of insulin across intestinal Caco-2 monolayers was studied and compared with chitosan, both in free-soluble form and in nanoparticulate form. The authors observed that the free chitosan solution showed better permeability than the nanoparticles. This observation was attributed to the reduction in the number of free positively charged amine groups in the case of particles, since the amine groups are "utilized" during formation of the nanoparticulate form [65].

It is generally understood that smaller (nano) particles are absorbed across the intestinal epithelium efficiently, whereas submicron particles are not. Another factor that determines particle uptake is the surface hydrophobicity/hydrophilicity and surface chemistry which also determines the charge of the particle. In conclusion, various groups have demonstrated the uptake of nanocapsules either by the transcellular or paracellular pathway [66]. It is well-known that nanoparticles are absorbed through the M cells (microfold cells) of Payer's patches, but it is now established that particle uptake also takes place through enterocytes [67–69]. A significant uptake of polystyrene nanoparticles in non-lymphoid intestinal tissue (enterocytes) has also been reported by various groups [68].

The use of polymeric nanocapsules for oral insulin delivery was first reported by Damge *et al.*, who observed a sustained reduction in blood glucose levels in streptozotocin-induced diabetic rats following oral insulin nanoparticle administration [70]. The prolonged hypoglycemic response that was elicited may be attributed to sustained insulin release from the polymer matrix following nanoparticle uptake via M cells and non-lymphoid tissue.

8.4.1.3. Synthetic and Natural Polymers

Poly(lactic-co-glycolic acid) (PLGA) is a polyester that has been widely investigated for oral peptide delivery applications. The main attraction of PLGA is its biodegradability and biocompatibility, and its approved clinical use by the US Food and Drug Administration. PLGA can be custom-synthesized to various molecular weights (generally around 30–80 kDa) and with various ratios of lactic and glycolic acid monomers (typically 3:1 to 1:1). However, there are many suppliers of PLGA polymers and fabrication of PLGA nanoparticles is straightforward using emulsion technology. A detailed review on the synthesis and characterization of PLGA nanoparticles exists [45]. However, it is reported that PLGA can adversely affect protein stability during the preparation and storage, primarily due to unfolding at the emulsion interface and acid-catalyzed degradation, respectively [71,72]. The latter arises because the hydrolysis of PLGA leads to the accumulation of the degradation products, lactic and glycolic acids, within the polymeric matrix. This leads to a significant reduction in pH of the immediate environment and consequently the denaturation of the encapsulated proteins.

The release of proteins from PLGA matrices is a complex subject. The sustained release arises primarily from the inability of hydrophilic macromolecules, such as polypeptides, to readily diffuse through a porous matrix of relatively hydrophobic polymer: the higher the lactic acid content of PLGA the higher its hydrophobicity [46]. In the manufacture of protein-encapsulated nanoparticles, it is also necessary to consider adsorption of the polypeptide to the nanoparticle surface (proteins are highly surface-active) [73]. Moreover, although a sustained release is generally desired, what is often observed is a so-called "burst release", wherein the majority of encapsulated polypeptide is rapidly released (followed by a sustained release over several days/weeks); this can lead to poor bioavailability when administered orally [74]. To overcome these drawbacks, various approaches have been explored. Davran *et al.* developed linear PLGA-poly (ethylene glycol) (PEG-PLGA) and star-branched β-cyclodextrin-PLGA (β-CD-PLGA) for nanoparticle preparations, which were used to successfully encapsulated insulin, with retention of structural stability [47]. The preparation process for these PLGA nanoparticles was reported to attenuate the burst release and so provide a more sustained release. Of the formulations developed using this technique, the β-CD-PLGA nanoparticles were reported to have a maximum loading efficiency of 95%, with a particle size of 120–350 nm displaying the slowest drug release profile. The stability of insulin in these nanoparticles was attributed to the β-CD segment of the copolymer.

Chitosan is a naturally occurring polymer which is biocompatible, bioresorbable, biodegradable and has mucoadhesive properties [48] (cf. Chapter 3, pages 52–54). The hydroxyl and, particularly, the amine groups of chitosan can be easily functionalized under mild reaction conditions to prepare modified chitosan derivatives with altered, desirable properties. Chitosan has been extensively investigated for the development of nanoparticles for oral insulin delivery [49–52,75]. Pan *et al.* demonstrated the efficacy of insulin-chitosan nanoparticles for lowering blood glucose levels over 15 hours at a dose of 21 IU/kg in alloxan-induced diabetic rats [49].

Chitosan copolymers have also been developed and investigated; Sarmento *et al.* reported that orally administered dextran sulphate-chitosan nanoparticles, encapsulating insulin, lowered blood glucose levels in diabetic rats by around 35%, with 50 and 100 IU/kg doses sustaining hypoglycemia over 24 h [51]. In our laboratory we have developed a novel derivative of chitosan, lauryl succinyl chitosan (LSC), which bears equal amounts of hydrophilic and hydrophobic moieties. Oral administration of insulin-loaded LSC particles to streptozotocin-induced diabetic rats lowered blood glucose levels by 35% (cf. 15% reduction for native chitosan-insulin nanoparticles), and this reduction was observed for 6 hours [52].

Sakuma *et al.* developed mucoadhesive polystyrene nanoparticles bearing surface hydrophilic polymeric chains [76]. The transit rates of nanoparticles having surface poly (*N*-isopropylacrylamide), poly (vinylamine), and poly (methacrylic acid) chains in the GI tract of rats was reduced. The rate of reduction in the transit time was correlated to improved adsorption of salmon calcitonin. It is possible that these hydrophilic polymer chains reduce transit time by association with the hydrophilic polymer chains of mucin in the GI tract.

8.4.2. Inorganic Nanoparticles

Calcium phosphate and ceramic nanoparticles are examples of widely used inorganic matrices used for various drug delivery applications. However, only limited work has been reported on nanosized oral peptide delivery applications [77–80]. BioSante Pharmaceuticals has developed a calcium phosphate-PEG-insulin-casein (CAPIC) oral insulin delivery system [77]. The authors observed that, following oral administration of 100 IU/kg insulin-CAPIC, blood glucose levels were reduced by 80% within the first 1 h of the treatment and remained reduced for 12 h. This was clinically comparable to 12.5 U/kg insulin administered by subcutaneous injection. However, one potential drawback of inorganic particles is that they do not degrade rapidly and/or fully *in vivo* and hence may accumulate in the body tissues, though further research is required to understand subsequent toxicological issues.

8.4.3. Liposomes and Lipid Nanoparticles

Liposomes are spherical phospholipid bilayers with a size that varies from 50 to 500 nm, which can encapsulate both water soluble and insoluble drug and proteins within its hydrophilic core or the bilayer respectively. As early as the 1970s, the use of liposomes for oral protein delivery was investigated using insulin as the model drug. However, while various groups reported on these oral insulin delivery systems, their observations of a hypoglycemic effect varied [81–83]. Weingarten *et al.* developed positively charged liposomes for insulin loading, and reported a protective action against insulin degradation in the presence of three digestive enzymes (pepsin, α-chymotrypsin and trypsin) [84]. The authors concluded that the binding of insulin to the phospholipids was primarily due to ionic interactions, rather than due to hydrophobic interaction, and that this adsorption to the external phospholipid bilayer was in such a manner as to conceal the peptide motifs from the enzymes [84]. They also suggested that the failure of oral insulin-liposome administration that had been observed in their previous work was due to the degradation (solubilization) of the liposomal membrane by bile salts in the intestinal lumen, rather than by the destruction of insulin by gastric enzymes [85]. Therefore, it could be suggested that simple liposomes are not sufficient as oral peptide delivery vehicles since they may be degraded by the presence of bile salts and lipases, and lack properties such as mucoadhesion. *In vitro* data acquired by Anderson *et al.* using targeted PEGylated liposomes furnished with folic acid for oral delivery were however promising, showing enhanced permeability of dextran across Caco-2 cell monolayers [86]. This study serves to highlight that extrapolating *in vitro* data to *in vivo* efficacy can be problematic. Research on oral liposomal delivery systems has moved forward with the development of polymer-modified liposomes.

Garcia-Fuentes *et al.* [87] developed PEG and chitosan coated lipid nanoparticles as oral delivery systems for salmon calcitonin (sCT). They studied the interaction of these surface-modified nanoparticles with Caco-2 cells to evaluate their efficacy for macromolecular delivery in rats. They observed that the association of these modified lipid nanoparticles with the Caco-2 cell monolayer was independent of the surface coating. The PEG-coated nanoparticles did not alter the transepithelial electrical resistance of Caco-2 cell monolayers, while the chitosan-coated nanoparticles showed a dose-dependent increase in the permeability of dextran (used as a marker) across the monolayers. In their *in vivo* experiments, chitosan-coated lipid nanoparticles loaded with sCT showed prolonged reduction of calcium levels in the serum, while the performance of PEG coated particles was not significantly different from the control (sCT solution). Thus the authors concluded that the favorable interaction of the chitosan-coated nanoparticles with intestinal mucosa together with their permeation enhancing characteristics might be responsible for the improved oral absorption of sCT.

Other studies have delivered sCT orally using "double liposomes" which consisted of small, inner liposomes of around 100 nm contained within larger liposomes. The inner liposome was derived from various lipids to create neutral and cationically charged liposomes, using two methods: Coatsome® and mechanochemical, both sized to 100 nm by extrusion [88]. The data showed that the double liposomes increased the oral bioavailability of sCT, and amongst the inner liposome preparations the group of rats treated with cationically charge liposomes showed the strongest hypocalcemic effects. The authors speculated that not only the size of liposomes but also the cationic charge plays an important role in the intestinal absorption of liposomal delivery systems. In all of the above reports, it is clear that the surface modification of liposomes is required for optimal efficacy if the proposed route of administration is oral.

8.5. STABILITY AND EVALUATION

8.5.1. Nanoparticles

Nanoparticles are characterized in terms of size, charge, shape, surface morphology, drug loading capacity and efficiency drug release profile. The techniques used in characterizing nanoparticles for evaluating a particle-based oral delivery system are summarized in Table 8.4. In order to successfully bring a nanoparticulate oral drug delivery system into medical use, assessment of the stability of these carriers is equally important. The material properties of the particles and the physicochemical characteristics of the formulation change should be monitored over storage. For example, the formulations will be usually be stored in the powder form (through lyophilization) and hence it will be important to monitor the dispersability of these particles in aqueous media, also measuring the resultant size, aggregation and drug release profiles.

TABLE 8.4 Summary of techniques use to characterize nanoparticles

Parameters	Techniques
Physicochemical characterization Particle size and size distribution	Photon correlation spectroscopy
Chemical composition	Energy Dispersive X-ray Spectroscopy (EDAX); elemental analysis
Surface charge	Zeta potential; capillary electrophoresis; laser Doppler anemometry
Surface hydrophobicity	Water contact angle measurement
Surface chemical analysis	Secondary ion mass spectrometry (SIMS) microscopy; X-ray photoelectron spectroscopy (XPS); Fourier transform infrared spectroscopy (FTIR)
Topology	Atomic force microscopy (AFM)
Morphology	Scanning electron microscopy (SEM)
In vitro performance Drug release studies	In simulated gastric and intestinal fluids (SGF and SIF), with and without enzymes
Blood contact properties	Whole blood clotting time; hemolysis; plasma protein adsorption; platelet aggregation/ activation; complement activation
Cell based assays	Cytotoxicity; phagocytosis; transepithelial electrical resistance (TEER); cell/fluorescence staining
Ex vivo performance Measure of the net ion transport across the intestinal epithelium	Ussing Chamber (cf. TEER)
Mucoadhesion	Various in-house techniques
Blood concentrations of drug over time	Drug/particle permeation across the *ex vivo* intestine; pharmacodynamic measurements
In vivo performance Toxicological studies	Systemic toxicity; oral toxicity; genotoxicity
Bioavailability and efficacy	Animal models of disease

8.5.2. Peptides and Proteins

The main peptide and protein drugs for which oral nanoparticulate formulations are being investigated include insulin, calcitonin, parathyroid hormone, octreotide, luteinizing hormone releasing hormone, cyclosporine A and a number of vaccines. As discussed above, the biological activity of proteins depends on maintaining their 3-dimensional structure, which requires consideration of secondary, tertiary and quaternary elements involving weak non-covalent interactions (other than disulphide bridges). In contrast, the biological activity of small peptides is generally not dependent on any secondary or tertiary conformation and these may therefore be more capable of withstanding encapsulation "stresses". Such stresses during nanoparticle development may include harsh conditions involving high pressures, organic/non-aqueous solvents, surfactants and mechanical forces (agitation, stirring and shearing) generating high oil/water interfacial areas that can alter the native structure. Therefore, careful consideration of pH, ionic strength, temperature must be made alongside minimization of the stresses mentioned above; i.e. to avoid protein unfolding, denaturation and subsequent aggregation within in the formulation. Insulin for example has a tendency to form hexamer aggregates, leading to an increase in its apparent molecular weight, which can minimize absorption across the intestine. Hogvaard *et al.* reported the use of dodecyl maltoside to minimize insulin aggregation, which also provided some protection against enzymatic degradation [89].

Protein loading into drug delivery systems may also lead to interaction and adsorption onto the carrier material (e.g. PLGA). This has disadvantages of altered release properties and loss of biological activity, through adsorption-induced protein unfolding. The adsorption of the protein onto the matrix surface is dependent on various factors such as the chemistry of the material, surface area, charge/pH, etc. The stability and biological activity of the loaded peptide or protein can be evaluated by various means including high performance liquid chromatography (HPLC, to assess aggregation), enzyme-linked immunosorbent assay (ELISA) and radioimmunoassay (RIA), to assess *in vitro* activity. Changes in peptide conformation can be easily observed by the use of circular dichroism, with changes in molecular weight observed by gel electrophoresis and mass spectrometry, and formation of dimers/trimers/etc., observed by analytical ultracentrifugation and gel filtration. Ultimately, if the product is for invasive administration then careful consideration must be given to the method of sterilization. Similarly, the lyophilization cycle must be optimized to avoid protein denaturation (see Chapter 1). Aggregation and denaturation not only attenuate biological activity but also lead to immunogenicity [90,91].

8.6. ORAL DELIVERY OF NANOPARTICLES: FURTHER CONSIDERATIONS

8.6.1. The Site of Drug Release

The intended intestinal site of release of the encapsulated peptide/protein requires consideration. With reference to insulin oral delivery systems, the stomach should be avoided since this site presents the smallest surface area available for absorption. A number of approaches are reported for reducing insulin release in the stomach, including the use of pH-sensitive polymers such as acrylates, or anionic polymers bearing carboxyl groups [52,57]. For example, chitosan is soluble in acidic media and the release of insulin from chitosan nanoparticles is higher at pH 1.2 than at pH 7.4 [57,92]. However, introducing carboxyl groups onto the chitosan nanoparticles (which are non-ionized at pH 1.2), produces a large change in the insulin release pattern: only 8.5% of the loaded insulin was released from lauryl succinyl chitosan particles in pH 1.2 buffer over the first 2 h, compared to 62.7% from native chitosan particles [57]. Sajeesh and Sharma also used polymethacrylic acid based nanoparticles for insulin delivery, suggesting that, at acidic pH, the non-ionized carboxylic acid groups prevent swelling of the matrix, in turn preventing insulin release and shielding the entrapped insulin from the gastric fluid [93]. Thus, pH-sensitive polymers can be used to minimize the loss of insulin in the stomach and thereby enhance its bioavailability.

8.6.2. The Presence of Food

Another factor that plays an important role in determining the efficiency of the drug delivery systems is the presence of food. Karsdal *et al.* studied the influence of food intake on the bioavailability and efficacy of oral calcitonin (sCT) [94]. They conducted a single-blind, randomized, partly placebo-controlled study in 36 healthy, post-menopausal, female volunteers aged 62–74 years. Though the formulation they investigated did not include nanoparticulate systems, their findings are relevant and interesting. The data suggested that oral doses of sCT were optimally absorbed in the fasted state, or prior to a meal. Fasting or preprandial dosing led to improved bioavailability and it was suggested by the authors that these results may aid the design strategy for oral peptide delivery.

8.6.3. Mucosal Barriers and Transit Time

The mucosal barrier is an important aspect which needs more research, though several reviews are highly informative in this regard [95–97]. Mucus is a viscoelastic, adhesive gel that forms a protective lining over internal epithelial surfaces such as the gastrointestinal tract, lung alveoli, eye and reproductive tract. It is composed of cross-linked and entangled mucin fibers secreted by the goblet cells in the epithelium and submucosal glands [98,99]. Mucins are well characterized by various groups, and form a heterogenous group of proteins which are rich in cysteine residues, such that the mucin monomers become cross-linked through disulphide bonds [95,100–101]. Mucin domains rich in serine and threonine residues are sites for glycosylation (O-glycan attachment); it is reported that 80% of the weight of mucin glycoproteins consists of carbohydrates such as N-acetylgalactosamine, N-acetylglucosamine, galactose, fucose and sialic acids [96].

It is reported that intestinal mucus is tenacious and entraps particulates, thereby preventing their direct contact with the intestinal epithelial cells via steric obstruction and/or adhesion. Gruber *et al.* observed that orally administered particulate drug carriers become entrapped within the mucosal layer of the intestine irrespective of their size. These mucus-entrapped particles then have very limited ability for systemic absorption and drug delivery [102]. It is reported that approximately 10 liters of mucus is secreted into the GI tract daily [95,96]. Hence a better understanding of the mucus layer thickness and clearance times at various mucosal surfaces is important to the development of oral drug delivery systems. These should be designed to maximize the gastrointestinal transit time and overcome the mucosal clearance mechanisms, which are reported to be about 4–5 hours in the GI tract [103].

For example, to prevent entrapment in mucin and subsequent clearance, mucoadhesive polymers and particulates have been developed to increase the transit time in the GI tract. These mucoadhesive polymers are designed so as to interact with and adhere to the mucosa, a process known as mucoadhesion [104–106]. It is now well established that mucoadhesive particles increase the GI transit time and improve drug absorption. A large proportion of oral delivery systems currently under investigation are focused on this design aspect.

Various approaches have been developed for improving mucoadhesivity such as introducing anionic, cationic or thiol groups, ligands such as lectins, etc. [61,76,107–109]. However, the major limitation to this approach is secreted mucin [96]; particles should be capable of penetrating the secreted mucin to reach the intestinal epithelium. Yoncheva *et al.* developed PEGylated nanoparticles based on poly(methyl vinyl ether-co-maleic anhydride) (PVM/MA) [110]. The authors reported that PEGylation of nanoparticles with PEG 2000 yielded a brush border, while PEGylation with PEG 1000 resulted in PEG being distributed either in the particle core or physically adsorbed to the surface. They observed that adsorption to mucin at pH 7.4 was significantly higher for the PEG 1000-derived nanoparticles than for PEG 2000-derived nanoparticles. Thus, the PEG 2000 brush border appeared to have decreased the interaction between the nanoparticle and mucin, facilitating penetration and achieving greater adhesive interactions with rat intestinal mucosa *in vivo* [110].

Lai *et al.* also hypothesized that coating particles with a high density of PEG molecules may reduce particle–mucus adhesive interactions [111]. They covalently modified the surfaces of small (100 nm) and large (200 and 500 nm) polystyrene nanoparticles with a high-density layer of PEG 2000, and studied particle diffusion in human mucus. Surprisingly, a larger fraction of the small PEG-coated nanoparticles became entrapped in the mucus compared to the larger nanoparticles. This was in agreement with the measured diffusion coefficients in mucus and water, which were 200-fold smaller for the small PEG-coated nanoparticles in mucus, but only 4- to 6-fold lower for the large PEG-coated nanoparticles. Therefore, Lai *et al.* demonstrated that large (500 nm) nanoparticles may still efficiently migrate through the mucin layer and may be useful in oral drug delivery, so long as they are properly coated [111].

8.6.4. Efflux by P-glycoprotein

P-glycoprotein (P-gp) is a member of the superfamily of ABC transporters which transport various molecules across cellular membranes, and is highly expressed in the intestinal epithelium. Efflux by P-gp can be a major limitation for the oral delivery of a number of drugs (particularly cytotoxics). To overcome this problem, various types of efflux pump inhibitors have been developed, but because of their side-effects, none of them has been approved in an oral delivery system [112]. It is suggested by various groups that, as an alternative, polymeric efflux pump inhibitors could be developed which have good inhibitory activity without being absorbed from the GI tract. Iqbal *et al.* developed a thiolated co-polymer of PEG grafted with polyethyleneimine (PEI) and evaluated its inhibition of P-gp efflux [113]. They observed that the thiolated PEG-g-PEI co-polymer increased transport of Rho-123 (a P-gp substrate) in the apical to basolateral direction, which reduced efflux in the opposite direction. Thus, the authors concluded that this novel polymer may be a useful tool for improving the intestinal uptake of P-gp substrates.

8.7. BIOAVAILABILITY

The aim of all the nanoparticulate and liposomal strategies discussed above is, of course, to achieve an increase in the oral bioavailability of peptides and proteins. This section therefore deals with what is a key determinant of whether or not a nanoparticulate strategy may be considered for translation to the clinic and commercialization. Despite original setbacks with both particulates and early liposomes, a general trend can be observed in the recent past which points towards a significant improvement in the reported bioavailability of peptides and proteins. There is some cause for optimism and selected nanoparticle formulations of insulin and calcitonin, shown in Table 8.5, illustrate this trend.

Chalasani *et al.* demonstrated a pharmacological availability of 26.5% for insulin orally delivered using vitamin B_{12} conjugated dextran nanoparticles [114]. They further optimized their formulation by minimizing the degree of cross-linking and studied the effect of vitamin B_{12} analogs with different linkages to the particle surface. These modifications only modestly improved the pharmacological activity to 29.4% [115]. The authors demonstrated that particle uptake occurred via vitamin B_{12} receptor mediated endocytosis, and that the matrix protected the entrapped insulin from enzymatic degradation.

TABLE 8.5 Therapeutic protein/peptide oral formulations and reported pharmacological/bioavailability

Peptide	Animal model	Nanoparticle material	Particle size (nm)	Oral dose	PA/BA (%)	Ref.
Insulin	Rat	Vitamin B_{12} coated dextran	192	20 IU/Kg	26.5%	[114]
	Rat	Aminoalkyl-Vitamin B_{12} coated dextran	150–300	20 IU/Kg		[115]
	Rat	FAO:PLGA (1:2) + 10% Fe_3O_4	< 1000	19.2 IU/rat	11.4	[50]
	Rat	SPC	200	20 IU/Kg	7.7%	[116]
	Rat	Chitosan-γ-PGA	218	30 IU/Kg	15.1%	[117]
	Rat	Poly(caprolactone)-Eudragit RS	358	50 IU/Kg	13.2	[118]
	Rat	Chitosan	25–400	21 IU/Kg	14.9%	[49]
	Rat	PLGA-HP55	169		6.27%	[119]
	Rat	Dextran Chitosan	500	50 IU/Kg 100 IU/Kg	5.6% 3.4%	[51]
Calcitonin	Rat	Sorbitol and PC proliposomes Proliposomes + TDC	~23	15 mg/Kg	1.9% 4.5%	[120]
	Rat	Cationic double liposome	~100 (inner liposome)	10 μg/Kg	3.5%	[88]

PA/BA = pharmacological availability/bioavailability of the dose administered; Eudragit RS = copolymer of acrylic and methacrylic acid esters with quaternary ammonium groups; FAO = fumaric acid oligomers; HP55 = hypromellose phthalate; TDC = taurodeoxycholate; PC = phosphatidylcholine; PGA = polyglutamic acid; SPC = soybean phosphatidylcholine; other abbreviations in main text.

Insulin-loaded chitosan nanoparticles (ICN), insulin-chitosan complex nanoparticles (ICCN) and enteric coated insulin-chitosan complex nanoparticles (EICCN) have also been developed as oral insulin delivery systems [50]. The particle sizes for the ICN, ICCN and EICCN nanoparticles were 265 ± 14,

284 ± 19 and 342 ± 23 nm, respectively, and the zeta potentials were 40.7 ± 0.7, 31.3 ± 0.4 and 34.1 ± 0.9 mV, respectively. The insulin entrapment efficiencies of the ICN, ICCN and EICCN nanoparticles were 66.8 ± 2.1, 81.3 ± 2.6 and 81.5 ± 3.1%, respectively, and the bioavailabilities of oral doses given to Wistar rats were 5.58 ± 0.7, 7.55 ± 0.9 and 8.33 ± 0.5% over 48 h, respectively. The authors established the usefulness of enteric coating in reducing the initial burst of insulin and enhancing the bioavailability of oral insulin preparations. However, it can be seen the bioavailabilities of these nanoparticles remain well below comparative doses given by subcutaneous injection.

Song *et al.* developed sCT-loaded "proliposomes" containing bile salts, and evaluated their potential for the oral delivery of calcitonin (sCT) [120]. Of the bile salts they investigated, it was shown that proliposomes containing sodium taurodeoxycholate (TDC) yielded the largest (7.1-fold) increase in bioavailability following duodenal administration to rats. Thus, the development of new polymers or surface-modified liposomes integrating multifunctional properties is encouraging. Mucoadhesive polymers exhibiting absorption enhancement as well as being inhibitors of proteolytic enzymes are similarly a potentially useful strategy in overcoming the intestinal barriers and enhancing the bioavailability of the protein/peptide.

8.8. BIOCOMPATIBILITY AND TOXICITY ISSUES

Nanoparticles obviously have a relatively high surface area and, along with potentially reactive surface chemistries, pose unique problems for biological systems. They need to be designed and developed in such a way that they are not cytotoxic, and are blood compatible. The effect of nanomaterials on the body has become a subject of intense interest in recent years, in part because there remains little or no guidance on the issue due to the lack of suitable and relevant data. There are still fundamental questions concerning nanomaterials that must be answered if nanotechnology is ultimately to have a significant impact beyond the laboratory, and move into the clinic.

How blood components, cells and tissues may interact with nanoparticles will determine the various consequences of (non) biocompatibility, and the obvious parameters which may be used to assess that response are nanoparticle size, shape, charge and surface chemistry. If an orally delivered nanoparticle reaches the systemic circulation, binding to plasma proteins (particularly albumin) will be a first step in determining its subsequent fate [121–123]. For example, the nature of bound plasma protein will determine the mechanism of clearance from the systemic circulation, or activation of other processes such as platelet activation, complement activation, etc. Thus, an ideal nanoparticle should be blood compatible and stable, remaining in the circulation undergoing erosion and degradation (e.g. for PLGA, followed by elimination of lactic and glycolic acids). The compatibility of degradation products in the blood is also a matter of concern, particularly for complex, novel co-polymers [124–125].

It is understood that nanoparticles may have a rapid intestinal uptake, but it is unknown whether prolonged administration can lead to accumulation and damage to the liver, kidney or lungs. Currently, few studies have been performed in this area. Reis *et al.* reported a toxicological assessment of alginate-chitosan nanoparticles for the oral delivery of insulin, evaluating biochemical, haematological and urine-based parameters, as well as organ and tissue histology [126]. They demonstrated that the insulin-loaded nanoparticles, orally administered for 15 days, did not result in mortality, morbidity or other abnormalities, and were therefore believed to be non-toxic. Similarly, a toxicity study on nanoparticles which had been self-assembled from chitosan-polyglutamic acid co-polymers showed these nanoparticles to be well tolerated following oral administration over 14 days to mice at doses 18 times higher than those used for therapeutic efficacy [117].

Since nanoparticles are thought to undergo absorption via enterocytes or M-cells, it would be logical to investigate how they may interact at these sites. Development of standardized assays using tracking technologies for labeled nanoparticles is still required however. Nevertheless, work towards imaging of nanoparticles using either simple light microscopy for dyed nanoparticles, or advanced coherent anti-Stokes Raman scattering (CARS) microscopy for non-labeled nanoparticles, is promising and such studies would immediately be in a position to exploit the wealth of information available on nanoparticle structure/absorption relationships [127–130].

8.9. CONCLUSIONS

Polymeric nanoparticles have advantageous features for the encapsulation and oral delivery of peptides and proteins, including controlled/sustained release properties and protection from endogenous enzymes. The focus of several research groups appears to be moving from *in vitro* cell culture models and data towards animal models of disease and the assessment of bioavailability. Much work is yet to be performed to determine the exact mechanism of nanoparticulate uptake and subsequent clearance, the associated potential for *in vivo* nano-toxicology, tissue specific targeting (whether or not this actually improves therapeutic efficacy) and modulation of GI transit; to date, no commercial formulations exist. Despite this, steady, albeit modest improvements in bioavailability look increasingly promising, and standardization of storage and dosage reconstitution should be relatively straightforward. It is hoped that emerging methods for the analysis of nano-systems and increasing commercial interest in nanotechnologies will provide strong drivers, such that in the near future a nano-based oral peptide/protein formulation will emerge successfully.

REFERENCES

[1] R. Langer, Where a pill won't reach, Scientific American 288 (2003) 50–57.
[2] R. Langer, Drug delivery and targeting, Nature 392 (1998) 5–10.

[3] O.C. Farokhzad, R. Langer, Nanomedicine: Developing smarter therapeutic and diagnostic modalities, Adv. Drug Deliv. Rev. 58 (2006) 1456–1459.
[4] K.A. Janes, P. Calvo, M.J. Alonso, Polysaccharide colloidal particles as delivery systems for macromolecules, Adv. Drug Deliv. Rev. 47 (2001) 83–97.
[5] F. Woodley, Enzymatic barriers for Gl peptide and protein delivery, Crit. Rev. Ther. Drug Carrier Syst. 11 (1994) 61–95.
[6] X.H. Zhou, Overcoming enzymatic and absorption barriers to non-parenterally administered protein and peptide drugs, J. Control. Release 29 (1994) 239–252.
[7] W.S. Snyder, M.J. Cook, E.S. Nassat, L.R. Karhausen, G.P. Howells, I.H. Tipton, International Commission on Radiological Partition 23, in: Report of the Taskgroup on Reference Man, Pergamon Press, New York, 1974, p. 139.
[8] S. Kerss, A. Allen, A. Garner, A simple method for measuring thickness of the mucus gel layer adherent to rat, frog and human gastric mucosa: Influence of feeding, prostaglandin, N-acetylcysteine and other agents, Clin. Sci. (Lond.) 63 (1982) 187–195.
[9] Y. Aoki, M. Morishita, K. Asai, B. Akikusa, S. Hosoda, K. Takayama, Region-dependent role of the mucous/glycocalyx layers in insulin permeation across rat small intestinal membrane, Pharm. Res. 22 (2005) 1854–1862.
[10] M. Morishita, I. Morishita, K. Takayama, Y. Machida, T. Nagai, Site-dependent effect of aprotinin, sodium caprate, Na2EDTA and sodium glycocholate on intestinal absorption of insulin, Biol. Pharm. Bull. 16 (1993) 68–72.
[11] A. Yamamoto, T. Taniguchi, K. Rikyuu, T. Tsuji, T. Fujita, M. Murakami, et al., Effect of various protease inhibitors on the intestinal absorption and degradation of insulin in rats, Pharm. Res. 11 (1994) 1496–1500.
[12] R.I. Mahato, A.S. Narang, L. Thoma, D.D. Miller, Emerging trends in oral delivery of peptide and proteins, Crit. Rev. Ther. Drug Carr. Syst. 20 (2003) 153–214.
[13] M. Hashimoto, K. Takada, Y. Kiso, S. Muranishi, Synthesis of palmitoyl derivatives of insulin and their biological activities, Pharm. Res. 6 (1989) 171–176.
[14] P. Calceti, S. Salmaso, G. Walker, A. Bernkop-Schnürch, Development and *in vivo* evaluation of an oral insulin-PEG delivery system, Eur. J. Pharm. Sci. 22 (2004) 315–323.
[15] I. Gomez-Orellana, D.R. Paton, Advances in oral delivery of proteins, Exp. Opin. Ther. Patents 9 (1999) 247–253.
[16] T. Uchiyama, T. Sugiyama, Y.S. Quan, A. Kotani, N. Okada, T. Fujita, et al., Enhanced permeability of insulin across the rat intestinal membrane by various absorption enhancers: Their intestinal mucosal toxicity and absorption enhancing mechanism of n-lauryl-b-D-maltopyranoside, J. Pharm. Pharmacol. 51 (1999) 1241–1250.
[17] A. Fasano, S. Uzzau, Modulation of intestinal tight junctions by zonula occludens toxin permits enteral administration of insulin and other macromolecules in an animal model, J. Clin. Invest 99 (1997) 1158–1164.
[18] N.N. Salama, N.D. Eddington, A. Fasano, Tight junction modulation and its relationship to drug delivery, Adv. Drug Deliv. Rev. 58 (2006) 15–28.
[19] E.M. Lane, C.M. O'Driscoll, O.I. Corrigan, Quantitative estimation of the effects of bile salts surfactant systems on insulin stability and permeability in the rat intestine using a mass balance model, J. Pharm. Pharmacol. 57 (2005) 169–175.
[20] N.F.H. Ho, J.S. Day, C.L. Barsuhn, P.S. Burton, T.J. Ranb, Biophysical model approaches to mechanistic transepithelial studies of peptides, J. Control. Release 11 (1990) 3–24.
[21] M. Gordberg, I. Gomez-Orellana, Challenge for the oral delivery of macromolecules, Nat. Rev. Drug Discov. 2 (2003) 289–295.

[22] E.S. Swenson, W.J. Curatolo, (C) Means to Enhance Penetration: (2) Intestinal permeability enhancement for proteins, peptides and other polar drugs: Mechanisms and potential toxicity, Adv. Drug Deliv. Rev. 8 (1992) 39–92.

[23] E.S. Swenson, W.B. Milisen, W. Curatolo, Intestinal permeability enhancement: Efficacy, acute local toxicity and reversibility, Pharm. Res. 11 (1994) 1132–1142.

[24] K.D. Hinds, S.W. Kim, Effects of PEG conjugation on insulin properties, Adv. Drug Deliv. Rev. 54 (2004) 505–530.

[25] D. Shah, W.-C. Shen, Transcellular delivery of an insulin-transferrin conjugate in enterocyte-like Caco-2 cells, J. Pharm. Sci. 85 (1996) 1306–1311.

[26] C.Q. Xia, J. Wang, W.-C. Shen, Hypoglycemic effect of insulin-transferrin conjugate in streptozotocin-induced diabetic rats, J. Pharmacol. Exp. Ther. 295 (2000) 594–600.

[27] N.J. Kavimandan, E. Losi, J.J. Wilson, J.S. Brodbelt, N.A. Peppas, Synthesis and characterization of insulin–transferrin conjugates, Bioconjugate Chem. 7 (2006) 1376–1384.

[28] C. Damge, C. Pinto, P. Maincent, Nanoparticle strategies for the oral delivery of insulin, Expert Opin. Drug Deliv. 5 (2008) 45–68.

[29] V.R. Babu, P. Patel, R.C. Mundargi, V. Rangaswamy, T.M. Aminabhavi, Developments in polymeric devices for oral insulin delivery, Expert Opin. Drug Deliv. 5 (2008) 403–415.

[30] C.B. Woitiski, R.A. Carvalho, A.J. Ribiero, R.J. Neufeld, F. Veiga, Strategies toward the improved oral delivery of insulin nanoparticles via gastrointestinal uptake and translocation, BioDrugs 22 (2008) 223–237.

[31] E.-S. Khafagy, M. Morishita, Y. Onuki, K. Takayama, Current challenges in non-invasive insulin delivery systems: A comparative review, Adv. Drug Deliv. Rev. 59 (2007) 1521–1546.

[32] C. Vauthier, K. Bouchemal, Methods for the preparation and manufacture of polymeric nanoparticles, Pharm. Res. 26 (2009) 1025–1058.

[33] P. Couvreur, C. Dubernet, F. Puisieux, Controlled drug delivery with nanoparticles: Current possibilities and future trends, Eur. J. Pharm. Biopharm. 41 (1995) 2–13.

[34] H. Vranckx, M. Demoustier, M. Deleers, A new nanocapsule formulation with hydrophilic core: Application to the oral administration of salmon calcitonin in rats, J. Pharm. Pharmacol. 42 (1996) 345–347.

[35] C. Pinto-Reis, R.J. Neufeld, A.J. Ribeiro, F. Veiga, Nanoencapsulation I. Methods for preparation of drug-loaded polymeric nanoparticles, Nanomedicine 2 (2006) 8–21.

[36] C. Damge, H. Vranckx, P. Balschmidt, P. Couvreur, poly(alkylcyanoacrylate) nanospheres for oral administration of insulin, J. Pharm. Sci. 86 (1997) 1403–1409.

[37] C. Damge, M. Aprahamian, W. Humbert, M. Pinget, Ileal uptake of polyalkylcyanoacrylate nanocapsules in the rat, J. Pharm. Pharmacol. 52 (2000) 1049–1056.

[38] M.S. Meisha, M.B. Sidhom, B. Fasipe, Oral and subcutaneous absorption of insulin poly (alkylcyanoacrylate) nanoparticles, Int. J. Pharm. 288 (2005) 289–293.

[39] S. Sajeesh, C.P. Sharma, Novel pH responsive polymethacrylic acid-chitosan-polyethylene glycol nanoparticles for oral peptide delivery, J. Biomed. Mater. Res. B Appl. Biomater. 76 (2006) 298–305.

[40] A.C. Foss, T. Goto, M. Morishita, N.A. Peppas, Development of acrylic-based copolymers for oral insulin delivery, Eur. J. Pharm. Biopharm. 6 (2004) 163–169.

[41] D.N. Robinson, N.A. Peppas, Preparation and characterization of pH-responsive poly (methacrylic acid-g-poly(ethylene glycol) nanospheres, Macromolecules 35 (2002) 3668–3674.

[42] M. Torres-Lugo, N.A. Peppas, Preparation and characterization of P(MAA-g-EG) nanospheres for protein delivery applications, J. Nanoparticle Res. 4 (2002) 73–81.

[43] E. Mathiowitz, J.S. Jacob, Y.S. Jong, G.P. Carino, D.E. Chickering, D.E. Chaturvedi, et al., Biologically erodable microspheres as potential oral delivery systems, Nature 386 (1997) 410–414.

[44] S. Furtado, D. Abramson, R. Burrill, G. Olivier, C. Gourd, E. Bubbers, et al., Oral delivery of insulin loaded poly(fumaric-co-sebacic) anhydride microspheres, Int. J. Pharm. 347 (2008) 149–155.

[45] C.E. Astete, C.M. Sabliov, Synthesis and characterization of PLGA nanoparticles, J. Biomater. Sci. Polymer. 17 (2006) 247–289.

[46] R.C. Mundargi, V.R. Babu, V. Rangaswamy, P. Patel, T.M. Aminabhavi, Nano/micro technologies for delivering macromolecular therapeutics using poly(d, l-lactide-co-glycolide) and its derivatives, J. Control. Release 125 (2008) 193–209.

[47] S. Davaran, Y. Omidi, M.R. Rashidi, M. Anzabi, A. Shayanfar, S. Ghyasvand, et al., Preparation and *in vitro* evaluation of linear and star-branched PLGA nanoparticles for insulin delivery, J. Bioactive Comp. Pol. 23 (2008) 115–131.

[48] L. Illum, Chitosan and its use as a pharmaceutical excipient, Pharm. Res. 15 (1998) 1326–1331.

[49] Y. Pan, Y.-J. Li, H.-Y. Zhao, Bioadhesive polysaccharide in protein delivery system: Chitosan nanoparticles improve the intestinal absorption of insulin *in vivo*, Int. J. Pharm. 249 (2002) 139–147.

[50] F. Cui, L. Zhang, J. Zheng, Y. Kawashima, A study of insulin-chitosan complex nanoparticles used for oral administration, STP Pharma. 14 (2004) 435–439.

[51] B. Sarmento, A. Ribeiro, F. Veiga, Oral bioavailability of insulin contained in polysaccharide nanoparticles, Biomacromolecules 8 (2007) 3054–3060.

[52] M.R. Rekha, C.P. Sharma, Synthesis and evaluation of lauryl succinyl chitosan particles towards oral insulin delivery and absorption, J. Control. Release 135 (2009) 144–151.

[53] S.H. Zhao, X.T. Wu, W.C. Guo, Y.M. Du, L. Yu, J. Tang, N-(2-hydroxyl) propyl-3-trimethyl ammonium chitosan chloride nanoparticle as a novel delivery system for Parathyroid Hormone-Related Protein 1-34, Int. J. Pharm. 393 (2010) 268–272.

[54] N. Zhang, J. Li, W. Jiang, C. Ren, J. Li, J. Xin, et al., Effective protection and controlled release of insulin by cationic β-cyclodextrin polymers from alginate/chitosan nanoparticles, Int. J. Pharm. 393 (2010) 212–218.

[55] N.A. Peppas, J.Z. Hilt, A. Khademhosseini, R. Langer, Hydrogels in biology and medicine: From molecular principles to bionanotechnology, Adv. Mater. 18 (2006) 1345–1360.

[56] D.A. Hutton, J.P. Pearson, A. Allen, S.N.E. Foster, Mucolysis of the colonic mucus barrier by faecal proteinases: Inhibition by interacting polyacrylate, Clin. Sci. (Lond.) 78 (1990) 265–271.

[57] H.L. Lueßen, J.C. Verhoef, G. Borchard, C.-M. Lehr, A.G. de Boer, H.E. Junginger, Mucoadhesive polymers in the peroral peptide drug delivery: II. Carbomer and polycarbophil are potent inhibitors of the intestinal proteolytic enzyme trypsin, Pharm. Res. 12 (1995) 1293–1298.

[58] F. Qian, F. Cui, C. Yin, Preparation, characterization and enzyme inhibition of methylmethacrylate copolymer nanoparticles with different hydrophilic polymeric chains, Eur. Polym. J. 42 (2006) 1653–1661.

[59] A. Bernkop-Schnürch, S.C. Thaler, Polycarbophil–cysteine conjugates as platforms for oral polypeptide delivery systems, J. Pharm. Sci. 89 (2000) 901–909.

[60] A. Bernkop-Schnürch, H. Zarti, G.F. Walker, Thiolation of polycarbophil enhances its inhibition of soluble and intestinal brush border membrane bound aminopeptidase-N, J. Pharm. Sci. 90 (2001) 1907–1914.

[61] G. Borchard, H.L. Lueβen, A.G. de Boer, J.C. Verhoef, C.-M. Lehr, H.E. Junginger, The potential of mucoadhesive polymers in enhancing intestinal peptide drug absorption. III: Effects of chitosan-glutamate and carbomer on epithelial tight junctions *in vitro*, J. Control. Release 39 (1996) 131–138.

[62] A.F. Kotzé, H.L. Lueßen, B.J. de Leeuw, A.G. de Boer, J.C. Verhoef, H.E. Junginger, Comparison of the effect of different chitosan salts and N-trimethyl chitosan chloride on the permeability of intestinal epithelial cells (Caco-2), J. Control. Release 51 (1998) 35–46.

[63] P. Artursson, T. Lindmark, S.S. Davis, L. Illum, Effect of chitosan on the permeability of monolayers of intestinal epithelial cells (Caco-2), Pharm. Res. 11 (1994) 1358–1361.

[64] E.K. Anderberg, C. Nystrom, P. Artursson, Epithelial transport of drugs in cell culture. VII. Effects of pharmaceutical surfactant excipients and bile acids on transepithelial permeability in monolayers of human intestinal epithelial (Caco-2) cells, J. Pharm. Sci. 81 (1992) 879–887.

[65] A.M.M. Sadeghi, F.A. Dorkoosh, M.R. Avadi, Permeation enhancer effect of chitosan and chitosan derivatives: Comparison of formulations as soluble polymers and nanoparticulate systems on insulin absorption in Caco-2 cells, Eur. J. Pharm. Biopharm. 70 (2008) 270–278.

[66] I. Behrens, A.I. Pena, M.J. Alonso, T. Kissel, Comparative uptake studies of bioadhesive and non-bioadhesive nanoparticles in human intestinal cell lines and rats: The effect of mucus on particle adsorption and transport, Pharm. Res. 19 (2002) 1185–1193.

[67] D.T. O'Hagan, The intestinal uptake of particles and the implications for drug and antigen delivery, J. Anal. 189 (1996) 477–482.

[68] A.T. Florence, A.M. Hillery, N. Hussain, P.U. Jani, Nanoparticles as carriers for oral peptide absorption: Studies on particle uptake and fate, J. Control. Release 36 (1995) 39–46.

[69] A. des Rieux, V. Fievez, M. Garinot, Y.-J. Schneider, V. Préat, Nanoparticles as potential oral delivery systems of proteins and vaccines: A mechanistic approach, J. Control. Release 10 (2006) 1–27.

[70] C. Damge, C. Michel, M. Aprahamian, P. Couvreur, New approach for oral administration of insulin with polyalkylcyanoacrylate nanocapsules as drug carrier, Diabetes 37 (1988) 246–251.

[71] L. Brannon-Peppas, Recent advances on the use of biodegradable microparticles and nanoparticles in controlled drug delivery, Int. J. Pharm. 116 (1995) 1–9.

[72] R.A. Jain, The manufacturing techniques of various drug loaded biodegradable poly (lactide-co-glycolide) (PLGA) devices, Biomaterials 21 (2000) 2475–2490.

[73] C.P. Reis, R.J. Neufeld, A.J. Ribeiro, F. Veiga, Nanoencapsulation II. Biomedical applications and current status of peptide and protein nanoparticulate delivery systems, Nanomedicine 2 (2006) 53–65.

[74] R. Yang, R. Gao, F. Li, H. He, X. Tang, The influence of lipid characteristics on the formation, *in vitro* release, and *in vivo* absorption of protein loaded SLN prepared by the double emulsion process, Drug Dev. Ind. Pharm. 37 (2011) 139–148.

[75] L. Yin, J. Ding, C. He, L. Cui, C. Tang, C. Yin, Drug permeability and mucoadhesion properties of thiolated trimethyl chitosan nanoparticles in oral insulin delivery, Biomaterials 30 (2009) 5691–5700.

[76] S. Sakuma, R. Sudo, N. Suzuki, H. Kikuchi, M. Akashi, M. Hayashi, Mucoadhesion of polystyrene nanoparticles having surface hydrophilic polymeric chains in the gastrointestinal tract, Int. J. Pharm. 177 (1999) 161–172.

[77] T. Morçöl, P. Nagappan, L. Nerenbaum, A. Mitchell, S.J.D. Bell, Calcium phosphate-PEG-insulin-casein (CAPIC) particles as oral delivery systems for insulin, Int. J. Pharm. 277 (2004) 91–97.

[78] W. Paul, C.P. Sharma, Tricalcium phosphate delayed release formulation for oral delivery of insulin: A proof-of-concept study, J. Pharm. Sci. 97 (2008) 875–882.
[79] R. Ramachandran, W. Paul, C.P. Sharma, Synthesis and characterization of PEGylated calcium phosphate nanoparticles for oral insulin, J. Biomed. Mater. Res. Part B: Appl. Biomater. 88B (2009) 41–48.
[80] W. Paul, C.P. Sharma, Fatty acid conjugated calcium phosphate nanoparticles for protein delivery, Int. J. Appl. Ceram. Technol. 7 (2010) 129–138.
[81] A. Hashimoto, J. Kawada, Effects of oral administration of positively charged insulin liposomes on alloxan diabetic rats: Preliminary study, Endocrinol. Jpn. 26 (1979) 337–344.
[82] H.M. Patel, B. Ryman, The gastro-intestinal absorption of liposomally entrapped insulin in normal rats, Biochem. Soc. Trans. 5 (1977) 1054–1055.
[83] J.F. Arrieta-Molero, K. Aleck, M.K. Sinha, C.M. Brownscheilde, L.J. Shapiro, M.A. Sperling, Orally administered liposome-entrapped insulin in diabetic animals. A critical assessment, Horm. Res. 16 (1982) 249–256.
[84] C. Weingarten, A. Moufti, J. Delattre, F. Puisieux, P. Couvreur, Protection of insulin from enzymatic degradation by its association to liposomes, Int. J. Pharm. 26 (1985) 251–257.
[85] C. Weingarten, A. Moufti, J.F. Desjeux, T.T. Luong, G. Durand, J.P. Devissaguet, et al., Oral ingestion of insulin liposomes: Effects of the administration route, Life Sci. 28 (1981) 2747–2752.
[86] K.E. Anderson, B.R. Stevenson, J.A. Rogers, Folic acid-PEO acid-PEO-labeled liposomes to improve gastrointestinal absorption of encapsulated agents, J. Control. Release 60 (1999) 189–198.
[87] M. Garcia-Fuentes, C. Prego, D. Torres, M.J. Alonso, A comparative study of the potential of solid triglyceride nanostructures coated with chitosan or poly (ethylene glycol) as carriers for oral calcitonin delivery, Eur. J. Pharm. Sci. 25 (2005) 133–143.
[88] K. Yamabe, Y. Kato, H. Onishi, Y. Machida, Potentiality of double liposomes containing salmon calcitonin as an oral dosage form, J. Control. Release 89 (2003) 429–436.
[89] L. Hovgaard, E.J. Mack, S.W. Kim, Insulin stabilization and GI absorption, J. Control. Release 19 (1992) 99–108.
[90] A.M. Hillery, Drug delivery, the basic concepts, in: A.M. Hillery, A.W. Lloyd, J. Swarbrick (Eds.), Drug Delivery and Targeting for Pharmacists and Pharmaceutical Scientists, Taylor & Francis, London, 2001, pp. 1–48.
[91] S. Frokjaer, D.E. Otzen, Protein drug stability: A formulation challenge, Nat. Rev. 4 (2005) 298–306.
[92] H.S. Lin, C.-T. Chen, H.-F. Liang, A.R. Kulkarni, P.-W. Lee, C.-H. Chen, H.-W. Sung, Novel nanoparticles for oral insulin delivery via the paracellular pathway, Nanotechnology 18 (2007) 105102.
[93] S. Sajeesh, C.P. Sharma, Cyclodextrin-insulin complex encapsulated polymethacrylic acid based nanoparticles for oral insulin delivery, Int. J. Pharm. 325 (2006) 147–154.
[94] M.A. Karsdal, I. Byrjalsen, M. Azira, M. Arnold, L. Choi, B.J. Rjis, et al., Influence of food intake on the bioavailability and efficacy of oral calcitonin, Br. J. Clin. Pharmacol. 67 (2009) 413–420.
[95] R.A. Cone, Barrier properties of mucus, Adv. Drug Deliv. Rev. 27 (2009) 75–85.
[96] S.K. Lai, Y.-Y. Wang, J. Hanes, Mucus-penetrating nanoparticles for drug and gene delivery to mucosal tissues, Adv. Drug Deliv. Rev. 61 (2009) 158–171.
[97] D.J. Thornton, J.K. Sheehan, From mucins to mucus: Toward a more coherent understanding of this essential barrier, Proc. Am. Thorac. Soc. 1 (2004) 54–61.

[98] C. Wickström, J.R. Davies, G.V. Eriksen, E.C.I. Veerman, I. Carlstedt, MUC5B is a major gel-forming, oligomeric mucin from human salivary gland, respiratory tract and endocervix: Identification of glycoforms and C-terminal cleavage, Biochem. J. 334 (1998) 685–693.
[99] H.W. Hovenberg, J.R. Davies, I. Carlstedt, Different mucins are produced by the surface epithelium and the submucosa in human trachea: Identification of MUC5AC as a major mucin from goblet cells, Biochem. J. 318 (1996) 319–324.
[100] J.K. Sheehan, I. Carlstedt, Hydrodynamic properties of human cervical-mucus glycoproteins in 6M-guanidinium chloride, Biochem. J. 217 (1984) 93–101.
[101] D.J. Thornton, D.F. Holmes, J.K. Sheehan, I. Carlstedt, Quantitation of mucus glycoproteins blotted onto nitrocellulose membranes, Anal. Biochem. 182 (1989) 160–164.
[102] P. Gruber, A. Rubinstein, V.H. Li, P. Bass, J.R. Robinson, Gastric emptying of nondigestible solids in the fasted dog, J. Pharm. Sci. 76 (1987) 117–122.
[103] C.-M. Lehr, F.G.J. Poelma, H.E. Junginger, J.J. Tukker, An estimate of turnover time of intestinal mucus gel layer in the rat *in situ* loop, Int. J. Pharm. 70 (1991) 235–240.
[104] S.A. Galindo-Rodriguez, E. Allemann, H. Fessi, E. Doelker, Polymeric nanoparticles for oral delivery of drugs and vaccines: A critical evaluation of *in vivo* studies, Crit. Rev. Ther. Drug Carr. Syst. 22 (2005) 419–464.
[105] G. Ponchel, J. Irache, Specific and non-specific bioadhesive particulate systems for oral delivery to the gastrointestinal tract, Adv. Drug Deliv. Rev. 34 (1998) 191–219.
[106] A. des Rieux, V. Fievez, M. Garinot, Y.J. Schneider, V. Préat, Nanoparticles as potential oral delivery systems of proteins and vaccines: A mechanistic approach, J. Control. Release 116 (2006) 1–27.
[107] H. Takeuchi, H. Yamamoto, Y. Kawashima, Mucoadhesive nanoparticulate systems for peptide drug delivery, Adv. Drug Del. Rev. 23 (2001) 39–54.
[108] F. Atyabi, F.A. Moghaddam, R. Dinarvand, M.J. Zohuriaan-Mehr, G. Ponchel, Thiolated chitosan coated poly hydroxyethyl methacrylate nanoparticles: Synthesis and characterization, Carb. Pol. 74 (2008) 59–67.
[109] A. Bernkop-Schnürch, Thiomers: A new generation of mucoadhesive polymers, Adv. Drug Del. Rev. 57 (2005) 1569–1582.
[110] K. Yoncheva, E. Lizarraga, J.M. Irache, Pegylated nanoparticles based on poly(methyl vinyl ether-co-maleic anhydride): Preparation and evaluation of their bioadhesive properties, Eur. J. Pharm. Sci. 24 (2005) 411–419.
[111] S.K. Lai, D.E. O'Hanlon, S. Harrold, S.T. Man, Y.Y. Wang, R. Cone, J. Hanes, Rapid transport of large polymeric nanoparticles in fresh undiluted human mucus, Proc. Natl. Acad. Sci. U.S.A. 104 (2007) 1482–1487.
[112] K. Yang, J. Wu, X. Li, Recent advances in research on P-glycoprotein inhibitors, BioScience Trends 2 (2008) 137–146.
[113] J. Iqbal, J. Hombach, B. Matuszczak, A. Bernkop-Schnürch, Design and *in vitro* evaluation of a novel polymeric P-glycoprotein (P-gp) inhibitor, J. Control. Release 147 (2010) 62–69.
[114] K.B. Chalasani, G.-J. Russell-Jones, S.K. Yandrapu, P.V. Diwan, S.K. Jain, A novel vitamin B12-nanosphere conjugate carrier system for peroral delivery of insulin, J. Control. Release 117 (2007) 421–429.
[115] K.B. Chalasani, G.-J. Russell-Jones, A.K. Jain, P.V. Diwan, S.K. Jain, Effective oral delivery of insulin in animal models using vitamin B12-coated dextran nanoparticles, J. Control. Release 122 (2007) 141–150.
[116] F. Cui, K. Shi, L. Zhang, Y. Kawashima, Biodegradable nanoparticles loaded with insulin–phospholipid complex for oral delivery: Preparation, *in vitro* characterization and *in vivo* evaluation, J. Control. Release 114 (2006) 242–250.

[117] K. Sonaje, Y.-H. Lin, J.-H. Juang, *In vivo* evaluation of safety and efficacy of self-assembled nanoparticles for oral insulin delivery, Biomaterials 30 (2009) 2329–2339.

[118] C. Damge, P. Maincent, N. Ubrich, Oral delivery of insulin associated to polymeric nanoparticles in diabetic rats, J. Control. Release 117 (2007) 271–280.

[119] F. Cui, A.-J. Tao, D.-M. Cun, L.Q. Zhang, K. Shi, Preparation of insulin loaded PLGA-Hp55 nanoparticles for oral delivery, J. Pharm. Sci. 96 (2006) 421–427.

[120] K.-H. Song, S.-J. Chung, C.-K. Shim, Enhanced intestinal absorption of salmon calcitonin (sCT) from proliposomes containing bile salts, J. Control. Release 106 (2005) 298–308.

[121] T. Blunk, D.F. Hochstrasser, J.-C. Sanchez, B.W. Müller, R.H. Müller, Colloidal carriers for intravenous targeting: Plasma protein adsorption patterns on surface modified latex particles evaluated by two-dimensional polyacrylamide gel electrophoresis, Electrophoresis 14 (1993) 1382–1387.

[122] M. Luck, B.R. Paulke, W. Schröder, T. Blunk, R.H. Müller, Analysis of plasma protein adsorption on polymeric nanoparticles with different surface chacteristics, J. Biomed. Mater. Res. 39 (1998) 478–485.

[123] D.E. Ovens III, N.A. Peppas, Opsonisation, biodistribution and pharmacokinetics of polymeric nanoparticles, Int. J. Pharm. 307 (2006) 93–102.

[124] J.B. Hall, M.A. Dobrovolskaia, A.K. Patri, Characterisation of nanoparticles for therapeutics, Nanomedicine 2 (2007) 789–803.

[125] D. Jacobson-Kram, T. McGovern, Toxicological overview of impurities in pharmaceutical products, Adv. Drug Deliv. Rev. 59 (2007) 38–42.

[126] C.P. Reis, I.V. Figueiredo, R.A. Carvalho, Toxicological assessment of orally delivered nanoparticulate insulin, Nanotoxicology 2 (2008) 205–217.

[127] J.F. Hillyer, R.M. Albrecht, Gastrointestinal persorption and tissue distribution of differently sized colloidal gold nanoparticles, J. Pharm. Sci. 90 (2001) 1927–1936.

[128] N. Hussain, V. Jaitley, A.T. Florence, Recent advances in the understanding of uptake of microparticulates across the gastrointestinal lymphatics, Adv. Drug Deliv. Rev. 50 (2001) 107–142.

[129] P. Jani, G.W. Halbert, J. Langridge, A.T. Florence, The uptake and translocation of latex nanospheres and microspheres after oral administration to rats, J. Pharm. Pharmacol. 41 (1989) 809–812.

[130] P. Jani, G.W. Halbert, J. Langridge, A.T. Florence, Nanoparticle uptake by the rat gastrointestinal mucosa: Quantitation and particle size dependency, J. Pharm. Pharmacol. 42 (1990) 821–826.

Chapter 9

Modulation of the Intestinal Tight Junctions Using Bacterial Enterotoxins

Christopher F. van der Walle [1] and Elke Schmidt [2]

[1] *University of Strathclyde, Strathclyde Institute of Pharmacy and Biomedical Sciences, 161 Cathedral Street, Glasgow, G4 0RE, Scotland,* [2] *Cancer Research UK Formulation Unit, University of Strathclyde, 27 Taylor St, Glasgow, G4 0NR, Scotland*

Chapter Outline

9.1. INTRODUCTION

Epithelial cell sheets limit the movement of solutes through the intercellular space by forming tight junctions (tjs) between adjacent epithelial cells, and therefore act as the major barrier between the internal and external environment of the body. With the largest surface area (*ca.* 200 m^2) in the body, the plasma membranes of the intestinal epithelium offer great potential for the absorption

Peptide and Protein Delivery. DOI: 10.1016/B978-0-12-384935-9.10009-4

of drugs. There are three transepithelial pathways for molecules to pass from the intestinal lumen to the bloodstream: the passive transcellular pathway, the carrier-mediated transcellular pathway and the paracellular pathway. Physicochemical properties, such as hydrophobicity, allow a molecule to passively partition from the intestinal lumen through the lipid bilayer into the cell. In contrast, hydrophilic properties prevent passive diffusion across the apical cell membrane. Some hydrophilic molecules, such as sugars and amino acids, are absorbed by specifically interacting with active transport systems on the cell membrane. Drug delivery via the paracellular pathway is less dependent on the physicochemical properties of the drug, and does not require a specific interaction with a transport system, and so is suitable for a large variety of molecules including peptides and proteins. Since opening of the tjs can cause the influx of other foreign substances, absorption enhancers acting via the paracellular pathway are required to modify the tight junctional structure *reversibly.* Early development of paracellular absorption enhancers was limited due to the lack of knowledge regarding the composition of the tjs and their regulatory role, resulting in unacceptable side-effects. Study of canonical enterotoxins (a class of exotoxin that acts on the intestinal epithelium) such as those released by *Clostridium perfringens* and *Vibrio cholerae* has increased our understanding to the point where we can exploit their mechanisms to facilitate peptide and protein delivery via the paracellular pathway [1,2].

9.2. TIGHT JUNCTION COMPOSITION

The tj complex is composed of the integral transmembrane proteins: claudin, occludin and junction adhesion molecules (JAMs), whose extracellular domains mediate cell—cell adhesion. Their cytosolic regions in turn associate with a set of adaptor molecules including ZO-1, ZO-2 and ZO-3, forming a link to the F-actin cell cytoskeleton. Occludin was the first protein identified in the tj [3]; its structure is composed of four hydrophobic transmembrane regions and two extracellular loops, with the amino- and carboxyl-termini facing the cytoplasm [4,5]. The C-terminal domain of occludin binds to ZO-1 and is also believed to be involved in the dynamic regulation of paracellular permeability. Both extracellular loops are essential for tj formation and sealing: the first loop undergoes homologous cell—cell interaction and the second facilitates a stable assembly [6]. Surprisingly, it was reported that occludin-deficient mice established well developed tjs [7], which led to the discovery of other integral components of the tj, namely claudin-1 and claudin-2 [8,9]. Today, our knowledge of the family of claudin proteins has expanded to 24 members, some of which have not yet been well characterized [10]. The claudin family of proteins are relatively small — between 20—27 kDa — and show four transmembrane domains, a significantly longer first extracellular loop of ~53 amino acids, a second extracellular loop of ~24 amino acids and a short carboxyl intracellular tail. Claudins interact in a homophilic or heterophilic manner [11], although the latter is limited to specific combinations which determines tj "tightness" and ion selectivity – attributed to the wide range of pI values of the extracellular domains [10]. JAMs are composed of two extracellular IgG-like domains and two N-glycosylation sites, and mediate both homotypic adhesion and monocyte transmigration [12]. Investigations of monocyte migration across epithelial monolayers suggest that JAMs provide an adhesive contact targeting monocytes to the intercellular pathway, or transduce a signal within the endothelial cell to open the tj space [13]. The occludin C-terminus interacts with the N-termini of the ZO proteins, though more precisely in case of ZO-1 it is the Gk domain and the acidic region that binds occludin [14]. Studies on recombinant ZO proteins identified their first PDZ domain to bind to the carboxyl terminal Tyr-Val (YV) sequence of the claudins [15].

The zonula occludens proteins (ZO-1, ZO-2, ZO-3) are members of the MAGUK (membrane associated guanylate kinase homologs) family of assembling multiprotein complexes. ZOs are peripherally associated membrane proteins, interacting together and anchoring membrane proteins like occludin, claudins and JAMs to the actin cytoskeleton. Their structure comprises three PDZ domains; a src-homolgy 3 domain (SH3), a guanylate kinase-like (Gk) domain and a carboxyl-end containing an acidic domain and a proline-rich region [16]. ZO-1 is the largest of the three ZO proteins, with a species-dependant mass of between 210 and 225 kDa. There is a specifically high abundance of ZO-1 in epithelial and endothelial cells, although it has also been found in non-epithelial cells. Two nuclear exporting signals have been associated with the PDZ-1 motif and the Gk domain, suggesting a participation in the regulation of transcription [17,18]. The SH3 domain mediates non-catalytic protein–protein interactions, such as coupling substrates to enzymes and thereby regulating enzymatic activities [19]. ZO-2 is a slightly smaller molecule than ZO-1 [20], with a molecular mass of 160 kDa, and similarly to ZO-1, it is present at the tj and adherens junctions of cells that lack tjs (e.g. fibroblast or cardiac muscle cells). ZO-3 was characterized as a 130 kDa protein with a surprisingly short carboxyl terminus. An extremely proline-rich region, potentially containing three SH3 binding domains, is found on the amino terminal between PDZ motifs two and three [21]. The function of the basic region inserted between PDZ domain one and two has yet to be established.

9.3. SIGNALING PATHWAYS REGULATING TJ FUNCTIONS

The major signaling pathways in epithelial barriers involve protein kinases A, C and G (PKA, PKC, PKG), Rho kinases, myosin light chain kinase (MLCK) and the mitogen activated protein kinase (MAPK) system (Figure 9.1) [22].

A wide range of pathological and physiological signals act through PKC, such as the depletion of extracellular Ca^{2+} or elevation of intracellular Ca^{2+} [23]. Ca^{2+} chelators, the first generation of absorption enhancers, also activate

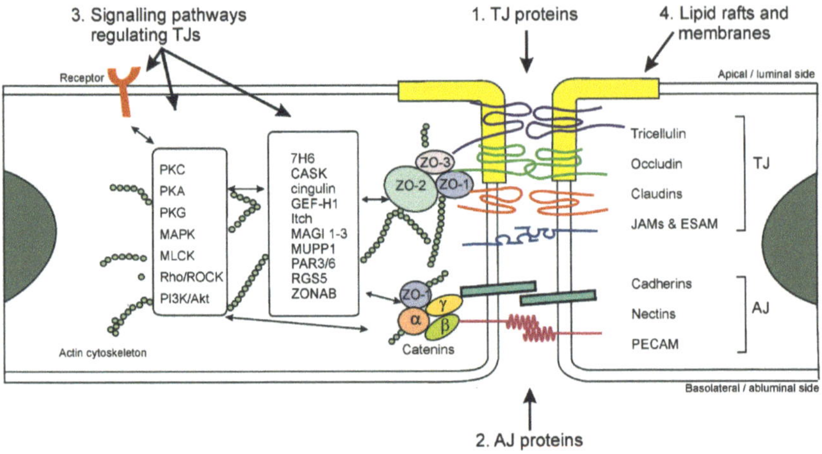

FIGURE 9.1 Schematic model of targets of tight junction modulators in epithelial and endothelial cells. Abbreviations: 7H6, cytoplasmic tight junction-associated protein; AJ, adherens junction; CASK, Ca^{2+}-dependent serine protein kinase; ESAM, endothelial selective adhesion molecule; GEF-H1, Guanine nucleotide exchange factor H1; Itch, E3 ubiquitin protein ligase; JAMs, junctional adhesion molecules; MAGI 1-3, membrane-associated guanylate guanylate kinase with inverted orientation of protein–protein interaction domains; MAPK, mitogen-activated protein kinase; MLCK, myosin light chain kinase; MUPP 1, multi-PDZ-protein 1; PAR3/6, partitioning defective proteins 3 and 6; PECAM, platelet-endothelial cell adhesion molecule; PI3K/Akt, phosphoinositide 3-kinase/Akt pathway; PKA, protein kinase A; PKC, protein kinase C; PKG, protein kinase G; RGS5, regulator of G-protein signalling 5; Rho/ROCK, Rho/Rho-associated protein kinase pathway; TJ, tight junction; ZO-1, zonula occludens 1; ZO-2, zonula occludens 2; ZO-3, zonula occludens 3; ZONAB, ZO-1-associated nucleic acid-binding protein. *Reprinted from reference [22] with permission from Elsevier.*

PKC. In addition, a range of pathological stimuli, such as oxidative stress, cytokines and vascular endothelial growth factor (VEGF), as well as toxins like HIV gp-120, Zonula occludens toxin (Zot), *Clostridium difficile* toxin A, *Escherichia Coli* OmpA all act via PKC to open tjs [22]. PKA regulation of the epithelial tj is contentious but has been linked to increased TEER, and a concomitant decrease in paracellular permeability in brain endothelial cells by elevating the intracellular cAMP levels [24]. In contrast, the elevation of intracellular cGMP levels through PKG activation of soluble guanylate cyclase facilitates the "opening" of the blood–brain barrier (BBB) [25].

Contraction of the actin-myosin belt can result in disassembly of the tight junction. Several cytokines, bacterial/viral pathogens, bile acids and the removal of extracellular Ca^{2+} activate MLCK, facilitating the phosphorylation of myosin light chain and contraction of the actin-myosin belt, resulting in the opening of the tj [26]. Three of the four groups of mitogen-activated protein kinases (MAPKs) regulate tj function: extracellular signal-regulated kinase (ERK), c-Jun N-terminal kinase (JNK) and p38 isoforms, and the disassembly of tjs can occur via all three pathways, or combinations thereof [27]. In this study, cross-talk between the fibronectin central cell binding domains [28] and ZO proteins was also implicated, although the functional significance in the intestine has yet to be established. The activation of the Rho family of proteins, Cdc42, Rac1 and RhoA, increases tj permeability. RhoA alters tj sealing through phosphorylation of occludin, claudin-5 and ZO-1 and -2 and also interacts with the PKC and MLCK pathways, indirectly altering tj function [29].

9.4. TIGHT JUNCTION MODULATORS

In the past, a lack of knowledge concerning the structure and physiology of the tj has prevented the development of pharmaceutically relevant absorption enhancers. The first generation of absorption enhancers was composed of a variety of categories such as surfactants, bile salts and their derivatives, Ca^{2+} chelators, medium chain length fatty acids/esters/amines and chitosans. Investigations on the effect of sodium dodecyl sulphate (SDS), an anionic surfactant, and polysorbate 80, a non-ionic surfactant, as well as the effects of bile salts (sodium taurocholate, sodium taurodeoxycholate and sodium taurodihydrofusidate (STDHF)) on epithelial permeability and integrity were performed with the human epithelial colorectal adenocarcinoma (Caco-2) cell culture model [30,31]. Although a concentration dependant increase in paracellular permeability was observed, a concomitant alteration of the cell morphology and cell membrane damage in case of SDS, STDHF and polysorbate 80 also occurred. Glycosylated bile acids with an additional amino or hydroxyl group, respectively, at the C3 position of the steroid structure were developed to reduce alteration of cell morphology and associated cellular damage [32]. In the class of zwitterionic surfactants, the naturally occurring acylcarnitines initially showed promise as absorption enhancers but this was later correlated with cytotoxicity [33,34].

Ethylenediaminetetraacetic acid (EDTA) is a tetravalent molecule that is capable of forming a stable complex with divalent ions, such as Ca^{2+}. Subsequent depletion of extracellular calcium activates PKC that mediates an expansion of the paracellular route [35]. As a mucoadhesive polymeric agent, chitosan and its derivatives induce a redistribution of F-actin and ZO-1 at the tj complex. Solubility issues are a drawback of chitosan but have been addressed by quaternizing the amino residues [30]. The only first generation absorption enhancer, which is clinically implemented in Japan, Sweden and Denmark, is the fatty acid sodium caprate [36]. This medium chain length fatty acid modulates intercellular permeability following the phospholipase C-signal transduction pathway, but also accounts for a reduction of cellular dehydrogenase activity and ATP levels.

The second generation of paracellular permeability enhancers is based on recent knowledge of tj composition and regulation. Initial approaches were based on disrupting the interaction of the extracellular loops of integral tj proteins, particularly occludin [37,38]. An unusual approach employed enzymatic cleavage: Der p1, a cysteine protease derived from *Dermatophagoides pteronyssinus*, cleaves the first extracellular loop of occludin [39]. Several studies on chick occludin identified that administration of the peptides that are homologous to the second extracellular loop of occludin resulted in manipulation of tj permeability [40]. Later, more specific motifs were discovered on the first extracellular loop of chicken occludin and also its human homolog [41]. These studies lead to the conclusion that peptide mimics of specific extracellular loop regions were capable of displacing various homologous or heterologous protein–protein interactions present at the tj [42]. Presently, various peptides of the extracellular domains of JAMs, occludin and claudins are being claimed to enhance mucosal delivery of therapeutic agents, in patent offices around the world [43].

Other approaches involve the modulation of the transcription of tj related proteins. The interference or degradation of mRNA of individual tj proteins was achieved through antisense oligonucleotides in the case of ZO-1, and small, interfering RNAs (siRNAs) in the case of occludin and claudin-4 [44,45]. Interestingly, the silencing of the occludin transcript had little impact on the paracellular permeability, excluding this method as a mode of permeation enhancement [41]. Of course, the major challenge with respect to strategies based on peptide and RNA technologies remains the effective delivery of these potential absorption enhancers themselves, although there are biologically stable analogs of peptides that could be envisaged.

Inspiration for macromolecular paracellular delivery can be found in the study of bacterial enterotoxins. These toxins are diverse in terms of their mode of action, ranging from rearrangement of the actin cytoskeleton to manipulation of specific tj proteins. The former mode of action is observed for the *Clostridium difficile* enterotoxins A and B, which inactivate the Rho family of proteins resulting in a degradation of the actin filaments and increased paracellular permeability [46,47]. The C-terminal domain of the *Clostridium perfringens* enterotoxin (C-CPE) binds to a cell receptor, which was later identified as claudin-4. Subsequently, claudin-4 was selectively removed from the tj with a concomitant disintegration of tj strands in a time and dose dependant manner [48]. *In vitro* and *ex vivo* studies have shown that C-CPE treatment increases the uptake of hydrophilic marker molecules [49,50]. Of particular interest is zonula occludens toxin (Zot), produced by *Vibrio cholerae* [51], which exhibits an ability to manipulate the tj sealing. Zot intervenes with an endogenous analog, zonulin, which regulates the intercellular permeability in a PKC-α-dependant manner [52]. The resulting effect on the peri-junctional actin ring has shown an increase in the uptake of various therapeutic agents, including insulin [53].

9.4.1. *Clostridium Perfringens* Enterotoxin (CPE)

CPE causes food-borne disease in humans associated with diarrheal and abdominal cramping symptoms. These symptoms are elicited following the release of CPE into the intestinal lumen, where it binds to its receptors on the surface of intestinal epithelial cells and triggers the formation of a large multiprotein membrane pore complex, ultimately resulting in cell lysis [54]. The process is very specific, since cells lacking expression of CPE receptors are completely unaffected by the toxin [55]. CPE is a single polypeptide of 35 kDa comprising an N-terminal cytotoxic domain and a C-terminal receptor-binding domain that binds to claudin-3 and -4 (Figure 9.2) [56,57]. The binding of CPE to a receptor on the cell membrane was first reported by Katahira *et al.* [58], but the biological function of the receptor remained unknown until it was matched to claudin-4. CPE cytotoxicity is mediated by the N-terminal domain, which forms small pores in the plasma membrane. Upon removal of this domain, the cytotoxic effect is removed, but receptor binding remains functional so long as the C-terminal 30 amino acids remain intact [59].

C-terminal CPE (C-CPE) binds to claudin-4 and the second extracellular loop of claudin-3 with high binding affinity [60]. Claudin-6, -7, -8 and -14 also bind C-CPE albeit with a lower binding affinity than with claudin-3, while claudin-5 and -10 do not bind C-CPE at all [60]. Many members of the claudin family show a distinct, organ-specific distribution pattern whereby the variation of expression profiles of claudin family members may be a critical factor for determining paracellular permeability in a specific tissue [61]. Claudin-1 plays a pivotal role in the epidermal barrier function and claudin-5 proves vital for the blood–brain barrier [62]. Targeting specific members of the claudin family may therefore allow for tissue-specific drug delivery. Structural analysis of C-CPE narrowed the claudin-4 modulating domain down to the C-terminal 16 amino acids of CPE, and highlighted the importance of the presence of tyrosine residues for functionality [48]. A later alanine scan identified Leu315 and Tyr306 as being critical and suggested that C-CPE and claudin-4 interact through formation of a hydrophobic cluster [50], which was supported by evidence showing that C-CPE mutants lacking Tyr306/310/312 and Leu315 lacked claudin-4 modulating activity. However, X-ray diffraction studies of C-CPE revealed a nine-strand β-sandwich structure over residues 194 to 319 showing similarities to other receptor binding domains of toxins of spore forming bacteria [63]. The receptor binding domain was localized to a large surface loop at strands 8 and 9, and the development of claudin modulators could be based on the sequences contained in this loop structure [64].

The absorption-enhancing effect of C-CPE was tested in an *in situ* loop assay using fluorescein isothiocyanate-dextran, and compared to sodium caprate [49]. C-CPE (184–319 aa) enhanced the absorption of dextrans up to 20 kDa in a dose dependant, non-cytotoxic manner, at a concentration of 0.1 mg/ml, compared to 40 mg/ml of sodium caprate for the same effect.

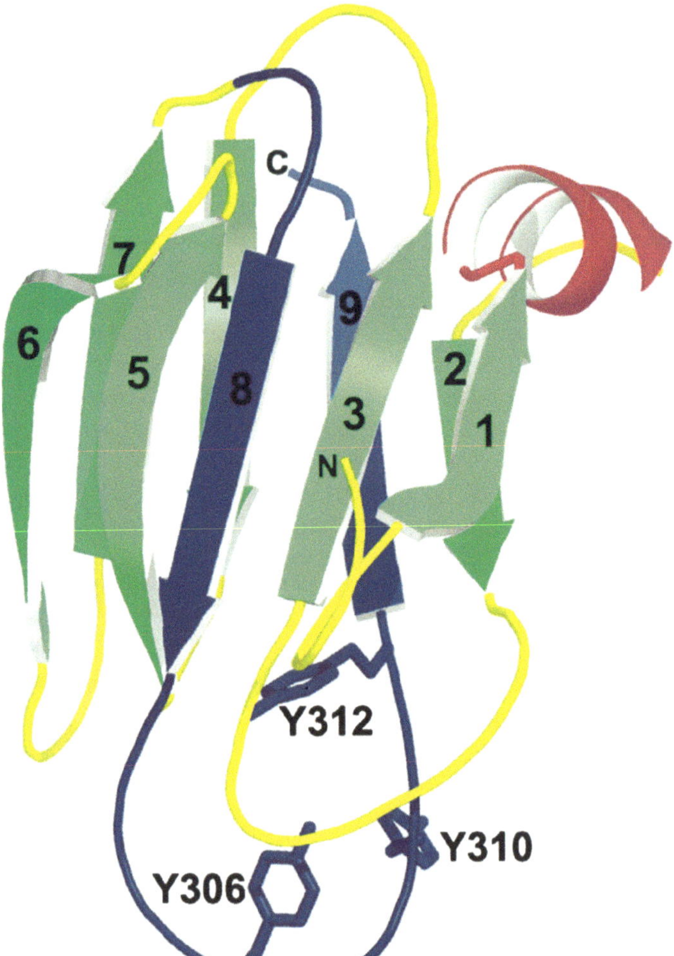

FIGURE 9.2 Structure of C-CPE-(194-319), showing the nine antiparallel β-strands and one α-helix. The binding site for claudin lies within the 30 C-terminal residues (dark blue), including strands β8 and β8 and the intervening surface loop. *Reprinted from reference [63] with permission from the American Society for Biochemistry and Molecular Biology.*

Furthermore, the absorption enhancing effect of C-CPE was observed in the jejunum but not in the colon, which would be considered advantageous on account of bacterial infiltration in the colon. Most recent studies assessed the absorption enhancing effect of C-CPE (184–319 aa) in jejunal, mucosal and pulmonary epithelium for a human parathyroid hormone derivative (hPTH [1–34]) [62]. Co-administration of C-CPE and hPTH resulted in a 2.5-fold absorption increase from nasal but not for jejunal or pulmonary tissue.

However, pre-incubation of the tissue with C-CPE increased jejunal absorption 7.5-fold, nasal absorption 5.6-fold and pulmonary absorption 2.4-fold. The experimental setup was limited due to the low aqueous solubility (0.3 mg/ml) of C-CPE.

Previous structural work on C-CPE (184–319 aa) by Van Itallie and co-workers established an increase in aqueous solubility (10 mg/ml) upon the removal of the 10 amino-terminal amino acids (C-CPE194) [63]. Based on these findings, Uchiba *et al.* prepared N-terminal truncates of C-CPE, and discovered that C-CPE184, C-CPE194 and C-CPE205 successfully bound claudin-4 and modulated tj integrity [62]. However, C-CPE derivatives lacking the first beta sheet (205–212), failed to maintain the C-CPE structure required for binding to claudin-4 and tj modulating activity. Based on these findings, C-CPE shows great potential as an absorption enhancer for peptide drugs, but since it is itself a polypeptide, the issue of antigenicity may arise. Nevertheless, the isolation of a common claudin-binding motif <XX(Y/W)(X)$_{3/4}$Y(Y/X) (L/I)XX> may lay the foundation for further design of claudin specific tj modulators [65].

9.4.2. Zonula Occludens Toxin (Zot)

Zot is an enterotoxin elaborated by *Vibrio cholera,* which was found more or less by accident during the development of a vaccine for cholera toxin (CTX) [66]. When the attenuated *V. cholerae* strain was administered to volunteers, more than half still developed a mild form of diarrhea. Comparison of the attenuated strain with the wild type led to the discovery that tissue exposed to the supernatant of either showed an overall decrease in the number of zonula occludens strands. Consequently, the factor that increased paracellular permeability and tissue conductance in rabbit ileum was found to be a 10–30 kDa peptide, which was named Zonula occludens toxin (Zot) [51]. Further studies to identify Zot were carried out, localizing the Zot encoding sequence immediately upstream of the *ctx* operon [67]. This led to a 1.3 kb open reading frame that, when translated, would theoretically result in a 44.8 kDa polypeptide containing 399 amino acids with a pI of 8.5. Zot and other genes on the chromosomal core region of *V. cholerae* are part of a filamentous bacteriophage named CTXΦ [68]. The analog of Zot in the filamentous bacteriophage is the pI protein, which led to the assumption of a dual function of Zot as a modulator of tjs and a putative NTPase [69,70]. The work by Koonin also identified the hydrophobic transmembrane membrane domain, and inferred N- and C-terminal regions of ZOT via hydropathy analysis (Figure 9.3) [69].

In vitro and *ex vivo* systems were examined to determine the molecular mechanisms of Zot, reporting an increase in polymerized F-actin mediated through a PKC α-isoform [52]. A hypothesis was established wherein Zot mimicked the effect of a putative physiological modulator of epithelial barrier function, by engaging to a specific surface cell receptor. In support of a putative

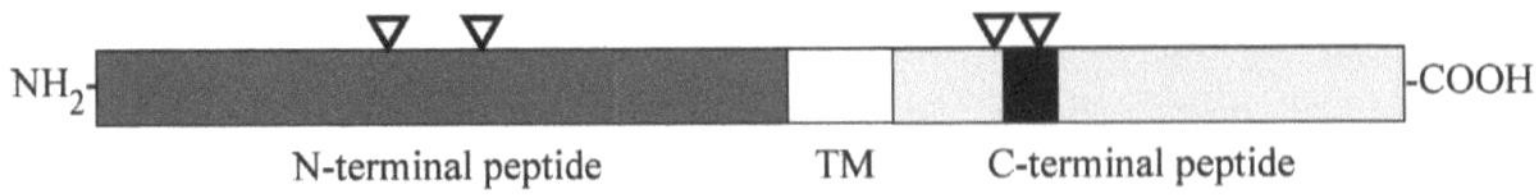


```
  1 SIFIHHGAPG SYKTSGALWL RLLPAIKSGR HIITNVRGLN LERMAKYLKM DVSDISIEFI
 61 DTDHPDGRLT MARFWHWARK DAFLFIDECG RIWPPRLTVT NLKALDTPPD LVAEDRPESF
121 EVAFDMHRHH GWDICLTTPN IAKVHNMIRE AAEIGYRHFN RATVGLGAKF TLTTHDAANS
181 GQMDSHALTR QVKKIPSPIF KMYASTTTGK ARDTMAGTAL WKDRKILFLF GMVFLMFSYS
241 FYGLHDNPIF TGGNDATIES EQSEPQSKAT VGNAVGSKAV APASFGFCIG RLCVQDGFVT
301 VGDERYRLVD NLDIPYRGLW ATGHHIYKDT LTVFFETESG SVPTELFASS YRYKVLPLPD
361 FNHFVVFDTF AAQALWVEVK RGLPIKTEND KKGLNSIF
```

FIGURE 9.3 Schematic of the Zot amino acid (aa) sequence; N-terminus coloured dark gray (aa 1-215), C-terminus coloured light gray (aa 245-399), with the transmembraneous region (TM) remaining clear (aa 225-244) and the putative active site is coloured black (aa 291-298); arrow heads indicate the position of cysteine residues. In the corresponding Zot amino acid sequence, the transmembraneous region is boxed and the active site underlined.

"Zot receptor", labeling of Zot-MBP confirmed engagement to a peripheral cell surface receptor [71]. One obstacle to the study of Zot was its poor expression in recombinant systems, and to increase yield and purity Zot was expressed as a recombinant protein fused to maltose binding protein (MBP) [72]. A high distribution of Zot-MBP binding coincided with a maximum in tissue conductance observed in the jejunum, decreasing down to the ileum and with virtually no binding in the colon [52]. The mature cells at the top of the villi seem to be preferentially targeted by Zot, rather than the crypt area.

Immobilization of Zot onto an affinity chromatography column to capture the putative receptor from lysates from tissue of human heart, brain and intestine, yielded two distinct proteins (45 and 55 kDa) with Zot-binding activity [73]. N-terminal sequencing of the 55 kDa binding partner confirmed an identical sequence with tubulin. The smaller, 45 kDa protein retained by the Zot affinity column, showed a 72% identity with the N-terminus of calprotectin. Radiolabeled Zot bound the purified Zot receptor isolated from the brain with a dissociation constant (K_d) of ca. 35 nM. Immunofluorescence and immunoblotting analysis confirmed Zot-MBP binding on the surface of Caco-2 and rat small intestinal epithelial (IEC6) but not human colon adenocarcinoma (T84) and Madin-Darby Canine Kidney (MDCK) epithelial cells [74]. Isolation of the Zot receptor from Caco-2 and IEC6 cell lines identified a 66 kDa protein with high similarity to α-1-chimaerin (46% identity, 69% similarity), a potential signal transduction molecule that contributes to the modulation of cytoskeletal organization. This appeared to be consistent with the hypothesis that Zot activated PKCα in a reversible, time and dose dependent manner, with the resulting protein phosphorylation leading to cytoskeletal rearrangements and an increase in paracellular permeability [52].

An entirely new direction was pursued with the co-administration of Zot and therapeutic proteins, such as insulin (5.7 kDa) and IgG (140 kDa), exploiting Zot's influence on small bowel permeability [53]. Since tjs govern the paracellular pathway in the small intestine, passage is generally restricted to molecules with molecular radii of around 11 Å, and is therefore not accessible to larger compounds (i.e. proteins). However, *in vitro* experiments on rabbit ileal mucosa resulted in an increased uptake of insulin and IgG of 72 and 52%, respectively. *In vivo* experiments in acute type 1 diabetic rats provided evidence for the bioactivity of insulin after oral administration without development of systemic side-effects, and normal histological findings [75]. Importantly, the action of Zot appears to be reversible and so it is assumed that a degree of control will be possible in any potential application as a permeation enhancer. A putative cleavage site at amino acid residue 290 of Zot was identified, resulting in a C-terminal 12 kDa Zot fragment which was biologically active [76]. The presence of a putative binding domain within the C-terminal fragment and its possible homology to an endogenous agonist, zonulin, were later described [77].

9.4.3. 'Delta G' (ΔG)

Several truncated Zot-MBP mutants were analyzed by immunoblotting to assess subsequent binding to IEC6 cells. Binding was reduced for Zot-MBP truncates deficient in a central region and completely impaired when the sequence from amino acid 118 to 299 was deleted, which was therefore implicated in Zot receptor binding. This putative binding domain was termed ΔG, and a key residue within this sequence was the Gly298, which when mutated to valine attenuated biological activity (Figure 9.4) [78]. Further experiments revealed that ΔG increased mannitol uptake *in vitro* and *in vivo,* but since ΔG was susceptible to enzymatic degradation *in vivo* protease inhibitors were required [79]. Since the Zot-mediated uptake of marker molecules was not impaired *in vivo*, it was postulated that the tertiary structure of Zot minimizes its enzymatic degradation [80].

9.4.4. Zonulin

Zonulin was identified as the endogenous analog to Zot and has been shown to modulate the intestinal tj in a similar manner to Zot (Table 9.1) [77]. Elevated levels of zonulin have been associated with autoimmune diseases, such as coeliac disease [81]. This is notable since a compromised intestinal tj can be hypothesized to allow antigens to pass the epithelial barrier, so triggering an autoimmune response in genetically susceptible individuals. The occurrence of autoimmune diseases has been associated with increased intestinal permeability in human, as well as in animal models [82,83]. These studies of human subjects affected by type 1 diabetes, and in diabetic prone rats, suggested an altered intestinal permeability in the prodromal stages of the disease which are associated with elevated levels of zonulin. Inhibition of this pathway by

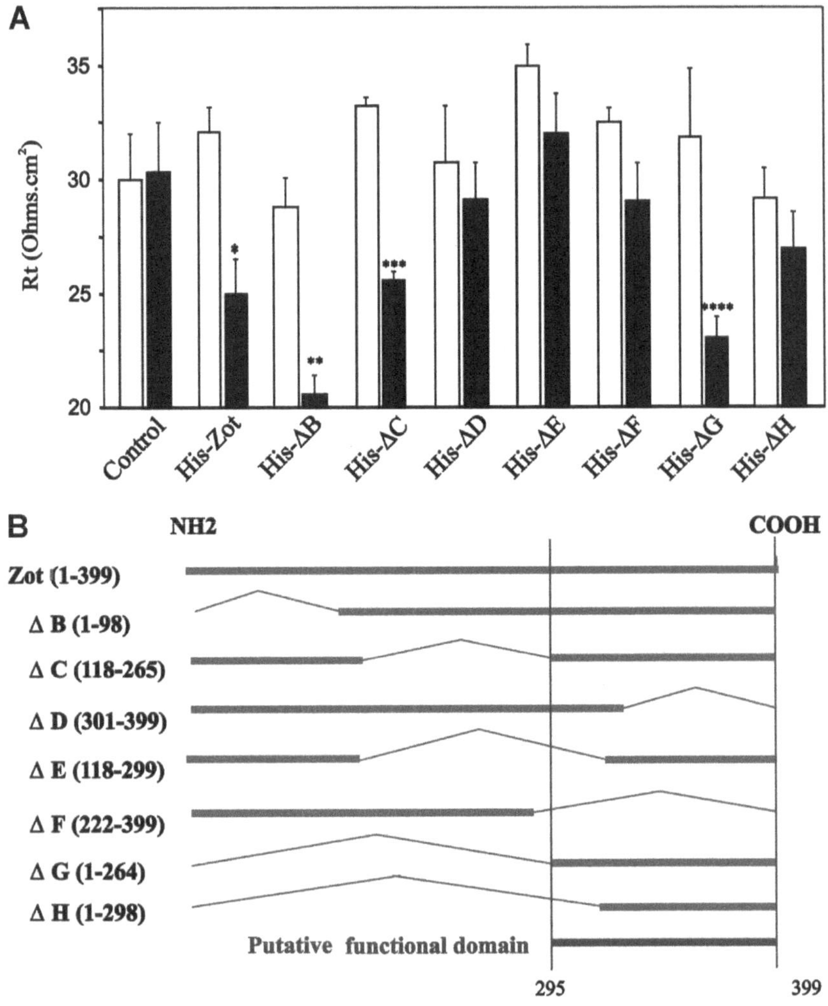

FIGURE 9.4 (A), effect of Zot and its deletion mutant derivatives on rabbit ileal Rt in Ussing chambers. Tissues were exposed to either 1×10^{-10} m of each protein or to a negative control (PBS), and variation in Rt between the baseline (■) and 90 min post-incubation (■) was monitored. All values were expressed as the means ± S.E. *, p between 0.03 and 0.0003 compared with either control, ΔD, ΔE, ΔF, or ΔH. **, $p < 0.01$ compared with control. ***, $p < 0.000006$ compared with control. ****, $p < 0.0001$ compared with control. (B), schematic description of Zot deletion mutants. The deletion for each protein is shown as a *broken filled line*. The putative Zot functional domain responsible for tj disassembly (delimited by the *two vertical lines*) maps in the toxin's carboxyl-terminal region (aa 295–399). Deleted amino acid residues are shown in *parentheses*. *Reprinted from reference [78] with permission from the American Society for Biochemistry and Molecular Biology.*

TABLE 9.1 Amino acid sequences (single letter code) within Zot and human zonulin, and compared to synthetic peptides. Putative regions of similarity are shown in bold

Protein/peptide	Source	Amino acid residues
Zot	*V. cholerae*	**FCIGRLCVQDG**FVT
Zonulin	Human adult intestine	EVQLVES**GGXL**
Zonulin	Human fetal intestine	MLQKAES**GGVLVQPG**XSNRL
Zonulin	Human brain	LTELEKALNXG**GGVG**HKY
Synthetic peptide (FZI/0 or AT-1001)	Synthetic	**GGVLVQPG**
AT-1002	Synthetic	**FCIGRL**

administering the zonulin inhibitor FZI/0 [78] may present an innovative tool for type 1 diabetes therapy, impeding the onset of increased intestinal permeability. Additionally, zonulin appears to modulate cell-mediated antigen presentation in human macrophages, subsequently changing the cytokine profile, which may then add to the switch from immune tolerance to autoimmunity [83]. Nevertheless, the initial trigger activating the increased zonulin secretion has yet to be identified.

9.4.5. The Zot Derived Peptides AT1001 and AT1002

In terms of a therapeutic application, the inhibition of the Zot/zonulin pathway may be beneficial in the treatment of diseases with elevated paracellular permeability [84]. However, to address the issues of the instability of ΔG and poor expression of full-length Zot, it would be desirable to have a small peptide agonist. An important step towards this goal was the synthesis of AT1001 (or FZI/0), an octapeptide which shares the putative binding domain of Zot [78]. AT1001, however, is an antagonist *in vivo* since it blocks the action of zonulin, leading to its clinical potential in celiac disease. A related hexameric synthetic peptide, named AT1002, which comprises the first six amino acids of ΔG (Table 9.1) was also found to increase the permeability of molecular weight markers via the paracellular space [85]. The development of the AT1002 peptide has also opened a new avenue of research in this field, based on homology with the protease activated receptor (PAR)-activating peptide (AP) motif [86].

AT1001 has been tested in a Phase I clinical trial for its potential treatment of celiac disease [84]. Design of the study parameters was influenced by the lack of validated markers for celiac disease progression and remission. Briefly, fasting subjects were dosed with either placebo or AT1001 followed by a sham gluten challenge (day 1) or a blinded gluten challenge (day 2), respectively. Subsequently, the intestinal permeability was determined by an orally administered probe solution containing sucrose, lactulose and mannitol. The degree of intestinal permeability was assessed as the lactulose to mannitol ratio from urinary samples. Building on the Phase I study, in 2009 a Phase IIb clinical trial with an enteric coated formulation of AT1001 (Larazotide Acetate) in celiac disease was completed in the USA by Alba Therapeutics Corporation (Baltimore). The primary outcome measure of this study investigated the villous height to crypt depth, the secondary outcome measures assessed the safety and tolerability of AT1001 (ClinicalTrials.gov Identifier: NCT00620451).

AT1002 represents the minimum Zot receptor binding domain, but its cysteine (at position 2) is prone to dimerization and structure-activity relationships have been explored, in which a number of other amino acids have been substituted for the Cys residue [87]. The synthetic peptides were tested *in vitro*, assessing the permeability of the marker Lucifer yellow, TEER and cytotoxicity. Compared to the dimeric form of AT1002, the reduced form increased permeability 52-fold, and substitution of allylglycine for Cys increased permeability 103-fold. These results suggest that the position 2 residue is an imperative structural unit for maintaining or enhancing the tight junction modulating activity. As mentioned, AT1002 comprises a protease activated receptor (PAR) 2 activator-like motif [86]. PARs belong to a family of G-protein coupled receptors that are activated by cleavage of their N-terminal domain by a proteolytic enzyme [88], unmasking the N-terminal sequence that subsequently acts as a tethered ligand that binds and activates the receptor itself. PAR-2 is found throughout the gastro-intestinal tract on several cell types, including enterocytes, mast cells, smooth muscle cells and endothelial cells [89]. Activation of PAR-2 occurs through trypsin, trypsin-like proteins or synthetic peptides, so called PAR-activating peptides (PAR-APs) that correspond to the amino acid sequence of the tethered ligand and are also able to selectively activate PARs [90]. PAR-2 involvement has been reported for different pathophysiological processes and led to colonic inflammation in mice and increased paracellular permeability [91]. Although the expression of PAR-2 has been reported on both apical and basolateral surfaces of enterocytes, evidence is lacking so far that direct activation of PAR-2 may lead to a dysfunction of the epithelial barrier [92].

Further investigation into the mechanism of AT1002 revealed a reversible tyrosine phosphorylation of ZO-1 concomitant with src kinase and MAP kinase activation [93]. Tyrosine phosphorylation of ZO-1 and occludin mediated by src kinase has previously been associated with increased tj permeability in oxidative stress [94]. Similarly, MAP kinase activation with subsequent opening of the tj has been associated with a variety of stimuli, such as alcohol, oxidative stress and cytokines [23]. However, the exact mechanisms of AT1002 induced src and MAP kinase activation are currently under investigation. Zonulin was shown to transactivate the epithelial growth factor receptor (EGFR) via PAR-2, resulting in a decrease in TEER in murine and human intestinal tissue [95]. This is not unusual; the role of PAR-2 in gut permeability has previously been described [96] and EGFR associated modulation of tj permeability has been reported for bile acids [97].

9.4.6. Comparison of Zot and Zot Derivatives With Other Absorption Enhancers

The Caco-2 intestinal cell culture model was first employed to assess the influence of Zot on the permeability of paracellular markers, such as mannitol (182 MW), PEG 4000 (4000 Da) and inulin (5000 Da). A concentration (0, 1, 2, 4 μg/ml) dependent increase in the permeability of molecules below 5000 Da was established, with prior incubation of Zot for 30 min found to enhance permeability further [98,99]. To compare the effect of Zot with the N-trimethyl chitosan chloride (TMC) absorption enhancers on the flux of marker molecules and therapeutic agents across the epithelial cell layer, data for apparent permeability coefficients (P_{app}) were used [98,100]. The results are compared in Table 9.2 and suggest that Zot increases the permeation of marker compounds (mannitol) to a greater extent than the TMC derivatives.

Zot increases the permeation of the poorly bioavailable peptide, cyclosporin, in a strongly dose-dependent manner. A 10-fold increase in cyclosporin permeation was observed for higher Zot concentrations and comparison against other poorly bioavailable compounds suggested a case-by-case assessment of

TABLE 9.2 Comparison of apparent permeability coefficients (P_{app}) of mannitol across Caco-2 cell monolayers treated with N-trimethyl chitosan chloride absorption enhancers (reprinted from reference [100] with permission from Elsevier) and Zot (reprinted from reference [98] with permission from Elsevier)

conc. TMCs % w/v	TMC40 P_{app} ($\times10^{-7}$ cm/s)	TMC60 P_{app} ($\times10^{-7}$ cm/s)	Conc. Zot (μg/ml)	Zot P_{app} ($\times10^{-7}$ cm/s)
0.05	7.3	18.1	0	7.6
0.1	8.2	18.5	1	16.4
0.25	9.1	18.9	2	76.0
0.5	17.0	23.1	4	95.7
1	19.0	27.3		

TABLE 9.3 Permeability coefficients (P_{app}) for compounds including peptide drugs across treated Caco-2 cells incubated with Zot (reprinted from reference [99] with permission from Springer)

Compound	P_{app} across a Caco-2 monolayer (control)	P_{app} in the presence of 1.0 μg/ml Zot	P_{app} in the presence of 2.0 μg/ml Zot
Acyclovir	2.85	3.00*	3.46*
Paclitaxel	10.3	20.5*	30.7*
Cyclosporin A	0.75	0.80	8.10*
Doxorubicin	2.44	3.48*	4.01*

** $p < 0.05$ when compared with control; units of P_{app} are $\times 10^{-6}$ cm/s.*

Zot-mediated permeation would be required (Table 9.3) [99]. Salama *et al.* investigated the effect of ΔG on the pharmacokinetic parameters of various classes of drug including the peptide cyclosporin. In directing the route of transit to the paracellular route, ΔG was shown to overcome the poor bioavailability of cyclosporin and also circumvent drug efflux via transporters such as P-glycoprotein [80]. As for Zot, the ΔG-mediated absorption process was dependant on many variables, such as drug molecular weight, lipophilicity, efflux transporter substrate, etc., resulting in limitations of absolute absorption enhancement. *In vivo* studies carried out to examine the effect of AT1002 on low molecular weight heparin (Ardeparin) showed a bimodal absorption profile that pointed to a significantly higher absorption compared to the control [101]. An early effect on the jejunal mucosa and a late effect on ileal mucosa were thought to be responsible for this adsorption profile. Plasma concentrations of cyclosporin were also significantly enhanced when co-administered with AT1002 (10 mg/kg), although the *in vitro* model did not show an enhanced absorption [85].

Permeability studies of paracellular marker molecules as well as chemotherapeutic agents across bovine brain microvessel endothelial cells, mimicking the BBB, were conducted with Zot [102]. Similarly, ΔG was tested *in vivo* to determine its effect on the distribution of marker and therapeutic molecules in the brains of male Sprague-Dawley rats [103]. Methotrexate, a very hydrophilic molecule (MW 454), and paclitaxel, a very lipophilic agent (MW 854), both show poor permeability across the BBB. ΔG, administered in combination with a protease inhibitor cocktail, increased the distribution of methotrexate 7-fold and that of paclitaxel 3-fold, suggesting that ΔG is a potential adjuvant for enhanced drug delivery to the brain.

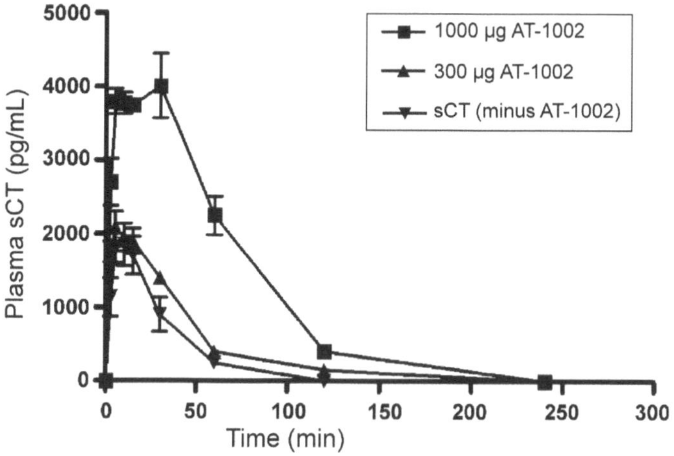

FIGURE 9.5 Effect of doses of AT-1002 (µg) on pulmonary absorption of sCT following liquid instillation of both AT-1002 and salmon calcitonin (sCT). Rats were instilled with saline or saline with 0, 300, or 1000 µg AT-1002 ($n = 6$) together with 10 µg sCT. Values are mean ± S.D. *Reprinted from reference [93] with permission from Elsevier.*

Modulation of the tj of the nasal mucosa may allow for systemic delivery of macromolecules, like peptides and proteins, or provide an adjuvant for mucosal vaccination. Song *et al.* further investigated AT1002 for nasal drug administration in combination with paracellular markers of different molecular size [85]. Promising results were obtained for AT1002 at 10 mg/kg; significantly enhancing inulin and PEG4000 uptake across the nasal mucosa of Sprague-Dawley rats. Furthermore, pulmonary co-administration of AT1002 with salmon calcitonin, a protein with naturally low bioavailability used to treat osteoporosis, increased bioavailability around 2–5-fold for low/high concentrations of AT1002 in male Sprague-Dawley rats (Figure 9.5) [93]. This result is particularly encouraging since it may point towards the future application of the Zot-derived peptides as absorption enhancers of protein drugs.

9.4.7. Application of Zot as an Immune Adjuvant

Compromising the barrier function of the epithelia may elicit mucosal immunization in conjunction with the delivery of soluble antigens, or IgA and IgG for local and systemic immunity, respectively [104]. Antigens could be directly targeted to the immune system at the site of entry, preventing pathogen adhesion and decreasing the chance of infection and disease. However, the immune response achieved by some mucosal vaccines is weak, lacking long lasting protective effects. Adjuvants which increase the immune response without having an antigenic effect themselves could resolve this problem. The functional characteristics of Zot make it a potential adjuvant by facilitating the delivery of antigens to the submucosa. Intranasal administration of antigen in combination with Zot has been shown to induce antigen-specific serum IgG as well as IgA in vaginal and intestinal secretions [105]. Long lasting effects were confirmed by high antigen-specific IgG titers that were maintained over a year [106]. Furthermore, Zot induced humoral and cell-mediated responses to tetanus toxoid, protecting C57BL/6 mice against a systemic challenge with tetanus toxin. The proven high efficacy is accompanied by a much lower immunogenicity than *Escherichia coli* heat-labile enterotoxin, an adjuvant for mucosal immunization associated with unfavorable side-effects. Since AT1001 blocked the Zot-mediated immune response, the modulation of tj permeability was suggested to be responsible for its role as a mucosal adjuvant.

9.5. FUTURE PERSPECTIVES

It is now established that careful engineering of bacterial enterotoxins, particularly CPE and Zot, can yield peptides which facilitate increased intestinal permeation of high MW compounds including otherwise poorly (oral) bioavailable peptides such as insulin and cyclosporin. Similar promise holds for peptide mimics of the extracellular loops of the claudins and occludin, though these peptides appear to be further from market potential at this stage. The susceptibility of some CPE- and Zot-derived peptides to enzymatic degradation does of course create something of a circular problem – how to deliver the permeation enhancers themselves. It is also important to distinguish between the various agonists and antagonists that have been developed during dissection of the molecular mechanisms underlying the action of Zot and zonulin. For example, Larazotide acetate (AT1001) is an antagonist and is intended to reverse paracellular permeability due to elevated levels of zonulin. It is, however, formulated as enteric coated multi-particulate beads for oral administration in gelatin capsules, and this may point to similar approaches for co-administration of therapeutic peptides with AT1002, a functional agonist. It is interesting to see research examining the effect of AT1002 on paracellular permeability of the airways epithelium, and its co-administration with the 32 amino acid peptide hormone calcitonin. Combining the potential of pulmonary delivery of peptides and proteins with AT1002 therefore holds promise. It may be that we will see the use of Zot-derived peptides as immune adjuvants in parallel with their development for peptide and protein delivery. It is further encouraging that comparisons of CPE and Zot with established permeation enhancers are favorable. The reversible action of Zot and possible tissue specificity of C-CPE are also promising with regard to future development along similar lines to those seen for AT1002. Localization of the Zot-receptor binding domain of Zot has led to parallels with the PAR-activating peptides and a better understanding of the signaling pathways involved in tj modulation. To stimulate the development of small molecule mimics, and circumvent issues associated with peptide adjuvants for protein delivery, a high resolution 3-dimensional structure of Zot would ideally be required. However, before this is possible, the poor expression of recombinant Zot proteins must be overcome.

REFERENCES

[1] M. Kondoh, K. Yagi, Tight junction modulators: Promising candidates for drug delivery, Curr. Med. Chem. 14 (2007) 2482–2488.

[2] M. Kondoh, K. Yagi, Progress in absorption enhancers based on tight junction, Expert Opin. Drug Deliv. 4 (2007) 275–286.

[3] M. Furuse, T. Hirase, M. Itoh, A. Nagafuchi, S. Yonemura, S. Tsukita, et al., Occludin: A novel integral membrane protein localizing at tight junctions, J. Cell Biol. 123 (1993) 1777–1788.

[4] M. Furuse, K. Fujimoto, N. Sato, T. Hirase, S. Tsukita, S. Tsukita, Overexpression of occludin, a tight junction-associated integral membrane protein, induces the formation of intracellular multilamellar bodies bearing tight junction-like structures, J. Cell Sci. 109 (Pt 2) (1996) 429–435.

[5] M. Furuse, M. Itoh, T. Hirase, A. Nagafuchi, S. Yonemura, S. Tsukita, Direct association of occludin with ZO-1 and its possible involvement in the localization of occludin at tight junctions, J. Cell Biol. 127 (1994) 1617–1626.

[6] G.J. Feldman, J.M. Mullin, M.P. Ryan, Occludin: Structure, function and regulation, Adv. Drug Deliv. Rev. 57 (2005) 883–917.

[7] M. Furuse, Knockout animals and natural mutations as experimental and diagnostic tool for studying tight junction functions *in vivo*, Biochim. Biophys. Acta 1788 (2009) 813–819.

[8] M. Furuse, K. Fujita, T. Hiiragi, K. Fujimoto, S. Tsukita, Claudin-1 and -2: Novel integral membrane proteins localizing at tight junctions with no sequence similarity to occludin, J. Cell Biol. 141 (1998) 1539–1550.

[9] M. Furuse, H. Sasaki, K. Fujimoto, S. Tsukita, A single gene product, claudin-1 or -2, reconstitutes tight junction strands and recruits occludin in fibroblasts, J. Cell Biol. 143 (1998) 391–401.

[10] E.E. Schneeberger, R.D. Lynch, The tight junction: A multifunctional complex, Am. J. Physiol. Cell Physiol. 286 (2004) C1213–C1228.

[11] S. Tsukita, M. Furuse, Occludin and claudins in tight-junction strands: Leading or supporting players? Trends Cell Biol. 9 (1999) 268–273.

[12] L.A. Williams, I. Martin-Padura, E. Dejana, N. Hogg, D.L. Simmons, Identification and characterisation of human junctional adhesion molecule (JAM), Mol. Immunol. 36 (1999) 1175–1188.

[13] I. Martin-Padura, S. Lostaglio, M. Schneemann, L. Williams, M. Romano, P. Fruscella, et al., Junctional adhesion molecule, a novel member of the immunoglobulin superfamily that distributes at intercellular junctions and modulates monocyte transmigration, J. Cell Biol. 142 (1998) 117–127.

[14] L. Gonzalez-Mariscal, A. Betanzos, A. Avila-Flores, MAGUK proteins: Structure and role in the tight junction, Semin. Cell Dev. Biol. 11 (2000) 315–324.

[15] M. Itoh, M. Furuse, K. Morita, K. Kubota, M. Saitou, S. Tsukita, Direct binding of three tight junction-associated MAGUKs, ZO-1, ZO-2, and ZO-3, with the COOH termini of claudins, J. Cell Biol. 147 (1999) 1351–1363.

[16] L. Gonzalez-Mariscal, A. Betanzos, P. Nava, B.E. Jaramillo, Tight junction proteins, Prog. Biophys. Mol. Biol. 81 (2003) 1–44.

[17] L. Gonzalez-Mariscal, A. Ponce, L. Alarcon, B.E. Jaramillo, The tight junction protein ZO-2 has several functional nuclear export signals, Exp. Cell Res. 312 (2006) 3323–3335.

[18] S. Islas, J. Vega, L. Ponce, L. Gonzalez-Mariscal, Nuclear localization of the tight junction protein ZO-2 in epithelial cells, Exp. Cell Res. 274 (2002) 138–148.

[19] M.R. Schiller, K. Chakrabarti, G.F. King, N.I. Schiller, B.A. Eipper, M.W. Maciejewski, Regulation of RhoGEF activity by intramolecular and intermolecular SH3 domain interactions, J. Biol. Chem. 281 (2006) 18774–18786.

[20] L.A. Jesaitis, D.A. Goodenough, Molecular characterization and tissue distribution of ZO-2, a tight junction protein homologous to ZO-1 and the *Drosophila* discs-large tumor suppressor protein, J. Cell Biol. 124 (1994) 949–961.

[21] J. Haskins, L. Gu, E.S. Wittchen, J. Hibbard, B.R. Stevenson, ZO-3, a novel member of the MAGUK protein family found at the tight junction, interacts with ZO-1 and occludin, J. Cell Biol. 141 (1998) 199–208.

[22] M.A. Deli, Potential use of tight junction modulators to reversibly open membranous barriers and improve drug delivery, Biochim. Biophys. Acta 1788 (2009) 892–910.

[23] L. Gonzalez-Mariscal, R. Tapia, D. Chamorro, Crosstalk of tight junction components with signaling pathways, Biochim. Biophys. Acta 1778 (2008) 729–756.

[24] I.A. Krizbai, M.A. Deli, Signalling pathways regulating the tight junction permeability in the blood–brain barrier, Cell. Mol. Biol. 49 (2003) 23–31.

[25] M.A. Deli, C.S. Abraham, Y. Kataoka, M. Niwa, Permeability studies on *in vitro* blood–brain barrier models: Physiology, pathology, and pharmacology, Cell Mol. Neurobiol. 25 (2005) 59–127.

[26] D. Yu, A.M. Marchiando, C.R. Weber, D.R. Raleigh, Y. Wang, L. Shen, et al., MLCK-dependent exchange and actin binding region-dependent anchoring of ZO-1 regulate tight junction barrier function, Proc. Natl. Acad. Sci. U.S.A. 107 (2010) 8237–8241.

[27] R. Yanai, J.A. Ko, N. Nomi, N. Morishige, T. Chikama, A. Hattori, et al., Upregulation of ZO-1 in cultured human corneal epithelial cells by a peptide (PHSRN) corresponding to the second cell-binding site of fibronectin, Invest. Ophthalmol. Vis. Sci. 50 (2009) 2757–2764.

[28] H. Altroff, C.F. van der Walle, J. Asselin, R. Fairless, I.D. Campbell, H.J. Mardon, The eighth FIII domain of human fibronectin promotes integrin alpha5beta1 binding via stabilization of the ninth FIII domain, J. Biol. Chem. 276 (2001) 38885–38892.

[29] M. Bruewer, A.M. Hopkins, M.E. Hobert, A. Nusrat, J.L. Madara, RhoA, Rac1, and Cdc42 exert distinct effects on epithelial barrier via selective structural and biochemical modulation of junctional proteins and F-actin, Am. J. Physiol. Cell Physiol. 287 (2004) C327–C335.

[30] M. Thanou, J.C. Verhoef, H.E. Junginger, Chitosan and its derivatives as intestinal absorption enhancers, Adv. Drug Deliv. Rev 50 (Suppl. 1) (2001) S91–101.

[31] K. Whitehead, N. Karr, S. Mitragotri, Safe and effective permeation enhancers for oral drug delivery, Pharm. Res. 25 (2008) 1782–1788.

[32] C.L. Bowe, L. Mokhtarzadeh, P. Venkatesan, S. Babu, H.R. Axelrod, M.J. Sofia, et al., Design of compounds that increase the absorption of polar molecules, Proc. Natl. Acad. Sci. U.S.A. 94 (1997) 12218–12223.

[33] E. Duizer, A.H. Penninks, W.H. Stenhuis, J.P. Groten, Comparison of permeability characteristics of the human colonic Caco-2 and rat small intestinal IEC-18 cell lines, J. Control. Release 49 (1997) 39–49.

[34] J.H. Hochman, J.A. Fix, E.L. LeCluyse, *In vitro* and *in vivo* analysis of the mechanism of absorption enhancement by palmitoylcarnitine, J. Pharmacol. Exp. Ther. 269 (1994) 813–822.

[35] M. Tomita, M. Hayashi, S. Awazu, Absorption-enhancing mechanism of EDTA, caprate, and decanoylcarnitine in Caco-2 cells, J. Pharm. Sci. 85 (1996) 608–611.

[36] T. Lindmark, N. Schipper, L. Lazorova, A.G. de Boer, P. Artursson, Absorption enhancement in intestinal epithelial Caco-2 monolayers by sodium caprate: Assessment of molecular weight dependence and demonstration of transport routes, J. Drug Target. 5 (1998) 215–223.

[37] R.S. Everett, M.K. Vanhook, N. Barozzi, I. Toth, L.G. Johnson, Specific modulation of airway epithelial tight junctions by apical application of an occludin peptide, Mol. Pharmacol. 69 (2006) 492–500.

[38] F. Lacaz-Vieira, M.M. Jaeger, P. Farshori, B. Kachar, Small synthetic peptides homologous to segments of the first external loop of occludin impair tight junction resealing, J. Membr. Biol. 168 (1999) 289–297.

[39] H. Wan, H.L. Winton, C. Soeller, D.C. Gruenert, P.J. Thompson, M.B. Cannell, et al., Quantitative structural and biochemical analyses of tight junction dynamics following exposure of epithelial cells to house dust mite allergen Der p 1, Clin. Exp. Allergy 30 (2000) 685–698.

[40] V. Wong, B.M. Gumbiner, A synthetic peptide corresponding to the extracellular domain of occludin perturbs the tight junction permeability barrier, J. Cell Biol. 136 (1997) 399–409.

[41] L. Gonzalez-Mariscal, P. Nava, S. Hernandez, Critical role of tight junctions in drug delivery across epithelial and endothelial cell layers, J. Membr. Biol. 207 (2005) 55–68.

[42] A. Nusrat, G.T. Brown, J. Tom, A. Drake, T.T. Bui, C. Quan, et al., Multiple protein interactions involving proposed extracellular loop domains of the tight junction protein occludin, Mol. Biol. Cell 16 (2005) 1725–1734.

[43] M. Kondoh, A. Takahashi, M. Fujii, K. Yagi, Y. Watanabe, A novel strategy for a drug delivery system using a claudin modulator, Biol. Pharm. Bull. 29 (2006) 1783–1789.

[44] K. Cui, B. Dutzar, Method for opening tight junctions. Patent Applicant: Nastech Pharmaceutical Company Inc. Publication info: US2007253956 (A1) – 2007–11–01, and US7696343 (B2) – 2010–04–13.

[45] A.S. Yu, K.M. McCarthy, S.A. Francis, J.M. McCormack, J. Lai, R.A. Rogers, et al., Knockdown of occludin expression leads to diverse phenotypic alterations in epithelial cells, Am. J. Physiol. Cell Physiol. 288 (2005) C1231–C1241.

[46] G. Hecht, C. Pothoulakis, J.T. LaMont, J.L. Madara, *Clostridium difficile* toxin A perturbs cytoskeletal structure and tight junction permeability of cultured human intestinal epithelial monolayers, J. Clin. Invest. 82 (1988) 1516–1524.

[47] I. Just, G. Fritz, K. Aktories, M. Giry, M.R. Popoff, P. Boquet, et al., *Clostridium difficile* toxin B acts on the GTP-binding protein Rho, J. Biol. Chem. 269 (1994) 10706–10712.

[48] A. Takahashi, M. Kondoh, A. Masuyama, M. Fujii, H. Mizuguchi, Y. Horiguchi, et al., Role of C-terminal regions of the C-terminal fragment of *Clostridium perfringens* enterotoxin in its interaction with claudin-4, J. Control. Release 108 (2005) 56–62.

[49] M. Kondoh, A. Masuyama, A. Takahashi, N. Asano, H. Mizuguchi, N. Koizumi, et al., A novel strategy for the enhancement of drug absorption using a claudin modulator, Mol. Pharmacol. 67 (2005) 749–756.

[50] A. Takahashi, E. Komiya, H. Kakutani, T. Yoshida, M. Fujii, Y. Horiguchi, et al., Domain mapping of a claudin-4 modulator, the C-terminal region of C-terminal fragment of *Clostridium perfringens* enterotoxin, by site-directed mutagenesis, Biochem. Pharmacol. 75 (2008) 1639–1648.

[51] A. Fasano, B. Baudry, D.W. Pumplin, S.S. Wasserman, B.D. Tall, J.M. Ketley, et al., *Vibrio cholerae* produces a second enterotoxin, which affects intestinal tight junctions, Proc. Natl. Acad. Sci. U.S.A. 88 (1991) 5242–5246.
[52] A. Fasano, C. Fiorentini, G. Donelli, S. Uzzau, J.B. Kaper, K. Margaretten, et al., Zonula occludens toxin modulates tight junctions through protein kinase C-dependent actin reorganization, *in vitro*, J. Clin. Invest. 96 (1995) 710–720.
[53] A. Fasano, S. Uzzau, Modulation of intestinal tight junctions by Zonula occludens toxin permits enteral administration of insulin and other macromolecules in an animal model, J. Clin. Invest. 99 (1997) 1158–1164.
[54] B.A. McClane, The complex interactions between *Clostridium perfringens* enterotoxin and epithelial tight junctions, Toxicon 39 (2001) 1781–1791.
[55] S.L. Kominsky, M. Vali, D. Korz, T.G. Gabig, S.A. Weitzman, P. Argani, et al., *Clostridium perfringens* enterotoxin elicits rapid and specific cytolysis of breast carcinoma cells mediated through tight junction proteins claudin 3 and 4, Am. J. Pathol. 164 (2004) 1627–1633.
[56] P.C. Hanna, E.U. Wieckowski, T.A. Mietzner, B.A. McClane, Mapping of functional regions of *Clostridium perfringens* type A enterotoxin, Infect. Immun. 60 (1992) 2110–2114.
[57] N. Sonoda, M. Furuse, H. Sasaki, S. Yonemura, J. Katahira, Y. Horiguchi, et al., *Clostridium perfringens* enterotoxin fragment removes specific claudins from tight junction strands: Evidence for direct involvement of claudins in tight junction barrier, J. Cell Biol. 147 (1999) 195–204.
[58] J. Katahira, N. Inoue, Y. Horiguchi, M. Matsuda, N. Sugimoto, Molecular cloning and functional characterization of the receptor for *Clostridium perfringens* enterotoxin, J. Cell Biol. 136 (1997) 1239–1247.
[59] P.C. Hanna, T.A. Mietzner, G.K. Schoolnik, B.A. McClane, Localization of the receptor-binding region of *Clostridium perfringens* enterotoxin utilizing cloned toxin fragments and synthetic peptides. The 30 C-terminal amino acids define a functional binding region, J. Biol. Chem. 266 (1991) 11037–11043.
[60] K. Fujita, J. Katahira, Y. Horiguchi, N. Sonoda, M. Furuse, S. Tsukita, *Clostridium perfringens* enterotoxin binds to the second extracellular loop of claudin-3, a tight junction integral membrane protein, FEBS Lett. 476 (2000) 258–261.
[61] A. Masuyama, M. Kondoh, H. Seguchi, A. Takahashi, M. Harada, M. Fujii, et al., Role of N-terminal amino acids in the absorption-enhancing effects of the c-terminal fragment of *Clostridium perfringens* enterotoxin, J. Pharmacol. Exp. Ther. 314 (2005) 789–795.
[62] H. Uchida, M. Kondoh, T. Hanada, A. Takahashi, T. Hamakubo, K. Yagi, A claudin-4 modulator enhances the mucosal absorption of a biologically active peptide, Biochem. Pharmacol. 79 (2010) 1437–1444.
[63] C.M. Van Itallie, L. Betts, J.G. Smedley 3rd, B.A. McClane, J.M. Anderson, Structure of the claudin-binding domain of *Clostridium perfringens* enterotoxin, J. Biol. Chem. 283 (2008) 268–274.
[64] K. Matsuhisa, M. Kondoh, A. Takahashi, K. Yagi, Tight junction modulator and drug delivery, Expert Opin. Drug Deliv. 6 (2009) 509–515.
[65] J. Ling, H. Liao, R. Clark, M.S. Wong, D.D. Lo, Structural constraints for the binding of short peptides to claudin-4 revealed by surface plasmon resonance, J. Biol. Chem. 283 (2008) 30585–30595.
[66] J.B. Kaper, H. Lockman, M.M. Baldini, M.M. Levine, Recombinant nontoxinogenic *Vibrio-cholerae* strains as attenuated cholera vaccine candidates, Nature 308 (1984) 655–658.

[67] B. Baudry, A. Fasano, J. Ketley, J.B. Kaper, Cloning of a gene (zot) encoding a new toxin produced by *Vibrio cholerae*, Infect. Immun. 60 (1992) 428–434.
[68] M.K. Waldor, J.J. Mekalanos, Lysogenic conversion by a filamentous phage encoding cholera toxin, Science 272 (1996) 1910–1914.
[69] E.V. Koonin, The second cholera toxin, Zot, and its plasmid-encoded and phage-encoded homologues constitute a group of putative ATPases with an altered purine NTP-binding motif, FEBS Lett. 312 (1992) 3–6.
[70] E. Schmidt, S.M. Kelly, C.F. van der Walle, Tight junction modulation and biochemical characterisation of the zonula occludens toxin C-and N-termini, FEBS Lett. 581 (2007) 2974–2980.
[71] A. Lee, N. White, C.F. van der Walle, The intestinal zonula occludens toxin (ZOT) receptor recognises non-native ZOT conformers and localises to the intercellular contacts, FEBS Lett. 555 (2003) 638–642.
[72] A. Fasano, S. Uzzau, C. Fiore, K. Margaretten, The enterotoxic effect of zonula occludens toxin on rabbit small intestine involves the paracellular pathway, Gastroenterology 112 (1997) 839–846.
[73] R. Lu, W. Wang, S. Uzzau, R. Vigorito, H.R. Zielke, A. Fasano, Affinity purification and partial characterization of the zonulin/zonula occludens toxin (Zot) receptor from human brain, J. Neurochem. 74 (2000) 320–326.
[74] S. Uzzau, R. Lu, W. Wang, C. Fiore, A. Fasano, Purification and preliminary characterization of the zonula occludens toxin receptor from human (CaCo2) and murine (IEC6) intestinal cell lines, FEMS Microbiol. Lett. 194 (2001) 1–5.
[75] A. Fasano, Novel approaches for oral delivery of macromolecules, J. Pharm. Sci. 87 (1998) 1351–1356.
[76] S. Uzzau, P. Cappuccinelli, A. Fasano, Expression of *Vibrio cholerae* zonula occludens toxin and analysis of its subcellular localization, Microb. Pathog. 27 (1999) 377–385.
[77] W. Wang, S. Uzzau, S.E. Goldblum, A. Fasano, Human zonulin, a potential modulator of intestinal tight junctions, J. Cell. Sci. 113 Pt 24 (2000) 4435–4440.
[78] M. Di Pierro, R. Lu, S. Uzzau, W. Wang, K. Margaretten, C. Pazzani, et al., Zonula occludens toxin structure-function analysis. Identification of the fragment biologically active on tight junctions and of the zonulin receptor binding domain, J. Biol. Chem. 276 (2001) 19160–19165.
[79] N.N. Salama, A. Fasano, R. Lu, N.D. Eddington, Effect of the biologically active fragment of zonula occludens toxin, delta G, on the intestinal paracellular transport and oral absorption of mannitol, Int. J. Pharm. 251 (2003) 113–121.
[80] N.N. Salama, A. Fasano, M. Thakar, N.D. Eddington, The impact of Delta G on the oral bioavailability of low bioavailable therapeutic agents, J. Pharmacol. Exp. Ther. 312 (2005) 199–205.
[81] A. Fasano, T. Not, W. Wang, S. Uzzau, I. Berti, A. Tommasini, et al., Zonulin, a newly discovered modulator of intestinal permeability, and its expression in coeliac disease, Lancet 355 (2000) 1518–1519.
[82] A. Sapone, L. de Magistris, M. Pietzak, M.G. Clemente, A. Tripathi, F. Cucca, et al., Zonulin upregulation is associated with increased gut permeability in subjects with type 1 diabetes and their relatives, Diabetes 55 (2006) 1443–1449.
[83] T. Watts, I. Berti, A. Sapone, T. Gerarduzzi, T. Not, R. Zielke, et al., Role of the intestinal tight junction modulator zonulin in the pathogenesis of type I diabetes in BB diabetic-prone rats, Proc. Natl. Acad. Sci. U.S.A. 102 (2005) 2916–2921.

[84] B.M. Paterson, K.M. Lammers, M.C. Arrieta, A. Fasano, J.B. Meddings, The safety, tolerance, pharmacokinetic and pharmacodynamic effects of single doses of AT-1001 in coeliac disease subjects: A proof of concept study, Aliment. Pharmacol. Ther. 26 (2007) 757–766.

[85] K.H. Song, A. Fasano, N.D. Eddington, Enhanced nasal absorption of hydrophilic markers after dosing with AT1002, a tight junction modulator, Eur. J. Pharm. Biopharm. 69 (2008) 231–237.

[86] M. Thakar, S. Vogel, J. Kao, A. Fasano, The vibrio cholerae-generated zonula occludens toxin (zot) N-terminal cleavage site contains a protease-activated receptor activating peptide (par-Ap) that retains biological activity on intestinal tight junctions, Gastroenterology 128 (2005) A541–T1742.

[87] M. Li, E. Oliver, K.M. Kitchens, J. Vere, S.S. Alkan, A.P. Tamiz, Structure-activity relationship studies of permeability modulating peptide AT-1002, Bioorg. Med. Chem. Lett. 18 (2008) 4584–4586.

[88] S. Nystedt, A.K. Larsson, H. Aberg, J. Sundelin, The mouse proteinase-activated receptor-2 cDNA and gene. Molecular cloning and functional expression, J. Biol. Chem. 270 (1995) 5950–5955.

[89] W. Kong, K. McConalogue, L.M. Khitin, M.D. Hollenberg, D.G. Payan, S.K. Bohm, et al., Luminal trypsin may regulate enterocytes through proteinase-activated receptor 2, Proc. Natl. Acad. Sci. U.S.A. 94 (1997) 8884–8889.

[90] C.U. Corvera, O. Dery, K. McConalogue, S.K. Bohm, L.M. Khitin, G.H. Caughey, et al., Mast cell tryptase regulates rat colonic myocytes through proteinase-activated receptor 2, J. Clin. Invest. 100 (1997) 1383–1393.

[91] N. Cenac, A.M. Coelho, C. Nguyen, S. Compton, P. Andrade-Gordon, W.K. MacNaughton, et al., Induction of intestinal inflammation in mouse by activation of proteinase-activated receptor-2, Am. J. Pathol. 161 (2002) 1903–1915.

[92] N. Cenac, A.C. Chin, R. Garcia-Villar, C. Salvador-Cartier, L. Ferrier, N. Vergnolle, et al., PAR2 activation alters colonic paracellular permeability in mice via IFN-gamma-dependent and -independent pathways, J. Physiol. 558 (2004) 913–925.

[93] S. Gopalakrishnan, N. Pandey, A.P. Tamiz, J. Vere, R. Carrasco, R. Somerville, et al., Mechanism of action of ZOT-derived peptide AT-1002, a tight junction regulator and absorption enhancer, Int. J. Pharm 365 (2009) 121–130.

[94] R.K. Rao, S. Basuroy, V.U. Rao, K.J. Karnaky Jr., A. Gupta, Tyrosine phosphorylation and dissociation of occludin-ZO-1 and E-cadherin-beta-catenin complexes from the cytoskeleton by oxidative stress, Biochem. J. 368 (2002) 471–481.

[95] A. Tripathi, K.M. Lammers, S. Goldblum, T. Shea-Donohue, S. Netzel-Arnett, M.S. Buzza, et al., Identification of human zonulin, a physiological modulator of tight junctions, as prehaptoglobin-2, Proc. Natl. Acad. Sci. U.S.A. 106 (2009) 16799–16804.

[96] L. Bueno, J. Fioramonti, Protease-activated receptor 2 and gut permeability: A review, Neurogastroenterol. Motil. 20 (2008) 580–587.

[97] F. Raimondi, P. Santoro, M.V. Barone, S. Pappacoda, M.L. Barretta, M. Nanayakkara, et al., Bile acids modulate tight junction structure and barrier function of Caco-2 monolayers via EGFR activation, Am. J. Physiol. Gastrointest. Liver Physiol. 294 (2008) G906–G913.

[98] D.S. Cox, H. Gao, S. Raje, K.R. Scott, N.D. Eddington, Enhancing the permeation of marker compounds and enaminone anticonvulsants across Caco-2 monolayers by modulating tight junctions using zonula occludens toxin, Eur. J. Pharm. Biopharm 52 (2001) 145–150.

[99] D.S. Cox, S. Raje, H. Gao, N.N. Salama, N.D. Eddington, Enhanced permeability of molecular weight markers and poorly bioavailable compounds across Caco-2 cell monolayers using the absorption enhancer, zonula occludens toxin, Pharm. Res. 19 (2002) 1680–1688.

[100] M.M. Thanou, A.F. Kotze, T. Scharringhausen, H.L. Luessen, A.G. de Boer, J.C. Verhoef, et al., Effect of degree of quaternization of N-trimethyl chitosan chloride for enhanced transport of hydrophilic compounds across intestinal caco-2 cell monolayers, J. Control. Release 64 (2000) 15–25.

[101] N.A. Motlekar, A. Fasano, M.S. Wachtel, B.B. Youan, Zonula occludens toxin synthetic peptide derivative AT1002 enhances *in vitro* and *in vivo* intestinal absorption of low molecular weight heparin, J. Drug Target. 14 (2006) 321–329.

[102] C.S. Karyekar, A. Fasano, S. Raje, R. Lu, T.C. Dowling, N.D. Eddington, Zonula occludens toxin increases the permeability of molecular weight markers and chemotherapeutic agents across the bovine brain microvessel endothelial cells, J. Pharm. Sci. 92 (2003) 414–423.

[103] D. Menon, C.S. Karyekar, A. Fasano, R. Lu, N.D. Eddington, Enhancement of brain distribution of anticancer agents using Delta G, the 12 kDa active fragment of ZOT, Int. J. Pharm. 306 (2005) 122–131.

[104] M.T. De Magistris, Zonula occludens toxin as a new promising adjuvant for mucosal vaccines, Vaccine 24 (Suppl. 2) (2006). S2-60-61.

[105] M. Marinaro, A. Di Tommaso, S. Uzzau, A. Fasano, M.T. De Magistris, Zonula occludens toxin is a powerful mucosal adjuvant for intranasally delivered antigens, Infect. Immun. 67 (1999) 1287–1291.

[106] M. Marinaro, A. Fasano, M.T. De Magistris, Zonula occludens toxin acts as an adjuvant through different mucosal routes and induces protective immune responses, Infect. Immun. 71 (2003) 1897–1902.

Chapter 10

Peptide and Protein Delivery with Cell-penetrating Peptides

Helin Räägel[1] and Margus Pooga[1,2]

[1] *Institute of Molecular and Cell Biology, University of Tartu, EE51010 Tartu, Estonia,* [2] *Estonian Biocentre, EE51010 Tartu, Estonia*

Chapter Outline

10.1. INTRODUCTION

The effective transport of hydrophilic molecules (i.e. proteins, peptides, etc.) across the cellular plasma membrane barrier has been an insurmountable obstacle for many decades. Because of this, numerous methods have been invented to attempt to overcome this hurdle; for instance, the use of helper molecules like polyethyleneimine (PEI) [1−3], encapsulating carrier systems such as liposomes (reviewed in [4,5]) or viral vectors ([5], reviewed in [6]), mechanical/physical membrane destabilizing techniques like electroporation [7,8] or microinjection [9,10] etc. Although these methods offer several advantages to the field of protein/peptide transport, they all possess significant drawbacks, for example, heterogeneous dispersion, limited *in vivo* application,

Peptide and Protein Delivery. DOI: 10.1016/B978-0-12-384935-9.10010-0

and above all immunogenicity issues in the case of viral carriers ([11], reviewed in [11,12]) . Therefore, the search for superior transport systems is ongoing.

In 1988, two independent research groups discovered that a trans-activating transcription factor, Tat, from the human immunodeficiency virus (HIV) was able to freely internalize into cultured cells [13,14]. Since this early discovery of proteins capable of inducing cell entry, much effort has been put into determining the requisite amino acid residues and the most effective sequences for translocation in order to develop an effective transport vector. This extensive research led to the emergence of the field of cell-penetrating peptides (CPPs), also called protein transduction domains (PTD) or Trojan peptides because of their ability to overcome the unassailable plasma membrane and carry attached cargo into the cell.

10.2. THE DIVERSE FAMILY OF CPPs

Cell-penetrating peptides comprise a large heterogeneous family of peptidic carriers. As mentioned above, the first CPPs were discovered from naturally occurring proteins, which were able to induce translocation across the cellular plasma membrane. In order to fine-tune the transporting properties of the vector, the quest for the most effective fragment was initiated. The explicit sequences **within the natural proteins** that possess the translocating ability were defined by evaluating the effective transport of the peptides after both truncation [15,16] and/or amino acid substitution analyses [16,17]. From these studies, the Tat peptide, comprising amino residues 48–60 from the full length Tat protein emerged, showing an elevated internalization efficacy over the parental protein.

Another widely used CPP from a natural origin comes from the third α-helix of the Antennapedia homeodomain from *Drosophila* (aka fruit fly) which was conveniently named Penetratin [18] because of its cell-penetrating activity. Additionally, CPPs like VP22 from the herpes simplex virus (HSV) envelope protein 22 [19], pVEC from the murine vascular endothelial cadherin [20] and others have stimulated interest in the scientific field of biomolecule transport.

Another subclass of CPPs incorporates peptides designed on the basis of naturally occurring sequences that have been derived from at least two different sources, forming a group of **chimeric cell-penetrating peptides.** The most well-known example from this ensemble is Transportan, which was designed by coupling the fragment of a neuropeptide galanin (amino acids 1–12) to the N-terminus of the wasp venom peptide mastoparan via a lysine residue (Lys) linker [21]. The Lys residue between the two distinct fragments acts like a joint dividing the peptide into two discrete clusters that can, to a certain extent, change the overall conformation of the peptide, making it more capable of interacting with membrane compounds. Yet, much like its progenitor galanin, Transportan was demonstrated to have a high affinity towards galanin receptors thus holding a potential to affect the activity of the downstream G-coupled proteins (GTPases) [21]. Therefore, in order to minimize the possible side-effects that could rise from receptor binding and/or misactivation of cellular GTPases, a series of deletion analogs of Transportan were designed and tested [22]. The best results were obtained with a peptide lacking the six N-terminal amino acid residues of the galanin fragment. This exhibited equal cellular uptake and lower cytotoxicity than the parent peptide. This truncated version of Transportan was designated as TP10. The list of available chimeric CPPs does not end here; several other examples exist in the field. The most commonly added motif is the nuclear localization signal (NLS), which assists with membrane binding due to its strong positive charge, while also acting as a transport signal inside the cells. An example of this type is Pep-1, which is comprised of a fusion peptide incorporating a fragment of the dimerization domain of the reverse transcriptase enzyme from the human immunodeficiency virus (HIV) and the NLS from the large T-antigen of the simian virus SV40 [23]. Similarly to Transportan and TP10, this peptide also contains a joint-like linker (here three amino acid residues in length) separating the highly hydrophobic region from the highly basic one, and giving it a somewhat flexible form.

A distinct subclass of **synthetic peptides** has emerged from the studies of the requisite amino acids for cell internalization, as these studies have demonstrated that positively charged amino acids play a positive role in the uptake of CPPs [15]. It is also known today that an arginine residue (Arg) with its bidentate side-chain structure – called a guanidinium head group – capable of forming two hydrogen bonds (instead of one) with the negatively charged plasma membrane components is therefore more potent in facilitating cell delivery than other positively charged amino acids like lysine (Lys), histidine (His) or ornithine [16,24]. This clearly proves that Arg plays a greater role in facilitating cellular uptake than either the mere charge or backbone structure of the peptide. Consequently, oligomers containing only Arg residues were designed and evaluated for their uptake efficiency. An optimizing assay revealed that oligoarginines with eight or nine arginine residues in their sequences exhibited the best internalization properties, being over 20-fold more efficient than the Tat peptide [16]. However, somewhat unexpectedly, the optimizing experiments revealed that increasing the number of Arg residues in the sequence above a certain number (usually 14 residues in length) elevated the cytotoxicity of the compound [24,25].

Another subclass of synthetic peptides arises from the analysis of delivery efficiency after alterations in the **secondary structure of the peptide**, showing that specific higher-order structures are preferred for membrane translocation. For example, the amphipathic α-helical structure exposes the hydrophilic amino acids to one side and the hydrophobic residues to the other side of the helix. This makes the peptide ideally suited for binding to membrane interfaces, since the polar surface can face the aqueous phase and the less polar surface will face the membrane interior. In nature, this structure is often seen in membrane proteins, such as ion channels or membrane pore complexes [26,27].

Because of its propensity to interact with membrane components, this amphipathic α-helix strategy has been incorporated into the optimization of the internalization efficiency of CPPs, and has resulted in the emergence of a new peptide carrier called MAP (model amphipathic peptide). MAP, designed using repetitive blocks of the amino acids lysine, leucine and alanine (Lys, Leu and Ala or KLA) comprising 18 residues in total that form the desired amphipathic α-helix [28], was demonstrated to efficiently internalize into mammalian cells [29]. Although MAP peptides interact strongly with the membrane, it is important to keep in mind that due to their natural pore-forming propensity and high hydrophobic moment, they may irreversibly destabilize or damage the lipid bilayers and thus exert cytotoxic effects [30].

Because of their recognized high potential as transport vectors, numerous new and reportedly even more effective or specific peptidic sequences are added to the pool of existing CPPs every year. Currently available computer-aided CPP prediction algorithms make it easier to detect or anticipate possible CPPs in existing amino acid sequences. However, care needs to be taken in trusting the predictions since prognosis and experimental data do not always correlate [31]. In spite of this, a couple of conceptual rules are highlighted in the review by Hansen *et al.* [31]. Briefly, firstly a high number of positive charges in the sequence tends to increase the interaction with the negatively charged plasma membrane. Furthermore, Arg residues should be preferred because of their electrochemical potency to form bidentate hydrogen bonds. Secondly, secondary amphipathic sequences lend a greater disposition to translocation into the lipid bilayer of the membrane, yet, they may also exert membranolytic effects giving rise to cytotoxicity. Thirdly, the hydrophobicity of the peptide also helps its translocation into the membrane lipid layer, but also leads to water insolubility and so makes it difficult to use. And finally, because of complications with synthesis and purification, shorter sequences are usually preferred.

10.3. DELIVERY OF CARGOES INTO CELLS

Since their discovery, CPPs have been shown to efficiently import a vast range of cargoes with various physicochemical properties, from peptides and proteins to DNA and nanoparticles, into different cells and tissues [23,32–34]. However, in this book, strong emphasis will be put on effective delivery of peptides and proteins.

Based on the principal purpose of the cargo molecule, the peptide and protein cargoes can be categorized into two distinct subclasses: reporter peptides/proteins (giving simply a positive read-out upon effective delivery) and functional peptides/proteins (allowing the detection of any alteration in the cellular processes). The most widely used reporter proteins in the field of CPPs are avidin or avidin-like (streptavidin, neutravidin) cargoes, tagged with attached reporter (fluorescence, gold, etc.) [35–37], β-galactosidase [38–40],

Cre-recombinase [41–44], green fluorescent protein [45], etc. All of these reporters have been successfully used in corroborating the transport properties of various CPPs and/or in specifying the intracellular localization of the CPP–cargo complexes. The application of functional protein cargoes will be discussed later in this chapter (under moving to *in vivo* systems).

10.4. PENETRATION VERSUS ENDOCYTOSIS

Essentially, the target in CPP technology is to create an effective transporter for a biologically functional cargo molecule. Yet, to reach that target, much research needs to be done first with the reporter cargoes, in order to select the most efficient delivery vector, and then to determine the route of uptake and the most productive transport pathway(s).

In spite of the extensive research towards defining the uptake mode of different CPPs, the subject is still hotly disputed and quite ambiguously understood. This often-encountered ambiguity probably arises from the different characteristics of the used cell lines, CPPs, attached cargoes, used linkers, methods of complex formation, etc. [46–49], making the generalization of the mechanisms of uptake and intracellular targeting a difficult, if not impossible, task.

The first question to consider when talking about CPP-mediated cargo delivery is the mechanism by which the CPPs (on their own or with the cargo) pass through the plasma membrane barrier. In other words, does the translocation of the complexes occur via direct penetration or an endocytic mechanism? Many past reports have claimed that the process is not of an endocytic nature, since uptake of the compounds was detected at low temperature (at or below 4°C), where invagination of the plasma membrane is arrested [21,50–52]. However, in earlier studies (especially before 2003), the treated cells were often fixed prior to microscopic analysis of the internalization and localization of CPP–cargo complexes. Unexpectedly, in the early 2000s, it was discovered that the fixation itself may potentially alter the localization and distribution of CPPs inside the cells [53], leading to a massive re-evaluation wave in the field of CPP-mediated transport [17,54]. Since then, live-cell imaging has been the suggested protocol for assessing the uptake efficacy and intracellular localization and/or trafficking of the CPP–cargo complexes.

Because of this tremendous scientific "stumble", new reports arguing for the direct passage of CPPs or especially for CPP–cargo complexes across the plasma membrane often encounter widespread skepticism. Although it is now commonly accepted that the transduction process, particularly in the case of CPP–cargo complexes, takes place via endocytosis, there is some evidence that certain CPPs could potentially still be capable of penetrating the plasma membrane and facilitating the uptake of the cargo directly into the cytoplasm of the cell [38,55]. As mentioned above, the size of the cargo may greatly influence the uptake route, since, for example, the Tat peptide fused to a peptide cargo was observed to rapidly internalize into cells and the nucleus, whilst the Tat peptide fused to a protein cargo was taken up by cells inside vesicular structures, indicating an endocytic process [48]. Nevertheless, it is important to recognize that even though a proportion of CPP–cargo complexes may infiltrate through the membrane with the help of their membrane destabilizing or interacting motifs, the majority of the complexes still make use of endocytic routes to gain entry to cells (observed as vesicular or punctate staining of the cytoplasm).

Whether the transduction of CPP–cargo complexes occurs via penetration or endocytosis, internalization begins with the **interaction of the compound with the cellular membrane**. The plasma membrane is extracellularly abundantly lined with different proteoglycans (i.e. heparin sulfate proteoglycans (HSPGs)) which, because of their negative charge and the far-reaching length of their side-chains, strongly interact with any positively charged molecule in the surrounding environment. Although some other cell surface "anchors", for example phospholipids [56–58], have also been suggested as participating in the primary attachment of the CPPs, HSPGs are the most widely agreed plasma membrane components that bridge the CPP–cargo complexes to the cellular membrane [59–62]. It has often been observed that CPPs have a tendency to adhere to the plasma membrane protrusions and the membrane curve at the base of the projection [36,63]. This phenomenon could result from the relatively motile nature of the protrusion compared to the other membrane areas and/or from the mere length of the outgrowth, giving it an advantage of extending further into the surrounding environment plus an elongated area of contact for the CPP complexes.

The plasma membrane constituents and components vary enormously between different cell types and tissues, yielding different uptake or intracellular trafficking of the internalized compounds. Additionally, since the plasma membrane components tend to accumulate in distinct microdomains (i.e. lipid rafts), it is clear that some regions are more engaged in the binding and/or internalization process of compounds than others. Furthermore, the role of different membrane areas can be of varying importance. For instance, it is highly probable that the penetration of CPPs and CPP–cargo complexes takes place at the liquid-disordered phase of the membrane, where the lipid bilayers are more fluid and flexible, so promoting the submersion of the assembled CPP complexes into the membrane. On the other hand, the endocytosis of the complexes via, for example, the caveolin-mediated pathway requires the clustering of caveolin and thus occurs at the highly ordered, caveolin-containing, lipid raft microdomains (reviewed in [64]). Due to the high relevance of endocytosis in CPP-mediated cargo delivery, this text will concentrate on the exact mechanisms behind the cell entry and the successive fate of the complexes.

After binding to the cell surface, CPP–cargo complexes may choose from a variety of different endocytic routes in order to reach the cell interior.

Endocytosis is a general term used to characterize the process of engulfing extracellular material by invagination of the plasma membrane to form a small vesicle, which after budding from the membrane migrates to the cell interior [65]. In fact, endocytosis can be subdivided into phagocytosis (occurring only in certain cell types, i.e. macrophages) and pinocytosis. The latter term segregates further into clathrin-mediated, caveolin-mediated, clathrin- and caveolin-independent endocytosis and macropinocytosis. A vast number of scientific papers have been published to precisely define the underlying endocytic pathway(s) that could be employed or even activated by the CPPs ([66], reviewed in [67]). The scientific basis of this research arises from using markers or inhibitors to either highlight or block specific endocytic routes, respectively. Another method is to make use of knockout cell lines, which are devoid of certain proteins. This results in the loss of cellular processes that require that specific protein. For example, in caveolin-1 knockout mice, caveolin-1 is not expressed, leading to the inhibition of the caveolin-mediated endocytic pathway. This makes these cells a near-ideal cell line to study the importance of caveolin in the uptake of different CPP complexes. Additionally, RNA interference technology has been widely applied. Here, short RNA sequences are implemented to downregulate the desired endogenous protein levels, but the outcome of this method is highly dependent on the use of small interfering RNA (siRNA). However, since the commonly used markers or inhibitors of different endocytic routes are not as specific as we would want them to be, and also because downregulation rarely achieves even 80% and may upregulate other endocytic pathways, the results of these studies must be interpreted with caution. Indeed, several parallel experiments need to be conducted in order to be able to draw valid conclusions. Despite the number of contradicting reports published in light of the search for the uptake route of CPP–cargo complexes, some conclusions can still be drawn.

The most thoroughly studied type of endocytosis is the **clathrin-mediated pathway** (CME) that is commonly used for internalization of receptors and proteins from the cell surface. Clathrin-mediated endocytosis initiates by binding the ligand to the receptor, stimulating the subsequent clustering of the receptor which thereon facilitates the recruitment of clathrin to the plasma membrane. The accumulation of clathrin to the cytosolic face of the membrane causes it to curve into the cell interior, and then bud off as a vesicle covered with a honeycomb-like clathrin lattice (reviewed in [68]). Since several cationic CPPs (i.e. nona-arginine and Tat), both on their own and also with protein cargo, can facilitate receptor internalization without displaying any specificity, it can be concluded that at least a part of them can internalize into the cells via clathrin-mediated endocytosis [69]. Further support for the partial use of this pathway comes from a study where the recycling of clathrin and the reformation of the clathrin-coated pits were inhibited by hyperosmolar conditions [70] resulting in a 20% or 10% decrease in the uptake of Tat-avidin and TP-avidin, respectively [35]. Also, some colocalization with Transferrin (a conventional marker for the clathrin-mediated pathway) was observed with Tat—Streptavidin complexes [71]. However, due to the slightly negative charge on the Transferrin molecule, the strongly positive avidin may bind extracellularly with the Transferrin causing a concomitant internalization and thus an overestimation of the actual colocalization between the two compounds. Therefore, it is fair to say that the clathrin-mediated pathway may, on one hand, contribute to the internalization of either Tat-, nona-arginine- or TP-bound cargo, yet, on the other, its participation is relatively scarce, especially in case of Transportan (TP).

As mentioned above, caveolin resides in the strictly ordered membrane microdomains called lipid rafts and, for this reason, the **caveolin-mediated pathway** originates only from these discrete plasma membrane regions. After budding from the plasma membrane, the small, flask-shaped, caveolin-1-positive vesicles fuse with the rosette-like caveosomes which can target some of its content to the Golgi complex, the endoplasmic reticulum or other intracellular compartments (reviewed in [72]). There are no specific live-cell imaging markers available for this pathway, although several toxins (i.e. the cholera toxin B subunit) and pathogens (i.e. SV40) are shown to at least partially make use of it as their gateway into the cell. Still, both of the aforementioned "markers" can also enter cells via other routes that are independent of caveolin [73,74], which lack of specificity makes the experimental data hard to evaluate. This phenomenon was observed, for instance, with Tat protein or peptide fused to a green fluorescent protein that colocalized better with the cholera toxin B subunit than with caveolin-1 itself [45]. This revealed that while caveolin-1 is important in the uptake of these constructs, other pathways must also be involved. Nevertheless, the role of caveolin-dependent endocytosis in the uptake of different CPP-cargo complexes has been described on several occasions [63,75].

Caveolin-1 is observed to be of considerable importance, especially for Transportan which to a great extent colocalizes intracellularly into caveolin-1-positive vesicles [63]. Furthermore, in the downregulation of caveolin-1 with siRNA, and also in caveolin-1 deficient knockout cells, the uptake of TP-protein cargo was reduced by more than half. Despite an early study demonstrating the localization of Tat fusion peptide with GFP to caveolin-1 positive endosomes [75], our studies of the Tat peptide indicated that it exploited the caveolin-mediated pathway to a lower extent, and colocalization of Tat-protein complexes with Caveolin-1 occurred more at the cell membrane and less inside the cytoplasmic vesicles [63]. These discrepancies probably come from the differences in the peptide constructs used, since in some cases the CPP sequence is fused to a glutathione-S-transferase (GST)-tag (necessary for the purification process after expressing the construct in the bacteria) but in others the CPP is fully synthesized by automated peptide synthesis, and needs no additional tag for further purification. However, to draw some parallels between different CPPs, it is reasonable to believe that, since oligoarginine (along with Tat) is a member of the arginine-rich CPP family, it may act similarly to the Tat peptide and thus employ analogous internalization pathway(s), also making some use of caveolin-dependent endocytosis.

Another well-characterized type of endocytosis that takes place without the participation of either caveolin or clathrin is **macropinocytosis**. Macropinocytic vesicles arise from the dynamic movement of the actin-rich cell surface outgrowths called membrane ruffles, which occasionally fall back to the plasma membrane. This process takes some of the extracellular contents entrapped between the protrusion and the plasma membrane into a vesicle that passes into the cell interior. Macropinosomes are heterogeneous in size, but tend to be larger than the vesicles formed by either clathrin- or caveolin-mediated endocytosis (reviewed in [76]). Many papers have highlighted the role of macropinocytosis in the entry of CPP-protein/peptide complexes [77,78], revealing that when this pathway is blocked by inhibitors (5-ethyl-isopropyl amiloride (EIPA) or amiloride), the internalization of the complexes is hindered by at least 30%. Another study, focusing on the transport characteristics of the Tat-Cre recombinase fusion protein, demonstrated that destabilization of lipid rafts by depletion or sequestration of the cholesterol residing there with β-cyclodextrin or nystatin, respectively, dramatically inhibited the uptake of Tat-Cre in a concentration-dependent manner [41]. However, since caveolins also inhabit these lipid rafts, the same report used a fluorescently labeled caveolin-1 to show that the two complexes are taken up by cells in different vesicular structures; the conclusion being that even though lipid rafts are central to the uptake of the Tat-Cre fusion protein, caveolin is not necessarily involved in the process. The relevance of macropinocytosis was further confirmed for a variety of arginine-rich peptides (i.e. Tat, oligoarginine, modified oligoarginines) [79], demonstrating that for GFP translocation, arginine-rich peptides use an actin-dependent macropinocytic pathway for cell entry. Additionally, shortly after internalization, a large proportion of Transportan in complex with streptavidin is detected in large, polymorphous, intracellular vesicles reminiscent of macropinosomes [36], adding supporting evidence for the theory of macropinocytic uptake. The underlying mechanism of the stimulation of macropinocytosis is, however, still undefined. There have been speculations that CPPs themselves can stimulate the membrane ruffling, and thus enhance the uptake of the cargo molecule via the macropinocytic pathway. Some reports suggest that the initial binding of the arginine-rich CPPs to the membrane-bound heparin sulfate proteoglycans (HSPGs) activates the GTPase Rac and the downstream actin polymerization pathway, leading to the membrane ruffling and the subsequently activated macropinocytic events [67]. The activation of Rac upon binding to the HSPGs has been discovered to take place for many different CPPs [80], confirming the assumptions that at least some CPPs are capable of stimulating the membrane fluctuations that help them cross the membrane barrier.

In addition to the above described endocytic routes, other, so called **clathrin-** and **caveolin-independent pathways**, facilitating the uptake of several receptors, molecules or pathogens, are active inside mammalian cells. For example, the vesicles referred to as clathrin-independent carriers (CLIC) or GPI-enriched endocytic compartments (GEEC), form as tubular structures from the plasma membrane through the activation of dynamin, several GTPases (i.e. RhoA) and the polymerization of actin filaments ([81], reviewed in [82]). Without ruling out the possibility that there are several distinct subtypes of this endocytic route, it is referred to as the CLIC/GEEC pathway. As the name infers, the CLICs/GEECs are used, for instance, for internalization of the glycosylphosphatidylinositol (GPI)-linked proteins [83]. However, a recent paper implies that the term "GPI-enriched endosomal compartment (GEEC)" is not entirely correct for these endosomes, since as yet there is no direct evidence that the density of GPI-linked proteins per unit area of membrane is any higher in the GEEC compartments than in the plasma membrane [84,85]. Moreover, the article also indicates that the GPI-linked proteins are sorted to the so-called GEECs chiefly by steric exclusion from the clathrin-coated pits. This means that they may enter the cells by multiple clathrin-independent endocytic pathways, including the typical caveolin-dependent one; i.e. the GPI-linked proteins cannot be exclusively used as markers for characterizing this specific pathway.

Analogously, in 2005, the uptake of SV40, which up to that point was thought to internalize into cells via the caveolin-mediated pathway, was discovered to at least partially move through an unconventional clathrin- and caveolin-independent pathway [74]. A similar tendency for the parallel use of the clathrin- and caveolin-independent pathway and the caveolin-mediated one was confirmed for the cholera toxin B subunit [86], ruling out the specificity of both peptides towards the caveolae, and hence also their use as definite markers for the latter pathway. To make matters worse, since accurate information about this pathway is lacking and the specificity of the known markers has been jeopardized, the role of the CLICs/GEECs in CPP-mediated peptide/protein delivery has not yet been analyzed to draw any decisive conclusions. At least one type of GPI-anchored protein is internalized via a clathrin-independent pathway, leading to the trafficking of the endocytosed material to an unorthodox subgroup of Rab5-independent tubular vesicles [83]. Thus, one cannot rule out the possibility of the use of this particular pathway in the Transportan-mediated protein transport, because the TP- and TP10-avidin complexes were demonstrated not to be targeted to the conventional Rab5-positive early endosomes [63].

Other markers, found to reside in the cholesterol-rich membrane microdomains and to be involved in a clathrin- and caveolin-independent pathway, are the **flotillin** (or reggie) family proteins [87,88]. Flotillin resides in punctate structures within the plasma membrane and in a specific population of endocytic intermediates facilitating, for instance, the internalization of the cholera toxin B subunit [88]. This again shows that the latter marker cannot be used to highlight any specific endocytic pathway, and the correlation seen between the uptake of the cholera toxin B subunit and any CPP-protein/peptide cargo complexes cannot be undeniably linked to a certain endocytic entry mechanism. For this reason, flotillin itself has been employed to characterize this subclass of clathrin- and caveolin-independent pathway. Unfortunately, the search for the association of Transportan-, TP10- and Tat-avidin complexes with flotillin-1 only led to a minor colocalization at the plasma membrane that diminished after cell entry of the complexes [63]. Therefore, the flotillin-mediated subclass of the clathrin- and caveolin-independent pathway can be discarded in the further scientific exploration for defining the specific endocytic gateway used by the CPP-protein/peptide cargo complexes.

What makes this search even more difficult is the recognition that, in the case of CPPs or CPP–cargo complexes, several different endocytic pathways can and probably are used in parallel (Figure 10.1) ([37,63,89], reviewed in [49]). Furthermore, downregulation or inhibition of specific routes may lead to the upregulation of other pathway(s) [90], and thus simply promote a switch from one endocytic route to another.

10.5. AFTER ENDOCYTIC ENTRY

After budding from the plasma membrane the CPP–cargo complexes that are confined in an endocytic vesicle are sorted and trafficked by the cellular machinery into the early endosomes (also called the early endosomal compartments). To date, two major intracellular pathways initiating at the plasma membrane and the early endosomes are known – the recycling and the endo-lysosomal pathway. For example, the receptors returning to the cell surface accumulate in the **recycling endosomes**, which are tubular portions of early endosomes located in the perinuclear Golgi region of the cell [91]. In certain cell types, the recycling endosomes are spatially segregated from the rest of the endosomes by clustering into the ring-like *trans*-Golgi structure [92], making it easier to analyze their role in the intracellular trafficking of, for instance, CPP–cargo complexes. It was found recently that neither Tat-, nona-arginine- nor TP-avidin complexes are targeted to the recycling endosomes [37], eliminating its importance in the trafficking of the CPP–protein cargo inside the cells. For this reason, little attention is given to the recycling pathway in this chapter. However, deeper interest will be given to the other – the endo-lysosomal pathway – which is considered to deliver the internalized content into degradative organelles; lysosomes.

The engulfed material that follows the classical **endo-lysosomal pathway** is sorted from the early endosomal compartment to the multivesicular bodies, and from there to late endosomes and lysosomes. The above mentioned subclasses of endosomes probably do not exist as distinct, stable organelles, but represent a particular stage of maturation and undergo continuous sorting of the

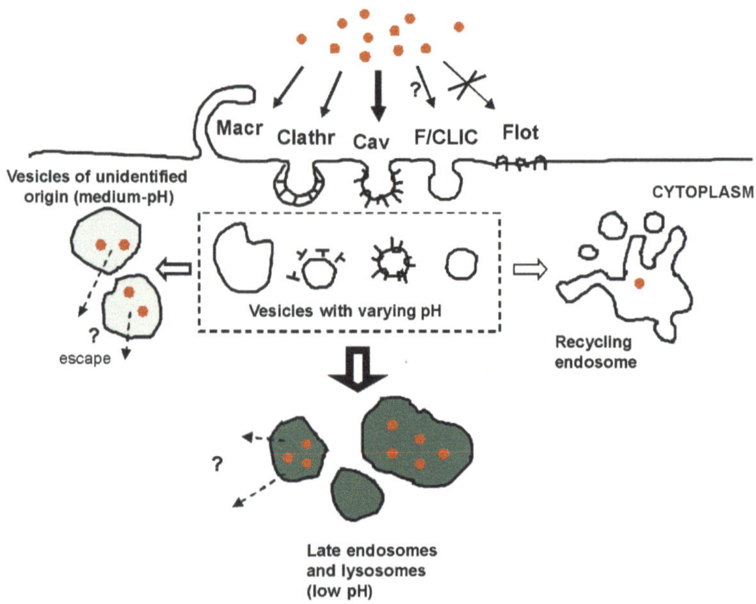

FIGURE 10.1 Endosomal entry and intracellular trafficking of CPP-protein complexes. CPP-protein complexes (red dots) may enter cells via different endocytic pathways (arrows outside the cell). The intracellular endocytic vesicles entrapping the complexes possess varying pH and may lead to either the recycling endosomes (minority of complexes, thin block arrow) or the endo-lysosomal pathway (majority of complexes, bold block arrow). However, some of the complexes are found also inside vesicles with a relatively neutral pH (medium block arrow), where the micro-environment may provide favorable conditions for the complexes to leak out into the cytosol. Arrow thickness indicates the extent of the endocytic pathways used in CPP-mediated protein uptake. Darkness of endocytic vesicles corresponds with their acidity. Macr – macropinocytosis; Clathr – clathrin-mediated endocytosis; Cav – caveolin-mediated endocytosis; F/CLIC – fluid phase/clathrin-independent carriers (involvement in CPP-mediated protein delivery has not yet been defined); Flot – flotillin-mediated endocytosis (plays no part in CPP-protein internalization). *Reprinted from reference [49] with permission from Elsevier.*

endocytosed material, separating the cargo and receptors by a pH shift and/or membrane differentiation events occurring at certain points during the development of the endosomal vesicle into a so-called ripe lysosome [91]. Several endocytic pathways ultimately target their content to acidic compartments. The accumulation of different CPP–cargo complexes into large, acidic lysosomes has been corroborated repeatedly [36,37,93,94]. Since more than 40 hydrolytic enzymes (including proteases, nucleases, glycosidases, lipases, phospholipases, phosphatases and sulphatases) are associated with the lysosomes, CPPs or their cargo are bound to be substrates for some of these enzymes. Therefore, that proportion of CPP–cargo complexes that are targeted to these low pH structures are probably destined for degradation. However, an extensive analysis of the pH of the intracellular vesicles containing either

TP-, nona-arginine- or Tat-avidin cargo complexes revealed three distinct subpopulations whose proportions depended on the used CPP, time and concentration of the CPP–cargo complexes [37]. Upon entry into cells, the complexes reside predominantly in smaller vesicles containing a low concentration of the complexes but having a variety of pH values, indicating that some of the material is rapidly subjected to degradation after internalization. With time, the fusion of the vesicles results in the accumulation of the complexes into fewer and fewer structures leading to the local high concentrations of CPPs attached to the cargo. These large, massively concentrated, complex-bearing vesicles can be divided into two groups – one with a low pH, representing the classical lysosomes, and the other with a relatively neutral pH, arising from an as yet unknown mechanism. Since the "effective" delivery of the intact bioactive cargo is typically to a target molecule that resides outside of the endocytic pathway, the transport vector must ensure, firstly, that the cargo is not degraded before its interaction with the target, and secondly, that the cargo escapes from the entrapping endosomes in order to reach its target. For this reason, the complexes inside highly acidic lysosomes can be discarded, due to their high potential of being digested before escape, leaving the "mysterious" near-neutral pH vesicles on the table for further investigation. Several research groups have used acidification inhibitors, also called lysosomotrophic agents (i.e. chloroquine, NH_4Cl), to enhance the proportion of this population of vesicles [95–97]. However, as mentioned above, at least TP, nona-arginine and Tat can mediate the uptake of the protein cargo into the medium pH vesicles, where the cargo molecule remains intact and functional.

Although the origin of these vesicles is not clear, some hypotheses have emerged. For example, the caveolin-mediated pathway has been indicated as participating in the formation of these neutral vesicular compartments, since not all contents of the caveosomes are intracellularly targeted for degradation, and are instead carried to long-lived, medium pH, caveosomal intermediates ([98], reviewed in [72]). This presumption is further corroborated by the fact that out of all the CPPs mentioned, Transportan delivered the largest amount of cargo to the near-neutral pH vesicles [37] and concurrently exhibited the strongest inhibition of uptake upon caveolin-1 removal [63]. On the other hand, some arginine-rich peptides (especially oligoarginines) have an inherent capacity to hinder the acidification of the vesicles they reside in, resulting in the generation of the near-neutral pH endosomes. Although the aforementioned explanations can be justifiably applied to complete the "big picture", the existence of other so far unknown mechanisms behind this phenomenon cannot be excluded.

Endosomal entry into cells represents a contradiction: on the one hand, endocytosis is a natural way for a cell to obtain molecules from the surrounding environment (thus being harmless) but on the other, internalized CPP–cargo complexes remain isolated from the cell interior and large amounts may be targeted for degradation. Therefore, in order to complete effective delivery, the transported peptide needs to somehow break free from this terminal pathway. The mechanism of CPP penetration of the vesicle membrane may be similar to that happening at the plasma membrane. Yet, there is one slight discrepancy that probably makes all the difference – the local concentration of the peptide. When applied extracellularly, the CPP–cargo complexes are distributed along the plasma membrane and, though aggregation does take place to some extent [99], the concentration of the peptide per unit area of the membrane is still rather low. During trafficking of endosomes harboring CPP–cargo complexes, the vesicles fuse with one another, so resulting in fewer and larger endocytic compartments with elevated concentrations of the complexes. This means that inside the vesicles the peptide number per unit area of the membrane is substantially increased. At high concentrations, the peptide accumulates and clusters at the membrane, interfering with its regular packing arrangement. In support of this, increasing the concentration of Transportan leads to a marked membrane leakage [30]. Additionally, it has been proposed that a higher oligoarginine peptide concentration induces membrane destabilization and pore formation (even at the plasma membrane) [57,100]. Therefore, it can be concluded that the liberation of the CPP–cargo complexes probably occurs via membrane destabilization and the subsequent slipping of the complexes out of the vesicles. It is not known whether the destabilization of the vesicle membrane brings about the complete breakage of the entire endosome or causes just some leakage through the membrane; however, from the cell's point of view, the latter mechanism would be less damaging. Moreover, the exact amount of the CPP–cargo complexes reaching the cytoplasm is not defined, yet, it is evident that it is occurring at least to some extent.

Escape from the endosomes is currently the limiting step in the efficiency of CPP-mediated cargo delivery and for this reason, scientists have developed ways to overcome this drawback. For example, some articles have demonstrated protein delivery using CPPs fused to the N-terminal part of the hemagglutinin protein (HA) of the influenza virus [41,101,102]. Because the HA epitope is a well-characterized, pH-sensitive, fusogenic peptide that destabilizes lipid membranes at low pH, the CPP–cargo complexes containing this motif can escape more efficiently from the vesicles, especially the ones undergoing a pH decrease. Since many of the vesicles entrapping the complexes are targeted to low pH structures, the addition of the HA tag can drastically enhance the cytoplasmic delivery of the endocytosed material. Other possible fusogenic compounds that may assist endosomal escape of the compounds are discussed in a recent review [103]. Additionally, as referred to earlier, inhibitors of endosomal acidification (i.e. chloroquine) are frequently used in cell culture to enhance the function of the internalized cargo protein [104] by: (1) protecting the complex from degradation and (2) simultaneously inducing leaky vesicles with already destabilized membranes for easier escape. However, *in vivo* the use of these chemical enhancers is limited.

An additional mechanism of cytosolic entry of arginine-rich CPPs was proposed to occur via the retrograde transport, in which the vesicles are carried from the early endosomes to the Golgi network where their contents may be released into the cytoplasm [105]. The chemical disruption of the *trans*-Golgi network by brefeldin-A inhibited the uptake of arginine-rich peptides. Moreover, the induction of retrograde transport resulted in an increase in the cytosolic fraction of fluorescently labeled Tat and oligoarginine, implicating this particular pathway during cell entry of these peptides. Another paper referred to the use of this pathway in enhancing the cytosolic delivery and subsequent direction to the nucleus [45], indicating that the retrograde transport of CPP–cargo complexes may play a role in their effective delivery. However, the accumulation of CPP–cargo complexes to the *trans*-Golgi network has been only rarely reported [36,37].

10.6. MOVING TO *IN VIVO* SYSTEMS

To proceed with the progression of the field of CPPs as transport vehicles for bioactive macromolecules *in vivo*, we must return to the beginning of this chapter and recall that they can be divided into two subclasses, based on the purpose of the delivered peptide/protein cargo – the reporter proteins/peptides and the functional proteins/peptides. So far, the utilization of reporter proteins/peptides has been discussed, to reveal the path(s) of cell entry of these CPP–cargo complexes. However, from this point on, the use of functional protein/peptide delivery, especially in the clinical/medical sense, will be discussed.

In cell culture, many reports have been published which demonstrate CPP-mediated delivery of functional proteins (e.g. the dominant negative form of Cdk2 that binds pRb and inhibits cell proliferation [106]) or peptides (e.g. the p53 and p21 peptides/fragments [107,108], a fragment from Bcl-X(L) [109], etc.). Yet, since an organism is a far more complex system for delivery, complications may arise from biodistribution, bioavailability, biodegradation, immunogenicity and so on – questions that need to be resolved before CPPs may become a supplement for existing and pioneering medicines.

10.6.1. Specificity

Firstly, in cellular systems, the delivery of protein/peptide by CPPs, usually reaching 100% of the cells, is highly efficacious, making CPPs better transport vectors than other techniques described at the beginning of this chapter. On moving to *in vivo* systems, however, this may not be a desirable feature unless systemic delivery is desired. The problem of selectivity arises especially, for example, in the case of anti-cancer treatments, where the drug needs to be targeted to the cancer cells without affecting the normal tissue surrounding the tumor. The issue of biodistribution has been under scrutiny since the early experiments with CPPs as transport vectors *in vivo*. For instance, it was apparent in early studies that the non-selective delivery of Tat-FITC [110] or Tat-fusion protein with β-galactosidase to various tissues (the heart, liver, spleen, brain, etc.) [110,111] represented simultaneously both an advantage and a disadvantage. It is a desired feature since, at least theoretically, CPPs can transport a bioactive cargo into different tissues and even cross the blood−brain barrier, but is an unwanted side-effect in a sense that modifications need to be implemented to accomplish targeted delivery and tissue-specificity.

When injected intravenously, the main site of accumulation of the CPP−cargo complexes will probably be the liver and spleen (due to their nature of being blood depots) and the kidneys (because of their function as filters) [112,113]. The non-specific accumulation of these chemicals in these organs may cause serious damage to the tissues, and in a worst case scenario lead to death of the whole organism, accentuating the necessity for precise targeting. Thus, in biomedical research, the goal of getting the functional protein cargo into the exact tissue or cell type required (after administration via injection, oral uptake or inhalation), where it actually performs its task, holds the real key to the development of new pharmaceuticals. Therefore, the main interest now is to find ways of enhancing the local concentration of the drug whilst at the same time minimizing systemic exposure. The specificity of CPP−cargo complexes can be achieved by many techniques. One of the easiest approaches is local administration to the site of interest, either by injection into the target tissue/area [114] or the tumor [115], intranasal or intratracheal administration for internalization into cells of the upper airways [116] or lung and tracheal tissue and nearby cells [117] (i.e. in the case of asthma or allergy) or simply rubbing the complexes onto an affected skin area [118]. In some cases, however, these techniques cannot be applied, requiring other ways of addressing their non-specificity to be developed.

Since cancer is one of the leading causes of death worldwide, affecting millions of people annually, there is a strong impetus for science to target this disease. The discovery of cancer-specific peptides, for instance, can be facilitated by phage display technology, where a library of bacteriophages exposing a different peptide sequence on their capsule are intravenously injected into animals bearing the tumor to be targeted. The phages displaying the "correct" sequence bind specifically to the tumor and are afterwards collected, amplified and their exposed sequence identified as the tumor targeting sequence. These so-called "tumor homing sequences" have been applied successfully *in vivo*, where effective accumulation of the vector alone [119] or vector−drug complexes [120] into the tumor was demonstrated. The same phage display technology can also be applied to achieve tissue-specific peptides.

An innovative approach that has been used to acquire tumor specificity involved attachment of a masking polyanionic sequence to a CPP via a cleavable linker that is dissected by specific metalloproteases which are over-expressed at the site of the tumor [121,122]. Tumors use metalloproteases to dissolve the extracellular matrix in order to expand and migrate into surrounding tissues (reviewed in [123]). Thus, this method utilizes the tumor's own biochemical signature in order to attack it. The polyanionic sequence masks the positive charges on the CPP, preventing its localization to non-tumorous tissues. When targeted at the site of tumor, the high concentration of activated metalloproteases leads to the cleavage of the linker and the unmasking of the CPP-sequence which then readily enters the cells, resulting in the accumulation of the CPP in the tumor. The main intention is to use technique as a tumorigenic marker for surgeons [124], making differentiation between normal and cancerous tissue clearer [125]. However, it could potentially also be used as an anti-cancer drug delivery system.

10.6.2. Bioavailability, Degradation and Toxicity

To be an effective transporter in an *in vivo* system, the vector needs to possess certain characteristics which to an extent contradict one another. On the one hand, the transporter needs to be stable enough to survive the degradation by the plasma or extracellular matrix enzymes, in order to be able to carry its cargo to the target site in the organism. On the other, it needs to be degraded by the organism intra- or extracellularly after completing its mission, to avoid any side-effects rising from the accumulation of non-degraded byproducts that may in time become toxic. Since the half-life of CPPs with L-amino acid composition is fairly low in the plasma [112,126], several modifications have been used to overcome this drawback; for example, replacing the L-amino acids with their D-counterparts [127,128] or synthesizing the so-called retro-inverso peptides [129] with a D-amino-acid sequence that has been inverted to better mimic the chirality or the side chain topology of the L-isomer. Since it has been shown that the backbone structure plays no role in CPP-facilitated entry to cells, these stability-enhancing methods do not decrease the uptake of CPPs [130]. Slower clearance from the blood, however, extensively increases the bioavailability of the CPP–cargo complexes due to low or non-existent degradation, but at the same time prevents the removal from the body of the CPP–cargo complexes that have already performed their "mission". The enhanced accumulation of target-specific CPPs (in for example cancer tissues) may prevent this issue from ever becoming problematic, since cancer cells are forced to undergo apoptosis during therapy.

10.6.3. Immunogenicity

Another issue that demands attention before moving towards clinical trials involving CPPs as transport vectors for bioactive macromolecules, is the problem with immunogenicity. Several viral carriers, especially adenoviruses, may elicit an immune response, which is both exhausting to the organism (allergic reactions and inflammation) and lethal to the carrying vector (due to secretion of anti-adenovirus neutralizing antibodies) ([11], reviewed in [131]).

The *in vivo* immunogenicity of different CPPs, however, is poorly described to date, but due to their relatively short half-life in the plasma, it is fairly unlikely that CPPs will promote an immune response. Still, the body's reaction to the more stable D- or retro inverso variants of CPPs cannot be estimated, and requires more investigation. Nevertheless, a simple yet ingenious solution to this problem has been proposed – if immunity to one particular CPP were to develop in long-term trials, the abundance of different CPP peptides would allow for replacement of the original peptide, thereby sidestepping the problem [132].

10.7. CONCLUSIONS AND FUTURE ASPECTS

To round up the information presented in this chapter, it can be concluded that CPPs have the potential to change the world of pharmaceutics as we know it today. The most commonly used CPPs: Tat, oligoarginine and Transportan, have all been demonstrated to facilitate the entry of protein/peptide cargoes into cells both *in vitro* and *in vivo*. In cellular systems, Transportan has displayed greater internalization properties than the aforementioned arginine-rich CPPs and their uptake efficiencies can be depicted as follows: Transportan $>$ oligoarginine $>$ Tat peptide. However, concomitant with their uptake efficiency, serious attention needs to be given to toxicity reports that show a successive series: Transportan $\geq$ oligoarginine $>$ Tat. When picking the "right" CPP sequence for cargo delivery, both of these aspects need to be considered, and when lower concentrations are used the Transportan or oligoarginine should be preferred. Additionally, since Transportan-cargo complexes are found in neutral intracellular vesicles to a greater extent, it is possible that this peptide is superior to the others in terms of effective delivery.

In vivo, however, the uptake efficiency, specificity and toxicity have not been extensively studied for the different CPPs. Nevertheless, it is evident that for targeted delivery, some extra motifs need to be added to the CPP sequence. A rather promising aspect for CPP-mediated delivery of peptides/proteins is that transcytosis in endothelial cells (that otherwise act as a barrier between the blood stream and underlying tissues) requires caveolin (reviewed in [133]). As was mentioned above, several CPPs but especially Transportan, exploit the caveolin pathway for cell entry, possibly giving Transportan a beneficial "edge" *in vivo*.

Great expectations lie in the field of CPPs for both biomedicine and surgery. Despite the number of obstacles and pending challenges still faced today, the growing number of examples of *in vivo* delivery (PKC inhibitors [114]; TP-PNA [134]; Tat-HSP70 for neuronal rescue [135,136]; Tat-Bcl-x(L) for improved neuronal precursor cell survival [137]) confirm that the problems can be overcome in one way or another. The existing proof that CPP technology can be effectively applied confirms their high potential in clinical applications in the future.

REFERENCES

[1] V.V. Didenko, H. Ngo, D.S. Baskin, Polyethyleneimine as a transmembrane carrier of fluorescently labeled proteins and antibodies, Anal. Biochem. 344 (2005) 168–173.

[2] J. Futami, M. Kitazoe, T. Maeda, E. Nukui, M. Sakaguchi, J. Kosaka, et al., Intracellular delivery of proteins into mammalian living cells by polyethylenimine-cationization, J. Biosci. Bioeng. 99 (2005) 95–103.

[3] M. Kitazoe, H. Murata, J. Futami, T. Maeda, M. Sakaguchi, M. Miyazaki, et al., Protein transduction assisted by polyethylenimine-cationized carrier proteins, J. Biochem. 137 (2005) 693–701.

[4] A. Chonn, P.R. Cullis, Recent advances in liposomal drug-delivery systems, Curr. Opin. Biotechnol. 6 (1995) 698–708.

[5] O. Zelphati, Y. Wang, S. Kitada, J.C. Reed, P.L. Felgner, J. Corbeil, Intracellular delivery of proteins with a new lipid-mediated delivery system, J. Biol. Chem. 276 (2001) 35103–35110.

[6] S. Boeckle, E. Wagner, Optimizing targeted gene delivery: Chemical modification of viral vectors and synthesis of artificial virus vector systems, AAPS J. 8 (2006) E731–E742.

[7] L.H. Raptis, H.L. Brownell, S.K. Liu, K.L. Firth, L.W. MacKenzie, C.D. Stiles, et al., Applications of electroporation of adherent cells *in situ*, on a partly conductive slide, Mol. Biotechnol. 4 (1995) 129–138.

[8] A.A. Deora, F. Diaz, R. Schreiner, E. Rodriguez-Boulan, Efficient electroporation of DNA and protein into confluent and differentiated epithelial cells in culture, Traffic 8 (2007) 1304–1312.

[9] B.D. Wright, M. Terasaki, J.M. Scholey, Roles of kinesin and kinesin-like proteins in sea urchin embryonic cell division: Evaluation using antibody microinjection, J. Cell Biol. 123 (1993) 681–689.

[10] C. Lappe-Siefke, C. Maas, M. Kneussel, Microinjection into cultured hippocampal neurons: A straightforward approach for controlled cellular delivery of nucleic acids, peptides and antibodies, J. Neurosci. Methods 175 (2008) 88–95.

[11] B.G. Harvey, N.R. Hackett, T. El-Sawy, T.K. Rosengart, E.A. Hirschowitz, M.D. Lieberman, et al., Variability of human systemic humoral immune responses to adenovirus gene transfer vectors administered to different organs, J. Virol. 73 (1999) 6729–6742.

[12] D.A. Muruve, The innate immune response to adenovirus vectors, Hum. Gene Ther. 15 (2004) 1157–1166.

[13] A.D. Frankel, C.O. Pabo, Cellular uptake of the tat protein from human immunodeficiency virus, Cell 55 (1988) 1189–1193.

[14] M. Green, P.M. Loewenstein, Autonomous functional domains of chemically synthesized human immunodeficiency virus tat trans-activator protein, Cell 55 (1988) 1179–1188.

[15] E. Vivés, P. Brodin, B. Lebleu, A truncated HIV-1 Tat protein basic domain rapidly translocates through the plasma membrane and accumulates in the cell nucleus, J. Biol. Chem. 272 (1997) 16010–16017.

[16] P.A. Wender, D.J. Mitchell, K. Pattabiraman, E.T. Pelkey, L. Steinman, J.B. Rothbard, The design, synthesis, and evaluation of molecules that enable or enhance cellular uptake: Peptoid molecular transporters, Proc. Natl. Acad. Sci. U.S.A. 97 (2000) 13003–13008.

[17] P.E. Thorén, D. Persson, P. Isakson, M. Goksör, A. Önfelt, B. Nordén, Uptake of analogs of penetratin, Tat(48-60) and oligoarginine in live cells, Biochem. Biophys. Res. Commun. 307 (2003) 100–107.

[18] D. Derossi, A.H. Joliot, G. Chassaing, A. Prochiantz, The third helix of the Antennapedia homeodomain translocates through biological membranes, J. Biol. Chem. 269 (1994) 10444–10450.
[19] G. Elliott, P. O'Hare, Intercellular trafficking and protein delivery by a herpesvirus structural protein, Cell 88 (1997) 223–233.
[20] A. Elmquist, M. Lindgren, T. Bartfai, Ü Langel, VE-cadherin-derived cell-penetrating peptide, pVEC, with carrier functions, Exp. Cell. Res. 269 (2001) 237–244.
[21] M. Pooga, M. Hällbrink, M. Zorko, Ü Langel, Cell penetration by transportan, FASEB J. 12 (1998) 67–77.
[22] U. Soomets, M. Lindgren, X. Gallet, M. Hällbrink, A. Elmquist, L. Balaspiri, et al., Deletion analogues of transportan, Biochim. Biophys. Acta 1467 (2000) 165–176.
[23] M.C. Morris, J. Depollier, J. Mery, F. Heitz, G. Divita, A peptide carrier for the delivery of biologically active proteins into mammalian cells, Nat. Biotechnol. 19 (2001) 1173–1176.
[24] D.J. Mitchell, D.T. Kim, L. Steinman, C.G. Fathman, J.B. Rothbard, Polyarginine enters cells more efficiently than other polycationic homopolymers, J. Pept. Res. 56 (2000) 318–325.
[25] I. Nakase, M. Niwa, T. Takeuchi, K. Sonomura, N. Kawabata, Y. Koike, et al., Cellular uptake of arginine-rich peptides: Roles for macropinocytosis and actin rearrangement, Mol. Ther. 10 (2004) 1011–1022.
[26] J.D. Lear, Z.R. Wasserman, W.F. DeGrado, Synthetic amphiphilic peptide models for protein ion channels, Science 240 (1988) 1177–1181.
[27] Y. Agawa, S. Lee, S. Ono, H. Aoyagi, M. Ohno, T. Taniguchi, et al., Interaction with phospholipid bilayers, ion channel formation, and antimicrobial activity of basic amphipathic alpha-helical model peptides of various chain lengths, J. Biol. Chem. 266 (1991) 20218–20222.
[28] V. Steiner, M. Schar, K.O. Bornsen, M. Mutter, Retention behaviour of a template-assembled synthetic protein and its amphiphilic building blocks on reversed-phase columns, J. Chromatogr. 586 (1991) 43–50.
[29] J. Oehlke, A. Scheller, B. Wiesner, E. Krause, M. Beyermann, E. Klauschenz, et al., Cellular uptake of an alpha-helical amphipathic model peptide with the potential to deliver polar compounds into the cell interior non-endocytically, Biochim. Biophys. Acta 1414 (1998) 127–139.
[30] M. Hällbrink, A. Florén, A. Elmquist, M. Pooga, T. Bartfai, Ü Langel, Cargo delivery kinetics of cell-penetrating peptides, Biochim. Biophys. Acta 1515 (2001) 101–109.
[31] M. Hansen, K. Kilk, Ü Langel, Predicting cell-penetrating peptides, Adv. Drug Deliv. Rev. 60 (2008) 572–579.
[32] S. Fawell, J. Seery, Y. Daikh, C. Moore, L.L. Chen, B. Pepinsky, et al., Tat-mediated delivery of heterologous proteins into cells, Proc. Natl. Acad. Sci. U.S.A. 91 (1994) 664–668.
[33] S.R. Schwarze, K.A. Hruska, S.F. Dowdy, Protein transduction: Unrestricted delivery into all cells? Trends Cell Biol. 10 (2000) 290–295.
[34] J.S. Wadia, S.F. Dowdy, Protein transduction technology, Curr. Opin. Biotechnol. 13 (2002) 52–56.
[35] P. Säälik, A. Elmquist, M. Hansen, K. Padari, K. Saar, K. Viht, et al., Protein cargo delivery properties of cell-penetrating peptides. A comparative study, Bioconjug. Chem. 15 (2004) 1246–1253.
[36] K. Padari, P. Säälik, M. Hansen, K. Koppel, R. Raid, Ü Langel, et al., Cell transduction pathways of transportans, Bioconjug. Chem. 16 (2005) 1399–1410.

[37] H. Räägel, P. Säälik, M. Hansen, Ü Langel, M. Pooga, CPP-protein constructs induce a population of non-acidic vesicles during trafficking through endo-lysosomal pathway, J. Control. Release 139 (2009) 108–117.
[38] S.T. Henriques, J. Costa, M.A. Castanho, Translocation of beta-galactosidase mediated by the cell-penetrating peptide pep-1 into lipid vesicles and human HeLa cells is driven by membrane electrostatic potential, Biochemistry 44 (2005) 10189–10198.
[39] K. Weller, S. Lauber, M. Lerch, A. Renaud, H.P. Merkle, O. Zerbe, Biophysical and biological studies of end-group-modified derivatives of Pep-1, Biochemistry 44 (2005) 15799–15811.
[40] H. Myrberg, M. Lindgren, Ü Langel, Protein delivery by the cell-penetrating peptide YTA2, Bioconjug. Chem. 18 (2007) 170–174.
[41] J.S. Wadia, R.V. Stan, S.F. Dowdy, Transducible TAT-HA fusogenic peptide enhances escape of TAT-fusion proteins after lipid raft macropinocytosis, Nat. Med. 10 (2004) 310–315.
[42] J.M. Gump, R.K. June, S.F. Dowdy, Revised role of glycosaminoglycans in TAT protein transduction domain-mediated cellular transduction, J. Biol. Chem. 285 (2010) 1500–1507.
[43] Y. Gitton, L. Tibaldi, E. Dupont, G. Levi, A. Joliot, Efficient CPP-mediated Cre protein delivery to developing and adult CNS tissues, BMC Biotechnol. 9 (2009) 40.
[44] P.A. Shaw, I.R. Catchpole, C.A. Goddard, W.H. Colledge, Comparison of protein transduction domains in mediating cell delivery of a secreted CRE protein, Biochemistry 47 (2008) 1157–1166.
[45] A. Fittipaldi, A. Ferrari, M. Zoppe, C. Arcangeli, V. Pellegrini, F. Beltram, et al., Cell membrane lipid rafts mediate caveolar endocytosis of HIV-1 Tat fusion proteins, J. Biol. Chem. 278 (2003) 34141–34149.
[46] J.R. Maiolo, M. Ferrer, E.A. Ottinger, Effects of cargo molecules on the cellular uptake of arginine-rich cell-penetrating peptides, Biochim. Biophys. Acta 1712 (2005) 161–172.
[47] S. El-Andaloussi, P. Järver, H.J. Johansson, Ü Langel, Cargo-dependent cytotoxicity and delivery efficacy of cell-penetrating peptides: A comparative study, Biochem. J. 407 (2007) 285–292.
[48] G. TÜnnemann, R.M. Martin, S. Haupt, C. Patsch, F. Edenhofer, M.C. Cardoso, Cargo-dependent mode of uptake and bioavailability of TAT-containing proteins and peptides in living cells, FASEB J. 20 (2006) 1775–1784.
[49] H. Räägel, P. Säälik, M. Pooga, Peptide-mediated protein delivery – Which pathways are penetrable? Biochim. Biophys. Acta 1798 (2010) 2240–2248.
[50] D. Derossi, S. Calvet, A. Trembleau, A. Brunissen, G. Chassaing, A. Prochiantz, Cell internalization of the third helix of the Antennapedia homeodomain is receptor-independent, J. Biol. Chem. 271 (1996) 18188–18193.
[51] M. Pooga, C. Kut, M. Kihlmark, M. Hällbrink, S. Fernaeus, R. Raid, et al., Cellular translocation of proteins by transportan, FASEB J. 15 (2001) 1451–1453.
[52] K. Kilk, M. Magzoub, M. Pooga, L.E. Eriksson, Ü Langel, A. Gräslund, Cellular internalization of a cargo complex with a novel peptide derived from the third helix of the islet-1 homeodomain. Comparison with the penetratin peptide, Bioconjug. Chem. 12 (2001) 911–916.
[53] M. Lundberg, M. Johansson, Positively charged DNA-binding proteins cause apparent cell membrane translocation, Biochem. Biophys. Res. Commun. 291 (2002) 367–371.
[54] J.P. Richard, K. Melikov, E. Vives, C. Ramos, B. Verbeure, M.J. Gait, et al., Cell-penetrating peptides. A reevaluation of the mechanism of cellular uptake, J. Biol. Chem. 278 (2003) 585–590.

[55] S. Deshayes, S. Gerbal-Chaloin, M.C. Morris, G. Aldrian-Herrada, P. Charnet, G. Divita, et al., On the mechanism of non-endosomal peptide-mediated cellular delivery of nucleic acids, Biochim. Biophys. Acta 1667 (2004) 141–147.
[56] A. Ziegler, Thermodynamic studies and binding mechanisms of cell-penetrating peptides with lipids and glycosaminoglycans, Adv. Drug Deliv. Rev. 60 (2008) 580–597.
[57] H.D. Herce, A.E. Garcia, J. Litt, R.S. Kane, P. Martin, N. Enrique, et al., Arginine-rich peptides destabilize the plasma membrane, consistent with a pore formation translocation mechanism of cell-penetrating peptides, Biophys. J. 97 (2009) 1917–1925.
[58] K. Cahill, Molecular electroporation and the transduction of oligoarginines, Phys. Biol. 7 (2009) 16001.
[59] S. Console, C. Marty, C. Garcia-Echeverria, R. Schwendener, K. Ballmer-Hofer, Antennapedia and HIV transactivator of transcription (TAT) protein transduction domains promote endocytosis of high molecular weight cargo upon binding to cell surface glycosaminoglycans, J. Biol. Chem. 278 (2003) 35109–35114.
[60] C. Marty, C. Meylan, H. Schott, K. Ballmer-Hofer, R.A. Schwendener, Enhanced heparan sulfate proteoglycan-mediated uptake of cell-penetrating peptide-modified liposomes, Cell Mol. Life Sci. 61 (2004) 1785–1794.
[61] E. Goncalves, E. Kitas, J. Seelig, Binding of oligoarginine to membrane lipids and heparan sulfate: Structural and thermodynamic characterization of a cell-penetrating peptide, Biochemistry 44 (2005) 2692–2702.
[62] G.M. Poon, J. Gariepy, Cell-surface proteoglycans as molecular portals for cationic peptide and polymer entry into cells, Biochem. Soc. Trans. 35 (2007) 788–793.
[63] P. Säälik, K. Padari, A. Niinep, A. Lorents, M. Hansen, E. Jokitalo, et al., Protein delivery with transportans is mediated by caveolae rather than flotillin-dependent pathways, Bioconjug. Chem. 20 (2009) 877–887.
[64] F. Galbiati, B. Razani, M.P. Lisanti, Emerging themes in lipid rafts and caveolae, Cell 106 (2001) 403–411.
[65] H. Lodish, A. Berk, S.L. Zipursky, P. Matsudaira, D. Baltimore, J. Darnell, Molecular Cell Biology, W.H. Freeman and Company, San Francisco, USA, 2000.
[66] I. Nakase, A. Tadokoro, N. Kawabata, T. Takeuchi, H. Katoh, K. Hiramoto, et al., Interaction of arginine-rich peptides with membrane-associated proteoglycans is crucial for induction of actin organization and macropinocytosis, Biochemistry 46 (2007) 492–501.
[67] S. Futaki, I. Nakase, A. Tadokoro, T. Takeuchi, A.T. Jones, Arginine-rich peptides and their internalization mechanisms, Biochem. Soc. Trans. 35 (2007) 784–787.
[68] T. Kirchhausen, Imaging endocytic clathrin structures in living cells, Trends Cell Biol. 19 (2009) 596–605.
[69] M. Fotin-Mleczek, S. Welte, O. Mader, F. Duchardt, R. Fischer, H. Hufnagel, et al., Cationic cell-penetrating peptides interfere with TNF signalling by induction of TNF receptor internalization, J. Cell Sci. 118 (2005) 3339–3351.
[70] J.E. Heuser, R.G. Anderson, Hypertonic media inhibit receptor-mediated endocytosis by blocking clathrin-coated pit formation, J. Cell Biol. 108 (1989) 389–400.
[71] J. Rinne, B. Albarran, J. Jylhävä, T.O. Ihalainen, P. Kankaanpää, V.P. Hytönen, et al., Internalization of novel non-viral vector TAT-streptavidin into human cells, BMC Biotechnol. 7 (2007) 1.
[72] I.R. Nabi, P.U. Le, Caveolae/raft-dependent endocytosis, J. Cell Biol. 161 (2003) 673–677.
[73] D.J. Barr, A.G. Ostermeyer-Fay, R.A. Matundan, D.A. Brown, Clathrin-independent endocytosis of ErbB2 in geldanamycin-treated human breast cancer cells, J. Cell Sci. 121 (2008) 3155–3166.

[74] E.M. Damm, L. Pelkmans, J. Kartenbeck, A. Mezzacasa, T. Kurzchalia, A. Helenius, Clathrin- and caveolin-1-independent endocytosis: Entry of simian virus 40 into cells devoid of caveolae, J. Cell Biol. 168 (2005) 477–488.
[75] A. Ferrari, V. Pellegrini, C. Arcangeli, A. Fittipaldi, M. Giacca, F. Beltram, Caveolae-mediated internalization of extracellular HIV-1 tat fusion proteins visualized in real time, Mol. Ther. 8 (2003) 284–294.
[76] G.J. Doherty, H.T. McMahon, Mechanisms of endocytosis, Annu. Rev. Biochem. 78 (2009) 857–902.
[77] J. Chen, G. Li, J. Lu, L. Chen, Y. Huang, H. Wu, et al., A novel type of PTD, common helix-loop-helix motif, could efficiently mediate protein transduction into mammalian cells, Biochem. Biophys. Res. Commun. 347 (2006) 931–940.
[78] S. Kameyama, M. Horie, T. Kikuchi, T. Omura, A. Tadokoro, T. Takeuchi, et al., Acid wash in determining cellular uptake of Fab/cell-permeating peptide conjugates, Biopolymers 88 (2007) 98–107.
[79] Y.W. Hou, M.H. Chan, H.R. Hsu, B.R. Liu, C.P. Chen, H.H. Chen, et al., Transdermal delivery of proteins mediated by non-covalently associated arginine-rich intracellular delivery peptides, Exp. Dermatol. 16 (2007) 999–1006.
[80] S. Gerbal-Chaloin, C. Gondeau, G. Aldrian-Herrada, F. Heitz, C. Gauthier-Rouviere, G. Divita, First step of the cell-penetrating peptide mechanism involves Rac1 GTPase-dependent actin-network remodelling, Biol. Cell 99 (2007) 223–238.
[81] R. Lundmark, G.J. Doherty, M.T. Howes, K. Cortese, Y. Vallis, R.G. Parton, et al., The GTPase-activating protein GRAF1 regulates the CLIC/GEEC endocytic pathway, Curr. Biol. 18 (2008) 1802–1808.
[82] G.J. Doherty, R. Lundmark, GRAF1-dependent endocytosis, Biochem. Soc. Trans. 37 (2009) 1061–1065.
[83] S. Sabharanjak, P. Sharma, R.G. Parton, S. Mayor, GPI-anchored proteins are delivered to recycling endosomes via a distinct cdc42-regulated, clathrin-independent pinocytic pathway, Dev. Cell 2 (2002) 411–423.
[84] P. Bhagatji, R. Leventis, J. Comeau, M. Refaei, J.R. Silvius, Steric and not structure-specific factors dictate the endocytic mechanism of glycosylphosphatidylinositol-anchored proteins, J. Cell Biol. 186 (2009) 615–628.
[85] B. Nichols, Endocytosis of lipid-anchored proteins: Excluding GEECs from the crowd, J. Cell Biol. 186 (2009) 457–459.
[86] M. Kirkham, A. Fujita, R. Chadda, S.J. Nixon, T.V. Kurzchalia, D.K. Sharma, et al., Ultrastructural identification of uncoated caveolin-independent early endocytic vehicles, J. Cell Biol. 168 (2005) 465–476.
[87] C.A. Stuermer, D.M. Lang, F. Kirsch, M. Wiechers, S.O. Deininger, H. Plattner, Glycosylphosphatidyl inositol-anchored proteins and fyn kinase assemble in noncaveolar plasma membrane microdomains defined by reggie-1 and -2, Mol. Biol. Cell 12 (2001) 3031–3045.
[88] O.O. Glebov, N.A. Bright, B.J. Nichols, Flotillin-1 defines a clathrin-independent endocytic pathway in mammalian cells, Nat. Cell. Biol. 8 (2006) 46–54.
[89] K. Melikov, L.V. Chernomordik, Arginine-rich cell penetrating peptides: From endosomal uptake to nuclear delivery, Cell Mol. Life Sci. 62 (2005) 2739–2749.
[90] C. Lamaze, S.L. Schmid, The emergence of clathrin-independent pinocytic pathways, Curr. Opin. Cell Biol. 7 (1995) 573–580.
[91] T.D. Pollard, W.C. Earnshaw, J. Lippincott-Schwartz, Cell Biology, Saunders Elsevier, USA, 2008.

[92] R. Misaki, T. Nakagawa, M. Fukuda, N. Taniguchi, T. Taguchi, Spatial segregation of degradation- and recycling-trafficking pathways in COS-1 cells, Biochem. Biophys. Res. Commun. 360 (2007) 580–585.

[93] S. Al-Taei, N.A. Penning, J.C. Simpson, S. Futaki, T. Takeuchi, I. Nakase, et al., Intracellular traffic and fate of protein transduction domains HIV-1 TAT peptide and octaarginine. Implications for their utilization as drug delivery vectors, Bioconjug. Chem. 17 (2006) 90–100.

[94] S.D. Laufer, T. Restle, Peptide-mediated cellular delivery of oligonucleotide-based therapeutics *in vitro*: Quantitative evaluation of overall efficacy employing easy to handle reporter systems, Curr. Pharm. Des. 14 (2008) 3637–3655.

[95] T.B. Potocky, A.K. Menon, S.H. Gellman, Cytoplasmic and nuclear delivery of a TAT-derived peptide and a beta-peptide after endocytic uptake into HeLa cells, J. Biol. Chem. 278 (2003) 50188–50194.

[96] N.J. Caron, S.P. Quenneville, J.P. Tremblay, Endosome disruption enhances the functional nuclear delivery of Tat-fusion proteins, Biochem. Biophys. Res. Commun. 319 (2004) 12–20.

[97] B. Groner, Peptides as Drugs: Discovery and Development, Wiley-VCH Verlag GmbH & Co, 2009.

[98] L. Pelkmans, J. Kartenbeck, A. Helenius, Caveolar endocytosis of simian virus 40 reveals a new two-step vesicular-transport pathway to the ER, Nat. Cell Biol. 3 (2001) 473–483.

[99] K. Padari, K. Koppel, A. Lorents, M. Hällbrink, M. Mano, M.C. Pedroso de Lima, et al., S4 (13)-PV cell-penetrating peptide forms nanoparticle-like structures to gain entry into cells, Bioconjug. Chem. 21 (2010) 774–783.

[100] S.M. Fuchs, R.T. Raines, Pathway for polyarginine entry into mammalian cells, Biochemistry 43 (2004) 2438–2444.

[101] H. Michiue, K. Tomizawa, F.Y. Wei, M. Matsushita, Y.F. Lu, T. Ichikawa, et al., The NH2 terminus of influenza virus hemagglutinin-2 subunit peptides enhances the antitumor potency of polyarginine-mediated p53 protein transduction, J. Biol. Chem. 280 (2005) 8285–8289.

[102] Y.J. Lee, A. Erazo-Oliveras, J.P. Pellois, Delivery of macromolecules into live cells by simple co-incubation with a peptide, Chem. Bio. Chem. 11 (2010) 325–330.

[103] A. El-Sayed, S. Futaki, H. Harashima, Delivery of macromolecules using arginine-rich cell-penetrating peptides: Ways to overcome endosomal entrapment, AAPS J. 11 (2009) 13–22.

[104] S.J. Kwon, K. Han, S. Jung, J.E. Lee, S. Park, Y.P. Cheon, et al., Transduction of the MPG-tagged fusion protein into mammalian cells and oocytes depends on amiloride-sensitive endocytic pathway, BMC Biotechnol. 9 (2009) 73.

[105] R. Fischer, K. Köhler, M. Fotin–Mleczek, R. Brock, A stepwise dissection of the intracellular fate of cationic cell-penetrating peptides, J. Biol. Chem. 279 (2004) 12625–12635.

[106] S.A. Ezhevsky, A. Ho, M. Becker-Hapak, P.K. Davis, S.F. Dowdy, Differential regulation of retinoblastoma tumor suppressor protein by G(1) cyclin-dependent kinase complexes *in vivo*, Mol. Cell Biol. 21 (2001) 4773–4784.

[107] G. Selivanova, V. Iotsova, I. Okan, M. Fritsche, M. Ström, B. Groner, et al., Restoration of the growth suppression function of mutant p53 by a synthetic peptide derived from the p53 C-terminal domain, Nat. Med. 3 (1997) 632–638.

[108] M. Bonfanti, S. Taverna, M. Salmona, M. D'Incalci, M. Broggini, p21WAF1-derived peptides linked to an internalization peptide inhibit human cancer cell growth, Cancer Res. 57 (1997) 1442–1446.

[109] E.P. Holinger, T. Chittenden, R.J. Lutz, Bak BH3 peptides antagonize Bcl-xL function and induce apoptosis through cytochrome c-independent activation of caspases, J. Biol. Chem. 274 (1999) 13298–13304.
[110] S.R. Schwarze, A. Ho, A. Vocero-Akbani, S.F. Dowdy, *In vivo* protein transduction: Delivery of a biologically active protein into the mouse, Science 285 (1999) 1569–1572.
[111] S.R. Cai, G. Xu, M. Becker-Hapak, M. Ma, S.F. Dowdy, H.L. McLeod, The kinetics and tissue distribution of protein transduction in mice, Eur. J. Pharm. Sci. 27 (2006) 311–319.
[112] H.J. Lee, W.M. Pardridge, Pharmacokinetics and delivery of tat and tat-protein conjugates to tissues *in vivo*, Bioconjug. Chem. 12 (2001) 995–999.
[113] Y. Lei, H. Tang, L. Yao, R. Yu, M. Feng, B. Zou, Applications of mesenchymal stem cells labeled with Tat peptide conjugated quantum dots to cell tracking in mouse body, Bioconjug. Chem. 19 (2008) 421–427.
[114] K. Inagaki, L. Chen, F. Ikeno, F.H. Lee, K.I. Imahashi, D.M. Bouley, et al., Inhibition of delta-protein kinase c protects against reperfusion injury of the ischemic heart *in vivo*, Circulation 108 (2003) 2304–2307.
[115] B. Cornelissen, K. McLarty, V. Kersemans, D.A. Scollard, R.M. Reilly, Properties of [(111) In]-labeled HIV-1 tat peptide radioimmunoconjugates in tumor-bearing mice following intravenous or intratumoral injection, Nucl. Med. Biol. 35 (2008) 101–110.
[116] C.T. McCusker, Y. Wang, J. Shan, M.W. Kinyanjui, A. Villeneuve, H. Michael, et al., Inhibition of experimental allergic airways disease by local application of a cell-penetrating dominant-negative STAT-6 peptide, J. Immunol. 179 (2007) 2556–2564.
[117] J.M. Choi, M.H. Ahn, W.J. Chae, Y.G. Jung, J.C. Park, H.M. Song, et al., Intranasal delivery of the cytoplasmic domain of CTLA-4 using a novel protein transduction domain prevents allergic inflammation, Nat. Med. 12 (2006) 574–579.
[118] L.H. Jin, J.H. Bahn, W.S. Eum, H.Y. Kwon, S.H. Jang, K.H. Han, et al., Transduction of human catalase mediated by an HIV-1 TAT protein basic domain and arginine-rich peptides into mammalian cells, Free Radic. Biol. Med. 31 (2001) 1509–1519.
[119] K.N. Sugahara, T. Teesalu, P.P. Karmali, V.R. Kotamraju, L. Agemy, O.M. Girard, et al., Tissue-penetrating delivery of compounds and nanoparticles into tumors, Cancer Cell 16 (2009) 510–520.
[120] K.N. Sugahara, T. Teesalu, P.P. Karmali, V.R. Kotamraju, L. Agemy, D.R. Greenwald, et al., Coadministration of a tumor-penetrating peptide enhances the efficacy of cancer drugs, Science 328 (2010) 1031–1035.
[121] T. Jiang, E.S. Olson, Q.T. Nguyen, M. Roy, P.A. Jennings, R.Y. Tsien, Tumor imaging by means of proteolytic activation of cell-penetrating peptides, Proc. Natl. Acad. Sci. U.S.A. 101 (2004) 17867–17872.
[122] E.S. Olson, T.A. Aguilera, T. Jiang, L.G. Ellies, Q.T. Nguyen, E.H. Wong, et al., *In vivo* characterization of activatable cell penetrating peptides for targeting protease activity in cancer, Integr. Biol. 1 (2009) 382–393.
[123] J.R. MacDougall, L.M. Matrisian, Contributions of tumor and stromal matrix metalloproteinases to tumor progression, invasion and metastasis, Cancer Metastasis Rev. 14 (1995) 351–362.
[124] E.S. Olson, T. Jiang, T.A. Aguilera, Q.T. Nguyen, L.G. Ellies, M. Scadeng, et al., Activatable cell penetrating peptides linked to nanoparticles as dual probes for *in vivo* fluorescence and MR imaging of proteases, Proc. Natl. Acad. Sci. U.S.A. 107 (2010) 4311–4316.
[125] Q.T. Nguyen, E.S. Olson, T.A. Aguilera, T. Jiang, M. Scadeng, L.G. Ellies, et al., Surgery with molecular fluorescence imaging using activatable cell-penetrating peptides decreases

residual cancer and improves survival, Proc. Natl. Acad. Sci. U.S.A. 107 (2010) 4317–4322.

[126] M. Aarts, Y. Liu, L. Liu, S. Besshoh, M. Arundine, J.W. Gurd, et al., Treatment of ischemic brain damage by perturbing NMDA receptor-PSD-95 protein interactions, Science 298 (2002) 846–850.

[127] A. Elmquist, Ü Langel, *In vitro* uptake and stability study of pVEC and its all-D analog, Biol. Chem. 384 (2003) 387–393.

[128] D.S. Youngblood, S.A. Hatlevig, J.N. Hassinger, P.L. Iversen, H.M. Moulton, Stability of cell-penetrating peptide-morpholino oligomer conjugates in human serum and in cells, Bioconjug. Chem. 18 (2007) 50–60.

[129] C. Bonny, A. Oberson, S. Negri, C. Sauser, D.F. Schorderet, Cell-permeable peptide inhibitors of JNK: Novel blockers of beta-cell death, Diabetes 50 (2001) 77–82.

[130] G. TÜnnemann, G. Ter-Avetisyan, R.M. Martin, M. Stockl, A. Herrmann, M.C. Cardoso, Live-cell analysis of cell penetration ability and toxicity of oligo-arginines, J. Pept. Sci. 14 (2008) 469–476.

[131] H. Sakurai, K. Kawabata, F. Sakurai, S. Nakagawa, H. Mizuguchi, Innate immune response induced by gene delivery vectors, Int. J. Pharm. 354 (2008) 9–15.

[132] E.L. Snyder, S.F. Dowdy, Cell penetrating peptides in drug delivery, Pharm. Res. 21 (2004) 389–393.

[133] P.G. Frank, S. Pavlides, M.P. Lisanti, Caveolae and transcytosis in endothelial cells: role in atherosclerosis, Cell Tissue Res. 335 (2009) 41–47.

[134] M. Pooga, U. Soomets, M. Hällbrink, A. Valkna, K. Saar, K. Rezaei, et al., Cell penetrating PNA constructs regulate galanin receptor levels and modify pain transmission *in vivo*, Nat. Biotechnol. 16 (1998) 857–861.

[135] F. Nagel, B.H. Falkenburger, L. Tonges, S. Kowsky, C. Poppelmeyer, J.B. Schulz, et al., Tat-Hsp70 protects dopaminergic neurons in midbrain cultures and in the substantia nigra in models of Parkinson's disease, J. Neurochem. 105 (2008) 853–864.

[136] T.R. Doeppner, F. Nagel, G.P. Dietz, J. Weise, L. Tonges, S. Schwarting, et al., TAT-Hsp70-mediated neuroprotection and increased survival of neuronal precursor cells after focal cerebral ischemia in mice, J. Cereb. Blood Flow Metab. 29 (2009) 1187–1196.

[137] T.R. Doeppner, G.P. Dietz, M. El Aanbouri, J. Gerber, O.W. Witte, M. Bähr, et al., TAT-Bcl-x(L) improves survival of neuronal precursor cells in the lesioned striatum after focal cerebral ischemia, Neurobiol. Dis. 34 (2009) 87–94.

Chapter 11

Peptide and Protein Bioconjugation: A Useful Tool to Improve the Biological Performance of Biotech Drugs

Stefano Salmaso [1] and Paolo Caliceti [2]

[1] *Department of Pharmaceutical Sciences, University of Padua, Via F. Marzolo 5, 35031, Padova – Italy,* [2] *Department of Pharmaceutical Sciences, University of Padua, Via F. Marzolo 5, 35031, Padova – Italy*

Chapter Outline

Peptide and Protein Delivery. DOI: 10.1016/B978-0-12-384935-9.10011-2

11.1. INTRODUCTION: WHY BIOCONJUGATION OF PEPTIDES AND PROTEINS?

The covalent conjugation of poly(ethylene glycol) (PEG) chains to proteins, known as PEGylation, has been successfully developed as a means of varying the *in vivo* properties of protein drugs. This technique offers less frequent administration to patients, greater convenience and improved efficacy [1].

PEG is a hydrophilic, biocompatible polymer approved by the main regulatory agencies, namely EMEA and FDA, for parenteral administration. The PEGylation of a protein can be achieved by random or selective conjugation protocols that yield the covalent attachment of one or more polymer chains to specific anchoring functions on the protein surface. Polymer conjugation induces size enlargement, charge and surface modifications and protein shielding, which ultimately result in improved solubility, stability [2,3], enhanced immunological profile [4], prolonged permanence in the body and altered tissue localization and cellular uptake. The beneficial effects of PEGylation usually compensate for the reduced biological activities of proteins modified by polymer conjugation.

11.2. *IN VIVO* BEHAVIOR OF PEGYLATED PROTEINS

11.2.1. Pharmacokinetic Profile of PEGylated Proteins

PEGylation has a strong effect on the pharmacokinetic profiles of proteins because it prolongs their presence in the bloodstream, increases their bioavailability and modifies their biodistribution profiles. Renal clearance represents one of the main routes of protein elimination from the bloodstream [5,6]. The glomerular capillary walls in the kidneys are organized into highly structured architectures with specialized barrier properties that control the ultrafiltration of hydrophilic macromolecules, namely proteins and polymers, and their reabsorption at the level of the proximal tubule. Both ultrafiltration and reabsorption depend on the composition, sizes and charges of the circulating molecules [7,8].

Proximal elements that are rich in sialoprotein and are located in the epithelium and the lamina rara interna of the glomerular basement membrane delay the filtration of circulating polyanions by repulsion. By contrast, distal components, namely the lamina rara externa of the glomerular basement membrane and the slit diaphragm, delay the clearance of cationic macromolecules. The passage of circulating polyanions is also retarded by the nonglomerular epithelium, which is negatively charged [8–10]. PEGylation protocols, which convert lysine and the terminal amino groups of amino acids into amidic bonds, result in a loss of the overall positive protein charge [1] that prolongs the body's exposure to proteins [11].

Proteins with molecular weights that are less than 65 kDa (the serum albumin size, which represents the ultrafiltration cut-off for globular proteins) are usually rapidly cleared through the kidney. Larger proteins, however, are cleared after being enzymatically digested into lower molecular weight fragments. The PEGylation of therapeutic proteins that have low molecular weights, or that have been downsized through engineering to enhance their biofunctional properties, can produce derivatives with high hydrodynamic sizes that overcome the glomerular ultrafiltration cut-off [12,13]. Due to its extended conformation, and the coordination of two to three water molecules per oxyethylene unit, PEG possesses a high hydrodynamic volume that is non-linearly correlated with molecular weight [14,15]. As a result, the ultrafiltration of PEGs with molecular weights below 8 kDa is not size-restricted. However, for PEGs between 8 and 30 kDa, the filtration rate depends on the molecular size. The 30 kDa PEG has a hydrodynamic size of 8 nm, which exceeds the 5 nm pores of the kidney glomerular membrane [16,17]. Nevertheless, by virtue of structural flexibility and extended conformation, PEGs with molecular weights that are significantly higher than 30 kDa are also cleared by kidney ultrafiltration. This clearing process occurs at a very slow rate through a pore diffusion process defined as "reptation" [12,18].

As a rule of thumb, the hydrodynamic volume and clearance of PEGylated proteins are dictated by the PEG molecular weight and structure, the number of attached polymer chains and the overall mass of the polymer forming the bioconjugate [19]. An 18 kDa PEG, having a hydrodynamic volume higher than that of albumin, is considered to have the minimum mass required to significantly reduce kidney ultrafiltration regardless of the molecular weight of the attached peptide [15]. Thus, a variety of proteins have been derivatized either with one PEG chain with a molecular weight above 18 kDa or with several PEG chains of lower molecular weight. The alpha- and beta-phase half-lives of interleukin-2 (IL-2) were found to increase as the number of attached 4 kDa PEG molecules increased, and the clearance of the derivative bearing three polymer chains and that obtained with the bioconjugate modified with one 10 kDa PEG chain overlapped [20]. This study confirmed that for diameters above 36 Å, which corresponds to the size of albumin, the kidney clearance is very slow regardless of the degree of PEGylation.

The modification of the single-chain Fv antibody fragment against tumor necrosis factor-α (anti-TNF-α scFv) by conjugation of single 5, 20 and 40 kDa PEG chains to cysteine yielded anti-TNF-α scFv derivatives with circulating half-lives of 2.57, 4.38 and 21.6 hours, respectively, while the non-PEGylated anti-TNF-α scFv had a half-life ($t^1/_2$) in blood of 0.15 hours [21]. The mono-PEGylation of the recombinant granulocyte colony stimulating factor (rh-G-CSF) with 5, 10, 20 and 40 kDa PEG increased the circulation half-life from about 10 minutes for the native protein to 25, 50, 98 and 180 minutes, respectively [22].

Comparative studies demonstrated that prolonged bloodstream presence after intravenous administration could be obtained using large and branched polymers [23]. The conjugation of interferon α-2a (INF α-2a) with a trimer-structured 43 kDa PEG prolonged the terminal half-life after intravenous injection in rats from 1.2 to about 37 hours [24]. INF α-2a modified with branched 40 kDa PEG (PEGASYS®) and interferon α-2b (INF α-2b) modified with linear 12 kDa PEG (PEGINTRON®) displayed significantly different pharmacokinetic profiles after subcutaneous administration [25].

Bovine lactoferrin mono-PEGylated with 40 kDa branched PEG led to an 8.7-fold increase in the half-life over that of the unmodified protein [26]. The serum half-life of the F(ab′)2-humanized segment of the anti-interleukin-8 antibody increased from 8.5 to 48 hours upon PEGylation [27].

Although the pharmacokinetic profile of bioconjugates is mainly dictated by the overall polymer mass of the derivative, studies performed with single-chain Fv showed that a few long PEG chains are usually more effective than many short PEG molecules in delaying the elimination half-life of proteins [28]. These results support the concept that PEGylation affects several mechanisms involved in protein clearance. In particular, it was hypothesized that the anchoring point for the polymer on the protein surface may be relevant both for enzymatic degradation and for immuno-recognition. This would thus affect the pharmacokinetic behavior more than the ultrafiltration process itself [29]. A study performed with growth hormones confirmed that the effect of the number of polymer chains on kidney filtration could not completely account for the observed slower clearance, suggesting that the polymer simultaneously alters various mechanisms involved in protein clearance [30].

PEGylation also affects the residence time in circulation of proteins that have a size that exceeds the kidney ultrafiltration limit. This effect was observed on a F(ab′)2 conjugate with one branched 40 kDa PEG chain. This derivative displayed a drug concentration/time area under the curve (AUC) in plasma that was 15.7-fold higher than that of the native protein. The effect was amplified when two branched PEG chains were conjugated to the antibody (the AUC value was 17.8-fold higher) [27].

After subcutaneous and intramuscular administration, PEGylated proteins showed increased AUC values and delayed concentration peak time (T_{max}) values compared to those of the native molecules, indicating an enhanced bioavailability. This enhancement is due to an increased resistance to inactivation in the injection site, delayed diffusive mechanisms through the biological membranes and a prolonged permanence in circulation.

For example, the AUC values in the blood after intramuscular and subcutaneous administration for random 5 kDa PEG conjugated superoxide dismutase (PEG-SOD) were 54 and 29%, respectively, as compared to the value of the intravenously administered PEGylated protein (Figure 11.1). However, the AUC value of the PEGylated enzyme was between 50- and 83-fold higher than that of the native enzyme, depending on the route of administration [31].

Pharmacokinetic studies of the subcutaneous injection of native or PEGylated INF α-2a in humans showed that the conjugation of one chain of the branched 40 kDa polymer shifted the T_{max} value from 12 to 80 hours, and that

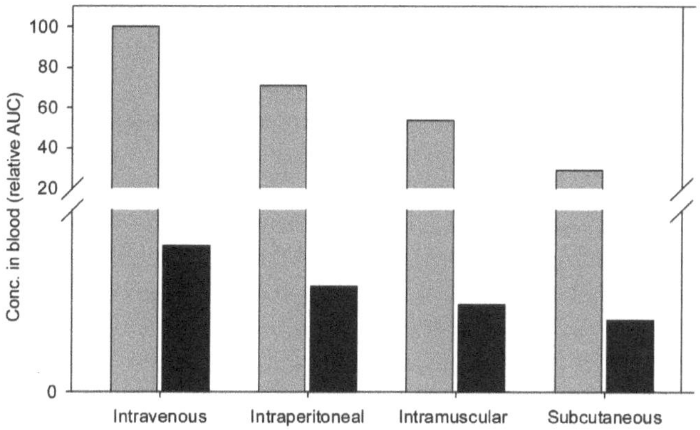

FIGURE 11.1 PEG-SOD (gray column) and SOD (black column) AUC in blood after administration via various routes.

the protein serum half-life ($t^1/_2$) and the plasma mean residence time (MRT) were 70- and 50-fold higher, respectively. Phase II/III hepatitis C clinical trials confirmed the improved pharmacological profile of the bioconjugate protein and its superior efficacy, as compared to those of the native protein [32].

Similarly to interferon-PEG, the PEGylation of rh-G-CSF was found to enhance pharmacokinetic and pharmacological performance. The values for T_{max} and the half-life of PEG-rh-G-CSF administered subcutaneously in humans were around 20–30 hours at all doses. These values are approximately three times higher than those observed for rh-G-CSF, the unPEGylated cytokine. The absolute neutrophil count (ANC) for PEG-rh-G-CSF was 6 to 8 times higher than that of the native protein, and the CD34+ cell count was more than 30 times higher at the highest tested doses. A weekly PEG-rh-G-CSF single subcutaneous dose was as efficient as daily rh-G-CSF doses in terms of ANC and CD34+ increase [33].

11.2.2. Biodistribution of PEGylated Proteins

After intravenous administration, PEGylated proteins usually display bicompartmental pharmacokinetic profiles, indicating that these bioconjugates can distribute in peripheral organs and tissues. The passage of a PEGylated protein from the blood to the peripheral compartment is dictated by features of the biological barriers and by the molecular size and structure of the conjugate [11].

Low molecular weight PEGs have shorter half-lives and larger distribution volumes (V_d) after intravenous administration than those of polymers with high molecular weights. It is noteworthy that small PEGs do not accumulate in the peripheral tissues, while large PEGs reside longer in the peripheral district.

The liver, the organ where the diffusion of molecules from the blood has limited restrictions, is involved in a size-dependent PEG-clearance process.

PEG molecules that are larger than 50 kDa can accumulate in the Küpffer cells, where they are stored indefinitely, while PEG chains with molecular weights below 50 kDa are mostly taken up by parenchymal cells (hepatocytes), by pinocytosis, at a rate that is inversely proportional to the polymer mass. Similar cellular uptakes occur in other organs that are rich in reticuloendothelial cells and that are characterized by active phagocytosis. Examples of such organs include the spleen, lymph nodes, lungs and kidneys [34]. PEG disposed in the liver can enter the enterohepatic cycle and be excreted into the bile.

Similarly to PEG, PEG–protein conjugates undergo unspecific internalization by Küpffer cells, hepatocytes and reticuloendothelial cells. However, specific receptor- or charge-mediated cellular uptakes are usually unpaired with the polymer on the protein surface. The limited cell internalization of PEGylated catalase in the liver (28% of the dose as compared to 70% for the native protein) was ascribed to the masking of both the protein glycosilic functions involved in the galactose-receptor-mediated uptake and to negative charges [35]. PEGylation is therefore regarded as a strategy to endow colloidal systems with stealth properties and reticuloendothelial system (RES) escape. Furthermore, studies performed with haemoglobin showed that PEGylation prevents the pairing of the protein with haptoglobin and the consequent RES uptake. These results indicate that the PEG molecules on the protein surface reduce the protein binding with the opsonins that promote phagocytosis [36].

PEGylated proteins undergo lymph node localization after intravenous administration, which decreases as the size of the bioconjugate increases. This property might be exploited for site-selective anti-cancer treatment, although PEGylation slows the lymph node disposition of proteins [37]. When PEG-interleukin-2 was intravenously injected into pigs, a 130-fold lower clearance from the plasma to the lymph nodes than that of the native protein was observed. The AUC_{lymph}/AUC_{plasma} ratio of native interleukin-2 was twice as high as that of the PEGylated derivative [38]. This lymph node uptake was attributed to the higher hydrophobicity of the native cytokine.

PEGylated proteins undergo passive distribution into solid tumors profiled by anomalous angiogenesis, leaky vasculature and low lymph drainage according to the mechanism known as "enhanced permeability and retention" (EPR), which is responsible for the preferential extravazation of macromolecules into the tumor tissue [39]. Accordingly, PEGylation has been exploited to achieve the passive tumor accumulation of PEGylated proteins for anti-cancer therapy. The intravenous administration of PEGylated xanthine oxidase in mice resulted in high tumor/blood ratios. The administration of xanthine promoted the decrease of the tumor mass as a consequence of the intratumor production of toxic oxygen free radicals generated from xanthine oxidation [40].

β-Glucuronidase, modified with three 5 kDa PEG chains and linked to a monoclonal antibody, showed an enhanced pharmacokinetic profile in nude mice as compared to the unPEGylated β-glucuronidase-antibody bioconjugate. The antibody-β-glucuronidase-PEG that was intravenously injected into nude mice bearing rat solid AS-30D hepatoma tumors accumulated efficiently in the tumor and led to a limited accumulation in the spleen, compared to the non-PEGylated construct. The antibody-β-glucuronidase-PEG resulted in complete regression of small solid tumors when mice were post-injected with a *p*-hydroxyaniline mustard glucuronide prodrug [41].

PEGylation was applied to anti-cancer cell antibodies to associate the EPR passive accumulation endowed by the polymer with the active targeting of xenographic tumors. Humanized A33 antibodies modified with PEG were found to accumulate in peripheral xenographic tumors with a delay that was several hours longer than that of the unmodified antibodies. The conjugation of several, high molecular weight (20 kDa) PEG chains yielded higher tumor disposition but lower biorecognition than the derivative obtained with several low molecular weight (5 kDa) PEG chains [42]. PEGylated Fab against the anti-carcinoembriogenic antigen accumulated efficiently in the tumor for a longer period of time and with a higher tumor/normal tissue ratio than the unmodified protein [43]. However, some concerns have been raised about the effective tumor cell targeting of PEGylated antibodies and their unspecific disposition in the interstitial space surrounding the tumor vasculature. In fact, a few studies report that deep penetration of the macromolecules into the tumor mass might be prevented by their large size and an elevated intratumor pressure [44].

11.2.3. Non-parenteral Routes of PEGylated Protein and Peptide Administration: Pulmonary, Nasal and Oral Delivery

The permeation of PEGylated proteins through biomembranes is limited by their high hydrophilicity and size, which interfere with the diffusion process, and also by enzymatic degradation. Nevertheless, several studies have demonstrated that polymer bioconjugation can significantly improve the peptide and protein bioavailability after trans-mucosal administration due to an increased stability towards proteolytic enzymes.

The lungs are considered to be a suitable route of administration for both the local and systemic delivery of many drugs, including proteins and peptides. Due to their large surface area, thin alveolar epithelium, elevated blood flow and lack of the first-hepatic-pass effect, pulmonary delivery can provide 10 to 200 times higher systemic peptide bioavailability than other non-invasive routes [45]. On the other hand, the presence of ubiquitous proteolytic enzymes and the phagocytic activity of macrophages, which characterize this particular physiologic barrier, limit the exploitation of this route of administration for biotech drugs.

α1 proteinase inhibitor (α1PI) is the only available therapeutic option for the treatment of hereditary emphysema caused by human leukocyte elastase (HLE) impairment [46]. The clinical use of this protein and its non-glycosylated recombinant form (rh-α1PI) is limited by a poor bioavailability after intravenous administration. To enhance the pharmacokinetic properties of rh-α1PI, the protein was site-specifically mono-PEGylated at the Cys^{232} with 5, 20 and 40 kDa maleimido-activated PEG. The deglycosylated rh-α1PI had shorter permanence in the bloodstream after intravenous injection as compared to the native glycosilated peptide, while the PEGylated inhibitor maintained the same activity of the native inhibitor. The blood half-life of the deglycosylated polypeptide was fully re-established by conjugation with 20 and 40 kDa PEG. After intranasal instillation to mice that had HLE-mediated lung damage, the 20 kDa PEG-conjugated rh-α1PI was found to persist longer in the bronchoalveolar fluid than rh-α1PI. Indeed, it prevented lung hemorrhage for up to 72 hours [47].

PEGylated calcitonin for pulmonary delivery was obtained by site-specific conjugation of 1, 2 and 5 kDa PEG to Lys^{18} according to a selective FMOC protection/deprotection method [48]. Calcitonin PEGylation with 1 and 2 kDa PEG reduced the protein biological activity by about 20%, while a 70% bioactivity reduction was obtained with 5 kDa PEG. These results indicate that the large, bulky PEG hampers target recognition. On the other hand, the protection of calcitonin from lung proteases was proportional to the PEG molecular weight; the conjugation of 5 kDa PEG prolonged the protein half-life survival by up to 1000 times [49]. Calcitonin is rapidly absorbed by the pulmonary route but circulates in the bloodstream for a very limited period of time ($t^{1}/_{2}$ = 34 minutes). The conjugation of 5 kDa PEG did not hamper the pulmonary absorption of calcitonin and prolonged the blood circulation half-life by up to 119 minutes. Pharmacodynamic studies performed by intrapulmonary instillation of native and PEGylated calcitonin in rats showed that all bioconjugates had a higher therapeutic efficacy than that of the native protein in terms of the total hypocalcaemic effect, with the 2 kDa PEG derivative giving the best result. The results of these studies indicate that the PEG molecular weight must be properly selected to balance the bioactivity decrease with the protection from proteases and the enhanced pharmacokinetic profile.

Glucagon-like peptide 1 (GLP-1) was also investigated for pulmonary delivery. GLP-1 has been proposed for type 2 diabetes and obesity treatment because it suppresses the appetite by stimulating the release of insulin after every meal and controls the haematic glucose levels. Nevertheless, the half-life of GLP-1 is only 2 minutes because it has a rapid glomerular extraction. GLP-1 was site-selectively PEGylated at Lys^{34} by a maleic anhydride blocking/deblocking method using 2, 5 and 10 kDa PEG. The PEGylated GLP-1 derivatives displayed an increased proteolytic stability in lung homogenates and enhanced the hypoglycaemic activity in type 2 diabetic mice that were intraperitoneally treated. Both increases were proportional to the increase in PEG molecular weight. GLP-1 was absorbed by the pulmonary route, but showed a blood half-life of only 8.9 minutes. PEGylation significantly improved its pulmonary absorption. The conjugation of 2, 5 and 10 kDa PEG, increased the protein half-life in the blood by 2.5, 4.6 and 9.1 times, respectively. A similar trend was observed for the AUC values. PEGylation also reduced the unwanted biodistribution of GLP-1 to the kidneys and the liver with evident therapeutic benefits [50]. In addition, PEGylation was found to reduce the proteolytic degradation of GLP-1 by the nasal mucosa homogenate as it increased the activity half-life by up to 11-fold. While the intranasally administered GLP-1 did not show any pharmacologic effect, the PEGylated derivatives induced hypoglycaemia that decreased with increasing PEG molecular weights [51].

Similarly to GLP-1, mono-PEGylated salmon calcitonin (sCT) showed increased resistance to nasal mucosal homogenate. After a two-hour incubation with the tissue homogenate, the PEGylated form maintained 90% of its starting activity, while less than 5% of the residual activity was obtained with native calcitonin [52]. The high resistance of the PEGylated protein was ascribed to the masking of the enzyme-sensitive protein sites by the polymer. The increased resistance to enzymatic degradation was reflected in the enhancement of the *in vivo* activity after intranasal instillation in rats. Calcitonin was absorbed by the nasal mucosa but gave transient hypocalcaemic effects, lasting only 30 minutes. However, PEGylated calcitonin was active for about 8 hours and promoted a calcium reduction that was up to 2.6 times higher than that of the native protein. In this case, the intranasal absorption of mono-PEG-sCTs was inversely related to the PEG molecular weight; the 2 kDa PEG derivative displayed the best results, while a 12 kDa PEG conjugation significantly reduced the protein pharmacological activity [53]. The systemic bioavailability of the intranasally administered $PEG_{2\text{-}kDa}$-calcitonin was approximately 91% of that of the intravenously injected sCT.

PEGylation was investigated as a means of enhancing the oral administration of proteins and peptides. Intestinal absorption of these biopharmaceuticals is strongly limited by low diffusion through the biomembranes, and also the harsh local conditions, i.e. the fluid composition and membrane-associated enzymatic activity that are involved in protein denaturation, degradation and final inactivation.

Lactoferrin is an antimicrobial, antiviral, immuno-modulatory, antioxidant, anti-inflammatory and analgesic agent that enhances the lipid metabolism and inhibits hepatitis-C viral cell entry. By virtue of its wide range of activity, it has been recently introduced into several anti-cancer clinical trials. Native lactoferrin can reach systemic circulation by intestinal absorption through the lymphatic system. However, its bioavailability is limited by its susceptibility to digestive juices. This phenomenon results in only a small amount of delivery to the intestinal tract. Random lactoferrin PEGylation with branched 20 kDa PEG halved the peptic digestion of the protein without altering its iron binding capacity. It also slightly reduced the *in vitro* biological activity. *In vivo*, after gastric administration to rats, a 10 times higher intestinal absorption of PEGylated lactoferrin was observed, compared to that of the native protein [54].

An oral insulin formulation was prepared by combining the protein PEGylation with mucoadhesive tablets. Insulin was mono-conjugated with a 0.75 kDa PEG to confer stability toward intestinal proteolytic enzymes and improve the absorption through the biomembrane [55]. It was then formulated into thiolated poly-acrylic acid matrices that were coated with poly-methacrylate to enhance the gastric stability [56]. The low molecular weight PEG on the insulin surface was sufficient to appreciably protect insulin from proteolysis when incubated with elastase and pepsin. The native insulin and PEG-insulin showed similar *in vivo* hypoglycaemic profiles after subcutaneous injection and permeation through an *ex vivo* intestine model. *In vivo* studies showed that the oral administration of PEG-insulin in solution had no pharmacologic activity. However, the formulated tablet displayed prolonged hypoglycaemic profiles [57].

11.2.4. Immunology

Before the advent of the biotechnology era, protein drugs were obtained by extraction from vegetal and animal cells, organs and tissues. Only a few human proteins were available. Urokinase, an enzyme extracted from urine, is one example. The structural features of xenoproteins make these products highly immunogenic when administered to humans. In addition, extraction processes may cause low immunogenic proteins to acquire high immunogenicity. The poor immunogenic properties of non-human proteins may be responsible for their inactivation, as is the case for adenosine-deaminase obtained from bovine sources. They can also lead to severe immunoreactions, as observed with asparaginase, uricase, avidin and streptavidin. Their use in medicine is therefore drastically limited. Nowadays, biotechnological processes allow the production of high quantities of safe and non-toxic recombinant human proteins. However, these macromolecular drugs are intrinsically immunogenic and their use entails the risk of an immunoresponse. Slight post-translational modifications and structural/conformational alterations caused by processing conditions may change self-proteins and recombinant human proteins into immunogenic or allergenic products [58]. Antibodies elicited against immunogenic proteins can neutralize their bioactivity and contribute to elimination, thus compromising the clinical effect or provoking severe adverse events related to cross-reactivity with autologous proteins [59].

The beneficial effects of PEGylation on the immunological properties of proteins have been widely demonstrated. PEGylation camouflages proteins from recognition by enzymes involved in protein processing and presentation, and also physically interferes with antibody or human leukocyte antigen-(HLA) epitope binding, which in turn results in the reduction or suppression of the native protein immunogenicity and antigenicity [23,60].

Several studies have demonstrated that the number and shape of polymer chains are more important than the polymer size in reducing protein immunogenicity. The immunogenicity and antigenicity of uricase was reduced to

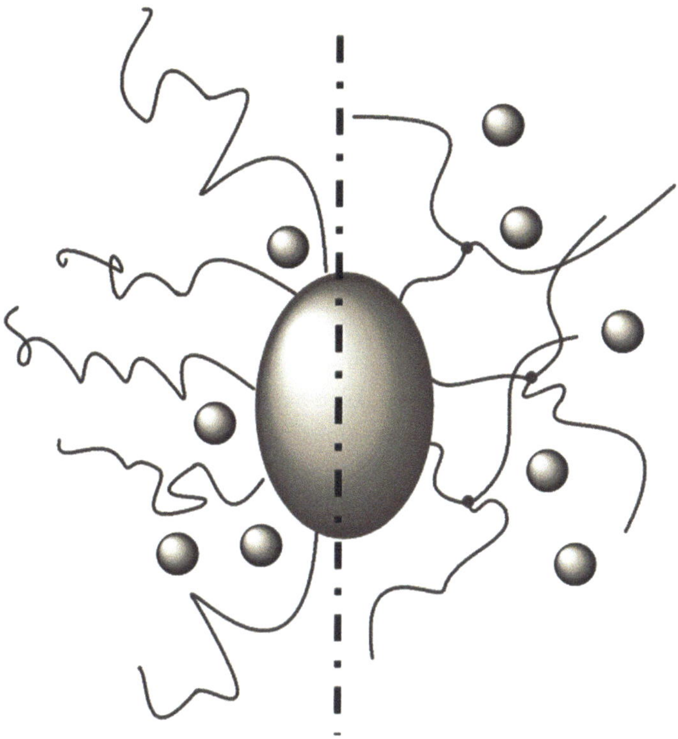

FIGURE 11.2 Structural representation of a protein conjugated to linear (left side) or branched (right side) PEG chains. Branched PEG molecules result in better protection against approaching antibodies, proteolytic enzymes and phagocytic cells than linear PEG chains.

a higher extent by conjugation with a branched 10 kDa PEG than by conjugation with a 5 kDa linear polymer. This is probably due to the fact that the former can mask the protein more efficiently than the latter from the immune system (Figure 11.2) [61,62].

The alteration in the immunogenic properties of proteins after PEGylation was found to significantly affect the pharmacokinetic behavior of the bioconjugates. PEGylation drastically reduced the clearance of humanized antibodies that were intravenously administered to mice, even though the antibody size is largely above the glomerular ultrafiltration cut-off limit. Forty-eight hours after administration, the blood levels of PEGylated antibodies were 165% higher than those corresponding to the unconjugated ones. The prolonged permanence of these large proteins in the bloodstream was ascribed to a lower protein immunorecognition conveyed by PEGylation [63]. Staphylokinase-PEG isomers, bearing the same amounts of PEG attached to different protein sites, displayed considerably different pharmacokinetic parameters: $t^1/_2$ ranged between 16 and 25 minutes (native protein $t^1/_2$ was 7 minutes), the AUC was between 9.3 and 27 g min ml^{-1} (that of the native protein was 4.6 g min ml^{-1}) and the clearance was between 1.1 and 0.4 ml min^{-1} (it was 2.2 ml min^{-1} for the native protein). These results demonstrate once again that glomerular ultrafiltration is not the only or the main clearance process for PEGylated proteins. The dissimilar pharmacokinetic behavior was in fact attributed to the ability of PEG chains to shield protein sites sensitive to immuno- or enzymatic recognition [64].

Although protein PEGylation is largely accompanied by a remarkable decrease in protein immunogenicity and antigenicity, and improvement of the therapeutic properties, PEGylation produces derivatives which have their own immunological profile. This results from the combination of the immunological properties of the native protein, the polymer and any other chemical entity introduced in the new bioconjugate, such as the protein–polymer linker.

PEG is usually considered as a non-immunogenic and non-antigenic material. Nevertheless, anti-PEG antibodies are in fact generated following either polymer or polymer–bioconjugate administration. The anti-PEG immuno-response is related to the molecular weight of the polymer and is more noticeable for low molecular weight chains. Specific anti-PEG antibodies are raised by a block of 6–7 repeated units of ethylene oxide [65]. Anti-PEG antibodies have been exploited as an analytical tool for detection of PEG conjugates in biological samples and are thought to condition the pharmacokinetic/pharmacodynamic behavior of PEGylated proteins in a relevant way [66,67].

The anti-PEG antibody titer increases when the polymer is conjugated to an immunogenic protein or in the presence of adjuvants. In such a case, the PEG immunogenicity is higher because the polymer molecular weight increases and the linear polymer chains are more immunogenic than the branched polymer chains. These results indicate that the polymer architecture and flexibility play a role in the polymer immunological properties [62,68].

β-Glucuronidase-PEG administered to mice elicited anti-β-glucuronidase IgG and anti-PEG IgM. The former recognized the protein fraction of the conjugate with an affinity that was 2.1 times lower than that observed for the native protein, and the latter displayed the ability to bind any other PEGylated protein. Anti-PEG IgM proved to be more efficient in removing the PEGylated enzyme from the blood than the anti-protein IgG. This is probably due to the multiple binding sites on the PEG chains [69].

Finally, conjugation may sometimes produce new epitopes as a consequence of partial protein denaturation due to polymer conjugation or the use of inappropriate spacers between the protein and PEG [70].

11.2.5. Toxicology Aspects of PEGylation

Although the amount of PEG administered with the PEGylated proteins is usually very low in chronic therapies (about 10 mg for six-month PEGASYS®

treatments and less than 2 mg for PEGINTRON® treatments), the potential toxicity of PEG has been a concern in the development of PEG–protein drugs. Over the past years, several investigations have been performed to examine the toxicity of PEG and PEGylated proteins and their effects on cell and organ biological function.

High doses of injected PEG–protein conjugates have been demonstrated to induce reversible renal tubular vacuolization, which is not associated with functional abnormalities [71].

Intravenous administration of high PEG levels has been shown to induce concentration and molecular-mass-dependent serum complement activation [72].

However, high PEG doses are used only in experimental veterinary settings for spinal cord injury and traumatic axonal brain injury investigations in which the formulations contain high polymer levels as a solubilizer/carrier. By contrast, the quantities of PEG administered in PEGylated therapeutics are tens-of-thousands-fold lower, and the risk of generating antibodies is rather remote.

11.3. FROM RANDOM TO SITE-SELECTIVE PEGylation

11.3.1. Random PEGylation

Proteins can be PEGylated by a wide range of methods [73], which include chemical and enzymatic protocols. Proper design of the conjugation strategy based on the full knowledge of the protein physicochemical properties, including the number, location and reactivity of the target functional groups in the protein structure, is of paramount importance for yielding derivatives with optimal therapeutic performance.

Chemical procedures require polymer activation for derivatization of selected amino acids in the protein structure that usually act as anchoring points.

Most of the bioconjugation approaches set up in the past decades were directed at the modification of the lysine and N-terminus amino groups of the protein because they are well represented in the protein structure, exposed on the protein surface to the bulk solvent and can be easily derivatized. Table 11.1 shows the main PEG for amino group derivatization obtained by activation of the terminal hydroxyl or carboxyl polymer group.

The literature reports a number of studies performed by protein PEGylation with amino-targeted activated PEGs. PEG-Adenosine deaminase (Adagen®) and PEG-asparaginase (Oncaspar®) are two bioconjugates approved by the FDA for the treatment of severe combined immunodeficiency disease (in the case of the former) and acute lymphoblastic leukemia and other lymphoid malignancies (in the case of the latter). These bioconjugates were obtained by extensive amino group derivatization with succinimidyl-succinate-activated 5 kDa PEG, which drastically reduced the immunogenic activity and prolonged the bloodstream residence time of these xenoproteins [74–76].

TABLE 11.1 Activated PEGs for amino group derivatization

Activated PEGs that maintain the charge of the native protein in the final conjugate	Activated PEGs that abolish the positive charge in the final conjugate with respect to the native protein	
PEG-aldehyde	PEG-succinimidyl ester	PEG-succinimidyl carbonate
PEG-dichlorotriazine		PEG-*p*-nitrophenyl carbonate
PEG-tresyl		PEG-2,3,5-trichlorophenyl carbonate
PEG-epoxide		PEG-benzotriazolyl carbonate

Amino group PEGylation usually quenches the positive charge resulting from the amino group protonation under physiological conditions. To avoid alteration of the overall charge of the biomolecule, which may reduce the bioactivity or impair the structural and biopharmaceutical properties of a protein, bioconjugation protocols that preserve the amino groups have been studied. Tresylated PEG yields straightforward PEGylated proteins without significant alteration of the isoelectric point, while aldehyde-PEG reacts with the protein amino groups to yield labile Schiff bases that can be stabilized by reduction with $NaCNBH_3$ and formation of a secondary amine.

Typically, amino group derivatization yields extensive PEGylation by a relatively simple, mild and fast reaction. The attachment of several linear or branched PEG chains with a molecular weight of 1.9–40 kDa has been found to provide efficient protein surface masking and size enlargement, with beneficial effects on protein stability, pharmacokinetic properties and immunological behavior [3,77,78]. This approach succeeded in the case of proteins that display their bioactivity by interaction with small substrates such as asparaginase, adenosine deaminase, superoxide dismutase and diamino oxidase. Nevertheless, random and extensive modifications rising from amino groups PEGylation often dramatically reduce biological activity of the protein by preventing the recognition of macromolecular substrates or receptors. Furthermore, polymer conjugation is not site-selective and cannot be easily controlled, resulting in the production of several isomers which are unsuitable for pharmaceutical applications because regulatory agencies require the extensive characterization of bioconjugates to be used for humans.

11.3.2. Site-selective PEGylation

Site-selective PEGylation strategies have been developed to obtain derivatives with defined structure and composition, and to counterbalance the reduction in biological activity with a significant enhancement of the physicochemical, immunological and biopharmaceutical properties of the proteins [79]. Site-directed PEGylation protocols exploit the selective reactivity of functional groups under controlled conditions, rare natural or artificial anchoring sites in the protein structure and enzyme-mediated bioconjugation. These methods produce derivatives with a well-defined structure, as a limited number of PEG molecules are anchored to selected protein points. Site-selective PEGylation is usually carried out using high molecular weight polymers to obtain derivatives with desired polymer/protein mass ratios. Mono- and site-specific PEGylated conjugates offer advantages in terms of regulatory acceptability, costs and therapeutic performances, when compared to multiple-PEGylated isomers formed when proteins are derivatized with low molecular weight PEG molecules.

11.3.2.1. Mono α-N Terminus PEGylation

One of the most successful site selective PEGylation methods exploits the difference in pKa between the ϵ-amino group of Lys and the α-N terminus [80].

Aldehyde-activated PEGs react at pH 5 under reductive conditions to preferentially yield α-N terminus mono-PEGylation, as shown in Figure 11.3 [81].

Amino terminal site-specific conjugation with aldehyde-activated PEG was used to prepare mono-PEGylated rh-G-CSF and a mono-PEGylated derivative of truncated recombinant human megakaryocyte growth and development factor (rHuMGDF) [82,83]. rh-G-CSF is a hematopoietic cytokine derivative with specific lineage effects on the neutrophilic granulocites. It is widely used clinically in anti-cancer therapies as it overcomes myelosuppression induced by chemotherapy and bone marrow transplantation [84,85]. rh-G-CSF was

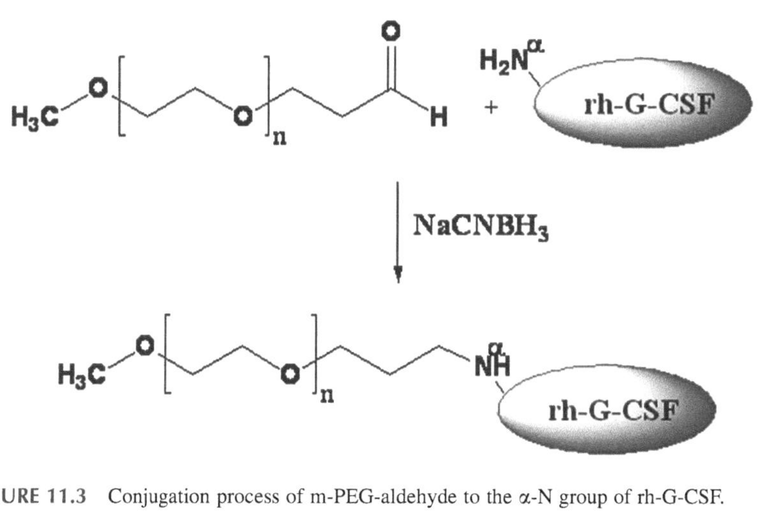

FIGURE 11.3 Conjugation process of m-PEG-aldehyde to the α-N group of rh-G-CSF.

reacted with aldehyde-activated PEGs of various molecular weights (12, 20, 25 and 30 kDa). About 90% mono-PEGylated products were isolated by ion-exchange chromatography and characterized by several analytical methods, including endoproteinase peptide mapping, MALDI TOF-MS and size-exclusion HPLC.

The pharmacological therapy consists of a daily dose to cancer patients after chemotherapy has begun, which must be continued until neutrophils reach a safe level. rh-G-CSF requires daily subcutaneous injections due to fast kidney- and neutrophil-mediated elimination from the circulation [86]. In contrast, the 20 kDa PEG amino terminal mono-PEGylated derivative (PEG-rh-G-CSF) displayed sustained release due to transmembrane diffusion, reduced renal clearance and increased stability towards neutrophil-mediated destruction. This allows for weekly administration with high enhancement of patient compliance [87]. Figure 11.4 shows the median rh-G-CSF blood concentration in patients treated with recombinant protein and different doses of PEG-rh-G-CSF.

A comparative study carried out using aldehyde-activated PEGs of different molecular weight (5, 10, 20 and 30 kDa) showed that the reaction with 5 kDa PEG led to large amounts of the diPEGylated derivative. This is probably due to a higher accessibility of the small polymer chain to the reactive amino groups

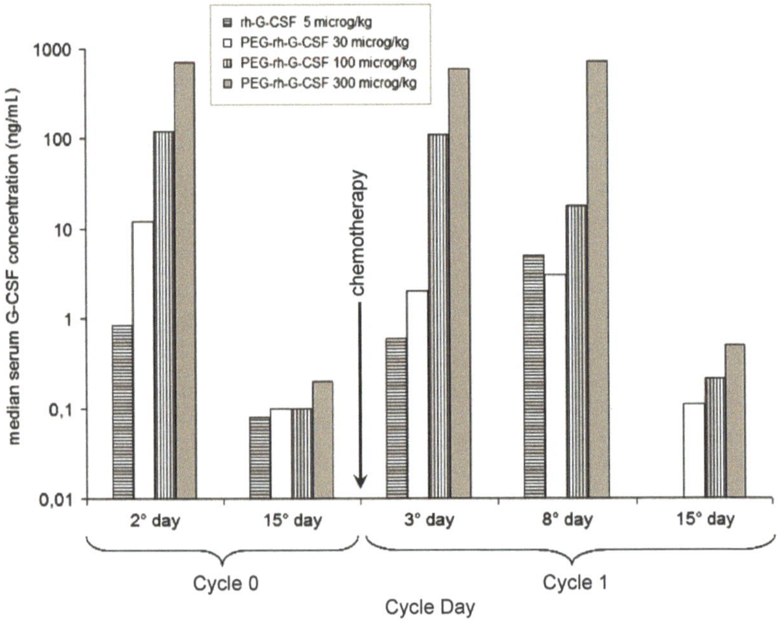

FIGURE 11.4 Median rh-G-CSF serum concentrations in patients receiving daily rh-G-CSF injections or single subcutaneous PEG-rh-G-CSF injections during each therapeutic cycle.

of the protein [22]. The *in vitro* activity of PEG-rh-G-CSF tested on a cytokine-dependent cell line decreased as the PEG molecular weight increased, by up to 72% with 20 and 30 kDa PEG. The conjugation of 5 and 30 kDa PEG increased the rh-G-CSF half-life by 15 and 170 minutes, respectively. Both 20 and 30 kDa PEG derivatives had high *in vivo* activities for up to 24 hours; the native cytokine was inactive after 60 minutes. The $PEG_{30\ kDa}$-rh-G-CSF showed a 60% higher bioavailability as compared to that of $PEG_{20\ kDa}$-rh-G-CSF, suggesting that the 30 kDa bioconjugate can be administered in a lower dose than $PEG_{20\ kDa}$-rh-G-CSF.

Similarly to rh-G-CSF, rHuMGDF was mono-PEGylated with a 20 kDa PEG-aldehyde [82]. MGDF is the natural hormone that is responsible for thrombopoiesis regulation. rHuMGDF, the N-terminal 163 amino acid fragment of MGDF, has a 5-fold higher specificity than the full-length protein. However, it lacks an adequate half-life for a therapeutic effect *in vivo*. The α-N terminus mono-PEGylation of rHu-MGDF prevents the non-enzymatic cyclization of the first two amino acids and the subsequent cleavage to form a des (Ser Pro)rHuMGDF, which is the main route of degradation [88]. Additionally, the intraperitoneal injection to mice of the PEGylated form rHuMGDF led to a platelet induction that was identical to that of the full length glycosylated protein. However, it showed a platelet response that was superior to that of both the glysosylated truncated form and the non-glycosylated full-length form [89].

Recently, a novel, reductive PEGylation protocol that uses an iridium catalyst as the reducing agent and sodium formate as the hydride source has been developed to better control the alkylation process [90]. This approach may also be viewed as an effort to develop "green chemistry" strategies since no harsh reagents are required.

11.3.2.2. Cysteine PEGylation

PEGylation protocols have been developed to modify those amino acids that are seldom represented in the protein structure to achieve site-directed PEGylations. One example is phenylglyoxal-activated PEG that can react with an arginine side group.

The thiol group of cysteine is an excellent anchoring site for obtaining derivatives with defined structure and composition. Cysteine is in fact an infrequent amino acid in the protein structure and, if natively present in the sequence, only one cysteine residue is usually available for conjugation since an even number of cysteines is normally involved in disulfide bonds. Additionally, Cys can be introduced by site-directed mutagenesis into a part of the peptide sequence that is not involved in the bio-function of the protein, thus allowing for site selective PEGylation.

Vinyl sulfone-, maleimide-, dithyopyridine- and iodoacetamide-activated PEGs have been developed for Cys modification. These reagents have different reactivity kinetics and their selectivity is dictated by coupling conditions.

The high stability of PEG-vinyl sulfone (PEG-VS) at pH 7–9 also allows for its use in multistep reactions [91]. At pH 8, PEG-VS can selectively modify the thiol groups at reaction rates that depend on their accessibility. Reaction of PEG-VS with reduced RNase at pH 8 resulted in coupling with all the thiol groups within one hour. The competitive reaction with amino groups was negligible over a period of 24 hours.

PEGylation of Cys has been used to enhance the biopharmaceutical properties of single chain Fv antibody fragments (scFv) by obtaining small multivalent tumor-targeting proteins, which have emerged for radio-imaging and therapy. The clinical use of these antibody fragments is hampered by their rapid blood clearance, inefficient tumor uptake and poor tumor residency [21,92]. Random polymer conjugation of the exposed lysines has been undertaken to improve the pharmacokinetic properties of scFv. It was found that such conjugation strongly reduced the bioactivity of the protein. Site-specific conjugation was investigated by insertion of Cys residues in the amino acid sequence. Four different anti-mucin-1 scFv (scFv-c) targeting epithelial cancer cells were prepared by introducing one cysteine residue in the close vicinity of the carboxyl terminus [93]. The mutated fragments were modified with mono-activated maleimide-PEG and di-activated (maleimide)$_2$-PEG with different sizes. The small structural variations of the four antibody fragments were found to influence the docking of the maleimide-PEG to the Cys and the low molecular weight PEGs resulted in higher conjugation yields. The di-activated (maleimide)$_2$-PEG produced divalent scFv-c with increased avidity and prolonged residence time in the target sites as compared to the monomeric scFv-c.

Trichosanthin, a type I ribosome-inactivating protein endowed with DNase-like activity for cleaving supercoiled double-stranded DNA [94], was shown to have a pro-apoptotic activity that inhibits tumor growth [95] and an anti-HIV activity [96]. The clinical use of this protein suffers from its immunogenicity and short half-life due to rapid kidney ultrafiltration and proteolytic degradation. Two trichosanthin muteins were designed *in silico* by replacing short amino acid sequences with sequences containing one cysteine in the most immunogenic sites according to parameters such as hydrophilicity, surface exposure, flexibility and formation of loops and turns [97]. The conjugation with 5 kDa PEG-maleimide maintained the DNase-like activity of the unPEGylated protein, but reduced the ribosome inactivation capacity by about 10-fold as compared to the wild type trichosanthin. Studies performed in mice showed that the PEGylation decreased the IgE value by 4-fold and IgG production by 40%. It also reduced the protein toxicity by about 50–80%. The proteolytic resistance resulted in a 4.5–6-fold longer residence time in the blood and a 2-fold slower clearance with a limited distribution in the liver.

Similar PEGylation strategies were adopted with other protein drugs, such as anti-HIV entry inhibitors, that suffer from antigenic potential, toxicity and rapid clearance [98]. Cyanovirin is a potent virucidal protein from the cyanobacterium

Nostoc elipsosporum that inactivates most HIV strains [100] by binding irreversibly to the gp120 HIV capside protein through its high-mannose oligomers, and by blocking the virus entry into lively cells (T-lymphocytes, monocytes and dendritic cells). Based on the 3-dimensional structure of the protein, glutamine 62, which is not included in the sequence forming the carbohydrate binding site, was replaced with Cys by site-directed mutagenesis [100,101]. *In vitro*, the biological activity of the mutein conjugated with a 20 kDa PEG-maleimide was very similar to that of the wild type protein. However, the 30 kDa PEG conjugate displayed a limited virucidal activity. A dose of 50 mg/kg cyanovirin induced mouse death within 24 hours of administration, while only limited toxicity was observed with administration of the same dose of PEG-cyanovirin. The immuno-response elicited by the PEGylated mutein was at least 30-fold lower than that observed for the unPEGylated mutein.

An unpaired cysteine in proteins is usually buried in hidden hydrophobic pockets of the tertiary structure that are not easily accessible to hydrophilic PEG for conjugation. This thus limits protein PEGylation under native conditions. Partial protein denaturation during polymer conjugation followed by refolding was found to increase the PEGylation yield. Unfortunately, this process can also induce irreversible conformational changes of the protein that may hamper its biological activity. Terminal hydrophobized PEG was successfully used for modification of buried Cys under non-denaturing conditions [102]. Linear 20 kDa PEG-NH_2 was modified with an 18-carbon alkyl chain and activated with maleimide. The alkyl spacer was designed according to the depth and width of the hydrophobic cleft which hosts cysteine 17 that was introduced into the rh-G-CSF used as protein model. While commercial PEG-maleimide was ineffective in the rh-G-CSF conjugation under non-denaturing conditions (16% conjugation yield), the PEG-C_{18}-maleimide, which can easily dock into the hydrophobic pocket in which Cys^{17} was located, resulted in a high protein PEGylation (58%). The PEGylated rh-G-CSF was stable to aggregation under native conditions and possessed similar biological activity to the commercial PEGylated rh-G-CSF Neulasta®.

11.3.2.3. Three-carbon Bridge PEGylation

Most therapeutic proteins have an even number of cysteines that couple to form disulfide bridges. Disulfide bridges are of paramount importance to the tertiary structure and stability of proteins [103], especially in the case of small proteins where disulfide bonds compensate for the lack of extensive hydrophobic intramolecular interactions [104].

Thiol-specific bis-alkylation reagents have been developed to achieve three-carbon bridge sulfur PEGylation after reduction of accessible disulfides [105]. The typical alkylating reagent (Figure 11.5A) comprises a substituted propenyl group at the end of PEG. This conjugation moiety includes an α,β-double bond, an α,β' sulfonyl group that undergoes elimination as sulfinic acid and an electron-withdrawing group (carbonyl) that is involved in the sulfinic acid

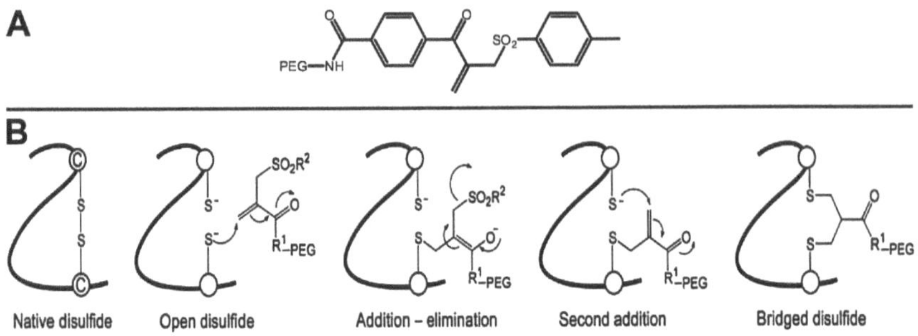

FIGURE 11.5 (A) The alkylating reagent employed to achieve three-carbon bridge sulfur PEGylation. (B) Mechanism for conjugating a three-carbon PEGylated bridge to a protein's native disulfide bond. *Reprinted adapted with permission from* Nat Chem Biol *2006, 2, 312–313. Copyright © by 2006 Nature Publishing Group.*

elimination, which is the rate-limiting step. The PEGylation mechanism is described in Figure 11.5B.

The accessibility of disulfide bonds on proteins can be ascertained by computational modeling, and the thermodynamic destabilization effect of the three-carbon bridge insertion can be predicted [106]. In the absence of denaturants, only exposed disulfide bonds, which are more involved in protein stability than in biological function [107], are involved in the PEGylation process. However, the cleaved disulfide bonds are reconstituted through the three-carbon bridge, resulting in maintained protein stability.

Bis-sulfone or mono-sulfone PEGs (Figure 11.6) have also been used to conjugate disulfide functions. At acidic pHs, a better performance was found for the mono-sulfone [108]. The bis-sulfone generates a reactive mono-sulfone, resulting in more specific reactions.

A three-carbon bridge PEGylation must be carried out under native conditions in order to avoid an increased distance between the two reduced cysteines, which will produce protein denaturation and inactivation.

Disulfide PEGylation of interferon α-2b produced a derivative with similar biological activity (~8%) than PEGINTRON®, the PEGylated interferon in clinical use [109]. Furthermore, the cytokine activity was not affected by the

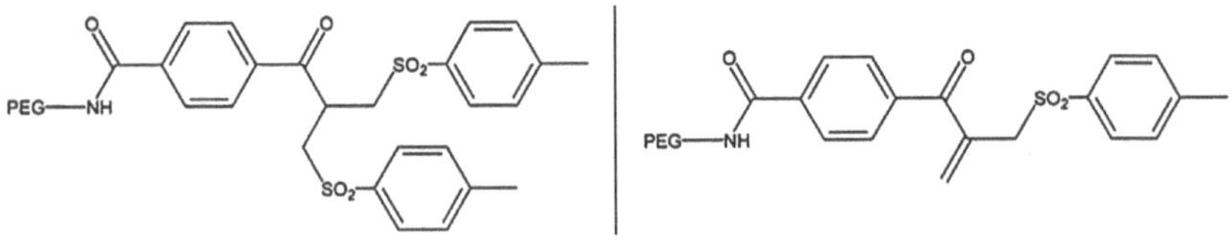

FIGURE 11.6 Bis-sulfone- (left) and mono-sulfone- (right) activated PEG for protein conjugation according to the "three-carbon bridge" procedure. *Reprinted adapted with permission from* Nat Chem Biol *2006, 2, 312–313. Copyright © by 2006 Nature Publishing Group.*

PEG molecular weight. This result is in contrast to that observed in the case of amine-directed PEGylation methods. Similar results were obtained by site-selective PEGylation of asparaginase, a tetrameric protein. Regardless of the PEG size, PEGylated asparaginase maintained its native enzyme activity.

11.3.2.4. PEGylation of Non-natural Tags

Site-specific PEGylation can be achieved by exploiting non-natural amino acids which have been genetically incorporated into proteins produced by prokaryotic and eukaryotic microorganisms [110−114].

The copper-mediated Huisgen [3+2] cycloaddition of an azide and an alkyne is orthogonal to all functional groups present in proteins and peptides and can thus be used for the selective PEGylation of proteins [115]. Thus, an azido-phenylalanine (Figure 11.7) was introduced in the superoxide dismutase (SOD) sequence using the *E. Coli* tyrosyl tRNA/tRNA-synthetase pair to produce the protein in yeast [116]. The mRNA coding for SOD was modified with a specific codon (TAG) to regulate the introduction of the non-natural amino acid.

Based on a crystallographic analysis, superoxide dismutase derivatives were designed in order to swap amino acids that are not involved in a catalytic activity with azide bearing, non-natural amino acids. The tryptophan at position 33 was selected as the best amino acid candidate and the mutated protein (33TAG-SOD) was obtained by fermentation of yeast incubated in a medium containing azido-phenylalanine. The azido-protein PEGylation was carried out by click-chemistry. More specifically, 33TAG-SOD was reacted with 5 or 20 kDa alkyne-terminating PEG in the presence of $CuSO_4$, a stabilizing ligand for Cu^{1+}, a Cu wire and tris(carboxyethyl)phosphine as the reducing agent, according to the scheme shown in Figure 11.8.

Fingerprint analysis, MALDI-TOF mass spectrometry and enzyme activity studies showed that click-chemistry can be properly exploited to produce mono- and site-specific PEGylated proteins with preserved biological activity. On the other hand, use of copper as the catalyst can induce protein structural modifications, unfolding and precipitation. Furthermore, copper is a toxic metal and should therefore be replaced by a non-toxic catalyst [117].

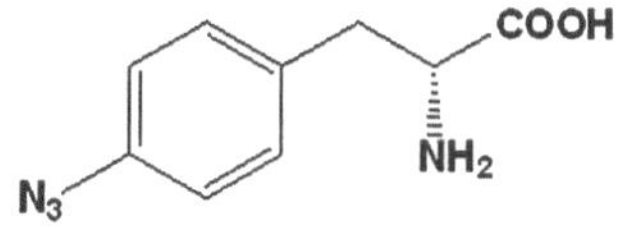

FIGURE 11.7 Azido phenylalanine used as a non-natural amino acid for protein PEGylation by cycloaddition.

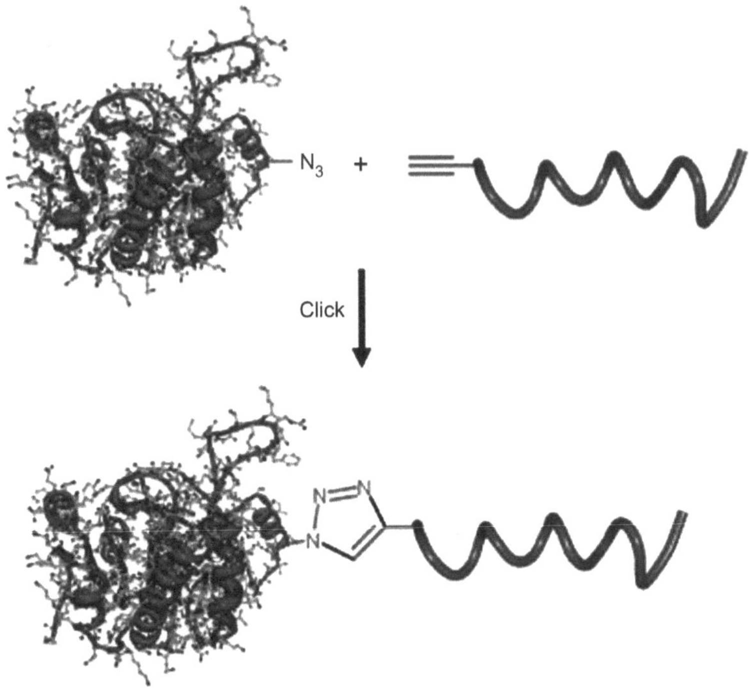

FIGURE 11.8 Site-specific PEGylation of 33TAG-SOD by [3+2] cycloaddition. *Reprinted with permission from* Macromol Rapid Comm *2008, 29, 1073–1089. Copyright © 2008 by WILEY-VCH Verlag GmbH & Co. KGaA, Weinheim.*

11.3.2.5. Direct Biosynthesis of PEGylated Proteins

Non-natural PEGylated amino acids have been introduced into the protein structure using a four-base codon in an *E. coli* cell-free translation system [118]. Succinimidyl carboxy-activated PEGs (4, 8 and 12 repeat units of ethylene glycol) were conjugated to p-aminophenylalanine-5′-O-phosphoryl-2′-deoxycytidylyl-(3′-5′)adenosine (aminophenylalanyl-pdCpA) and lysil-5′-O-phosphoryl-2′-deoxycytidylyl-(3′-5′)adenosine (lysyl-pdCpA). The PEGylated aminoacyl-pdCpAs were then ligated with a yeast phenylalanine tRNA that included a CCCG four-base anticodon and were added to an *E. coli* cell-free ribosomal translation system [119]. The PEGylated p-aminophenylalanine-tRNA and lysine-tRNA were exploited to produce streptavidin muteins by an *E. coli* cell-free translation reaction. PEGylated-phenylalanine was successfully introduced in the protein sequence according to a PEG-size-dependent efficiency; a PEG with four repeat units led to an incorporation efficiency of about 46% and 80% depending on the mutein. By contrast, PEGylated lysine was not efficiently incorporated. Studies performed with branched, high molecular weight PEGs did not yield PEGylated proteins, indicating that large, bulky polymers interfere with the ribosomal machinery.

11.3.2.6. Enzyme-mediated PEGylation

Recently, new enzymatic methods have been explored as a novel and mild PEGylation technique. Due to the substrate specificity of enzymes, such methods are intrinsically able to achieve site-directed polymer conjugation.

Transglutaminase (TGase), which is a widespread enzyme with many physiologic functions, was the first enzyme exploited for the production of PEGylated proteins [120–122]. This enzyme catalyses the acyl transfer reaction between the γ-carboxyamide group of a protein-bound glutamine (Gln) residue (acyl donor) and a variety of unbranched primary amines [123]. The specificity of the TGase was found to depend mainly on the surface accessibility of the Gln rather than on a specific amino acid sequence [124,125].

The TGase-mediated IL-2 PEGylation yielded the stoichiometric addition of one PEG molecule to Gln^{74}, which is located in an exposed protein loop [126]. Gln^{74} was a preferred substrate even though other 6 glutamines were available along the hormone sequence.

The PEGylation of the recombinant human growth hormone (rh-GH), a protein containing 13 glutamine residues, resulted in Gln^{40} and Gln^{141} modifications, with Gln^{40} being the preferred PEGylation site [127]. Electrospray mass spectrometry and tandem MS analysis of the bioconjugates, combined with proteolytic studies and structural dynamics, which indicate the degree of mobility along the polypeptide chain of a protein (B-factor, Figure 11.9), showed that PEGylation was selective for the highly flexible chain regions that encompass glutamine residues.

TGase-mediated PEGylation of rh-G-CSF showed that enzymatic conjugation was not affected by molecular weight of the polymer. Preliminary

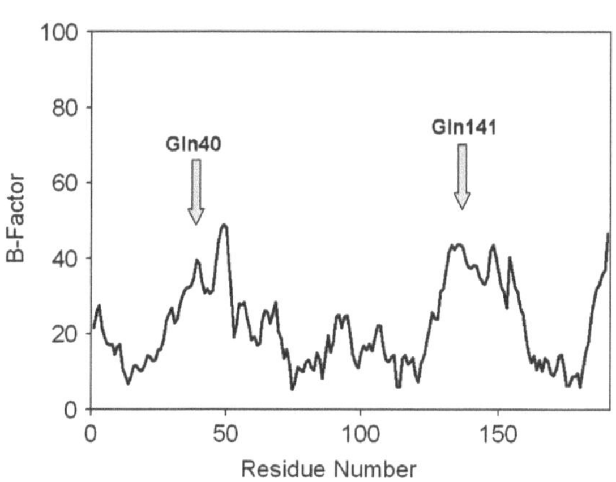

FIGURE 11.9 Profile of the backbone B-factor along the 191-residue polypeptide chain of rh-GH, as derived using StrucTools (<http://helixweb.nih.gov/structbio/basic.html>) from the Worldwide Protein Data Bank (PDB), entry 3HHR. A discontinuity of the B-factor profile indicates the protein disorder of the corresponding chain segment. Arrows indicate the sites of TGase-mediated PEGylation.

3-dimensional structural analysis, molecular dynamics simulations and protein–protein docking calculations identified a single potential PEGylation site out of 17 glutamines in the hormone molecule, Gln^{134}, which was confirmed by analysis of the PEGylated protein [128]. Therefore, *in silico* design combined with protein engineering may provide a tool for producing tailored bioconjugates.

A two-step enzymatic PEGylation method, glycoPEGylation, has been developed to PEGylate non-glycosylated recombinant proteins at the natively glycosylated protein sites. The method involves:

(i) Polypeptide N-acetylgalactosamine (GalNAc) transferase-mediated O-glycosylation of serine and threonine residues; and
(ii) Sialyltransferase-mediated sialic-PEG conjugation to the newly introduced GalNAc residue.

The enzymatic reactions allow the conjugation of sialic-PEG [129].

rh-G-CSF and IFN α2b site-directed O-glycosylation was obtained by exploiting a human polypeptide N-acetylgalactosaminyl transferase isoform (GalNac-T2). The enzyme glycosylates the amino acids that are constitutively glycosylated, i.e. threonine 133 and threonine 106 on rh-G-CSF and IFN-α2b, respectively. Sialyltransferase was then used to transfer cytidine monophosphate-sialic acid-PEG_{20kDa} to the GalNAc residue of rh-G-CSF and IFN-α2b. The glycosylation and PEGylation site may be tuned by using the proper N-acetylgalactosaminyl transferase isoform and sialyltransferase. In rats, the glycoPEGylated rh-G-CSF showed a 10-fold lower plasma clearance and 1.8-fold higher AUC value than the commercial rh-G-CSF-PEG. *In vitro* studies showed that glycoPEGylation slightly affects the rh-G-CSF bioactivity. The capacity to increase the white blood cell count *in vivo* was almost twice as high as that obtained with the unPEGylated cytokine. Promising results were also found with glycoPEGylated IFN-α2b. Indeed, this cytokine displayed an *in vivo* anti-viral activity at a dose that was 26-fold lower than that necessary in the case of the unPEGylated interferon.

GlycoPEGylation was used to modify the coagulation factor VIIa (FVIIa), a trypsin-like serine protease that is responsible for initiating blood coagulation when it associates with a tissue factor (TF). Although this protein has potential application in prophylactic treatment of hemophilia, a short *in vivo* half-life (2–4 hours) limits its use [130,131]. In order to ameliorate pharmacokinetic activity without impairing biological activity, FVIIa was site-specifically PEGylated by reacting a 2–40 kDa sialic acid-PEG with the two bis-antennary N-glycans at positions 145 and 322 [132]. PEGylation slightly reduced the binding constant of FVIIa for TF by decreasing the association rate according to a PEG-size-dependent behavior. The PEGylation at position 322 reduced the biological activity by 36%, while PEGylation at position 145 had no effect on the protein bioactivity. The factor X (FX) activation rate was not affected by the polymer conjugation of FVIIa, but had a beneficial effect on the deactivation rate of FVIIa by Antithrombin III (AT III) as it was reduced. Furthermore, glycoPEGylation reduced FVIIa removal from the bloodstream by endothelial cells via the specific receptors TF and endothelial cell protein C receptor with consequent prolongation of the *in vivo* half-life [133].

11.4. RESTORING PHARMACODYNAMIC PROPERTIES BY REVERSIBLE PEGYLATION

Reversible PEGylation has been performed to obtain prodrug-like protein bioconjugates that restore biological activity by polymer release. Releasable PEGylation exploits "self-immolative" chemical linkers that reversibly tether PEG to the protein [134], and that thus offer the biopharmaceutical advantages of extensive polymer conjugation with a high *in vivo* activity. Although the linkers used for reversible PEGylation have not shown any toxicity in animal studies, a complete analysis of their *in vivo* properties needs to be performed in the future [135].

Reversible PEGylation techniques that generate native proteins *in vivo* can be exploited to achieve polymer conjugation of the abundant and reactive amino groups without site-selective modification [136]. This approach requires the specific design of the linker for PEG conjugation. The PEG conjugate contains a trigger segment, which is responsible for the protein–polymer linker degradation. The PEG release rate in the blood can occur on a timescale that varies from hours to days depending on the linker structure.

PEGylation through ester bonds is frequently employed to obtain reversible polymer conjugation. Alternatively, the use of substituted benzyl linkers allows for the modulation of PEG release kinetic *in vivo* and reconstitution of the parent protein (Figure 11.10A) [137]. Bis-N-2-hydroxyethylglycinamide (bicin) linkers (Figure 11.10B) undergo hydrolysis according to a deamination mechanism. The PEG release rate can be modulated using different linkers between the polymer and the bicin moiety [138,139].

The immunotoxin obtained by conjugating *Pseudomonas exotoxin A* with the anti-mesothelin binding domain (Fv) antibody developed for tumor targeting suffers from a non-specific binding to normal tissues, formation of neutralizing antibodies and a limited blood residency. Preliminary studies showed that irreversible PEGylation completely abolished the immunorecognition and the cytotoxic activity of the immunotoxin. Therefore, immunotoxin was randomly PEGylated using the bicin releasable linker [140]. The polymer conjugation reduced the toxicity of the immunotoxin by about 30%. *In vivo* studies performed by intravenous administration of unPEGylated and PEGylated immunotoxin to nude mice bearing A431-K5 human cancer cells that express mesothelin showed that the polymer increased the plasma half-life by a factor of ten and led to complete tumor regression after one injection at a dose that was lethal in the case of the unPEGylated derivative.

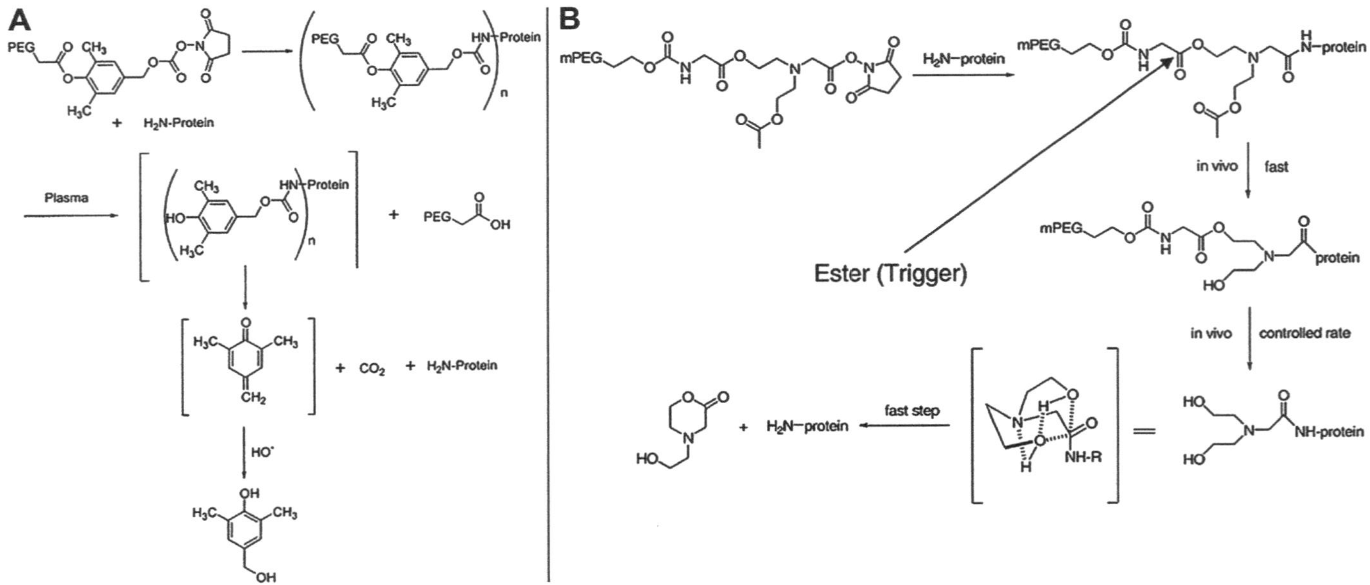

FIGURE 11.10 (A) Conjugation of a model protein through a benzyl linker and PEG-release mechanism by enzymatic cleavage followed by a classical 1,6-benzyl elimination reaction restoring the native protein. (B) Protein PEGylation through a bicinconic linker and PEG-release mechanism yielding the native protein. *Reprinted adapted with permission from* Bioconjugate Chem. *2001, 12, 163–169, Copyright © 2001 by American Chemical Society and from* Bioconjugate Chem. *2006, 17, 341–351. Copyright © 2006 by American Chemical Society.*

Exedin-4, a peptide containing 27 amino acids, is a glucagon-like peptide receptor agonist approved for the treatment of type 2 diabetes. Being a small peptide, it is rapidly cleared from the blood. Because permanent PEGylation reduces its biological potency by 100-fold, exedin-4 was reversibly PEGylated using the sulfo-fluorenylmethoxycarbonyl (FMS) derivative of the amino-reversible fluorenylmethyloxycarbonyl (Fmoc) protecting group as the linker. PEGylation occurred predominantly at the terminal amino group [141]. The peptide regenerated at a constant rate by a β-elimination reaction at position 9 of the fluorenyl moiety, as described in Figure 11.11. The native polypeptide was fully released in 48 hours when incubated in serum or buffer at pH 8.5.

In vivo, exedin-4-FMS-PEG induced delayed (8−12 hours from administration) and prolonged (12 hours) hypoglycaemia when compared to the native peptide.

The atrial natriuretic peptide (ANP) that generates hypotension *in vivo* has potential application for improving the hemodynamics in patients with chronic heart failure. However, its short half-life is detrimental for therapeutic use [142]. ANP is rapidly eliminated by receptor uptake and proteolysis. A long-lasting version of this hormone was obtained by site-selective N-terminus PEGylation with $PEG_{30\ kDa}$-FMS [143]. *In vitro* cell studies showed that the PEGylated derivative lacked receptor affinity. However, binding was restored in the presence of BSA. ANP was constantly released from the PEGylated derivative at a 7% daily rate. While native ANP induced a short reduction of the arterial and systolic pressure, the *in vivo* $PEG_{30\ kDa}$-FMS-ANP activity lasted for up to 10 hours.

Reversible PEGylation can be pursued by exploiting the polymer conjugation with oxidized glycosidic moieties [144]. Since most of the pharmaceutical recombinant proteins lack a glycosylation site, a protocol to chemically glycate proteins has been developed [145]. Albumin was Cys^{34} site-directed conjugated with maleimido-alkyl-maltosyl or maleimido-alkyl-galactosyl derivatives. The glycosidic moieties underwent selective oxidation with periodate to generate aldehyde functions that were coupled to PEG-hydrazide. The reaction produced conjugates bearing two and three PEG chains for the BSA-alkyl-galactosyl and BSA-alkyl-maltosyl derivatives, respectively.

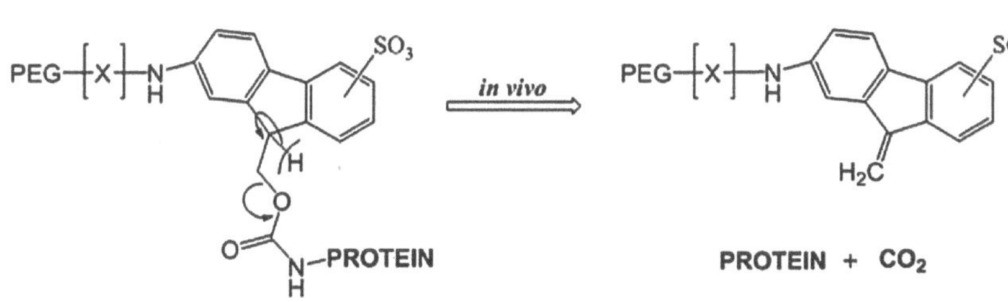

FIGURE 11.11 PEG release mechanism from a PEG-FMS-protein bioconjugate.

The polymer protein hydrazone bond was found to hydrolyze according to a pH-dependent kinetic behavior, being stable at neutral pH and faster under mildly acidic and basic conditions.

11.5. GROWING-FROM PEGylation

Most PEGylation protocols developed so far can be classified as "grafting-to" techniques because the preformed polymers are straightforwardly conjugated to proteins. Recently, novel "growing-from" approaches that involve the transformation of proteins into macroinitiators for subsequent living radical polymerization have been investigated. This latter approach has three main advantages over the classical "grafting-to" methods: high bioconjugation efficiency due to reduced steric hindrance; simple purification of the final bioconjugate and easy control of the polymer properties, such as the composition, molecular shape and chain length.

"Growing-from" has been used to conjugate PEG-like, comb-shaped polymers by reacting PEG-methacrylate (PEG-MA) monomers. This technique is very versatile and can be adapted to a variety of methacrylate monomers [146].

Polymer growth on the protein surface occurs by living radical polymerization. More specifically, it occurs by atom transfer radical polymerization (ATRP) and reversible addition-fragmentation transfer (RAFT), which are characterized by a lifetime increase of the propagating radical polymerization on the timescale of the polymerization reaction. The sharp kinetic control of the polymerization process results in a product having well-defined molecular weight, homogeneity and polydispersity [147–150]. ATRP and RAFT differ by a few operative conditions but yield products with comparable features.

The α-chymotrypsin (α-CT) macroinitiator was obtained by lysine derivatization with 2-bromoisobutyryl bromide. Polymerization was carried out using a 1.1 kDa PEG-MA. The resulting α-CT-polyPEG-MA was nearly homogenous in terms of polymer chain number and molecular weight and the protein retained 86% of the native protein activity [151]. On the other hand, the classical "grafting-to" amino PEGylation performed with PEG-succinimidyl propionate (PEG-SPA) yielded a highly polydisperse α-CT-PEG with dramatically reduced protein bioactivity. α-CTs polymerized with monomethoxy-polyethyleneglycol-methacrylate (PEG-MA) or *N*-2-hydroxypropylmethacrylamide (HPMA) by ATRP showed very different stability profiles in serum [152].

"Growing-from" ATRP was carried out using a biotin-initiator-complexed streptavidin that yielded four polyPEG-MA chains on the four biotin recognition sites of streptavidin [153].

Bovine serum albumin (BSA) "growing-from" site-directed derivatization was carried out using a pyridyl-disulfide-bearing RAFT agent. By conjugation with the unpaired cysteine of BSA, the agent switched the protein into a macroinitiator that was then polymerized with N-isopropylacrylamide (NIPAAm)

[154]. The polymerization reaction yielded monodisperse bioconjugates and did not affect the biological esterase-like activity of BSA.

Myoglobin was activated at its N-terminus with 2-(aminooxy)ethyl 2-bromo-2-methylpropanoate (ABM) and then polymerized with PEG-MA by ATRP. The procedure yielded a defined product in terms of molecular weight and polydispersity, and neither altered the tertiary structure of the protein nor its peroxidase-like activity. *In vivo*, the myoglobin-polyPEG-MA showed a 40-fold longer distribution phase and a 6-fold increased terminal elimination phase compared to the unPEGylated protein [155].

"Growing-from" rh-GH polymerization was carried out by fictionalization of seven protein amino groups with succinimidyl-tetra(ethylene glycol)-Br. PEG-MA polymerization was carried out by ATRP under mild conditions in buffer at 4°C. The procedure resulted in the controlled growth of 19.5 kDa polymer chains with low polydispersity and the maintenance of the protein structural conformation. The biological activity of the rh-GH was fully retained, as confirmed by *in vivo* studies. The daily administration of the bioconjugate to rats raised the same biological response after subcutaneous injection as did the native rh-GH. The *in vivo* response to twice-a-week administrations was superior to that obtained with the native protein [156].

Hybrid, comb-shaped materials, such as rhodamine-methacrylate/PEG-MA, were co-polymerized on bovine serum albumin surfaces by ATRP to introduce a probe along the copolymer chain for analytical purposes [157]. Furthermore, cleavable comb-shaped polymers can guarantee the release of polymer modules with easier elimination of the bioconjugate from the bloodstream and partial protein activity regeneration.

11.6. STIMULI RESPONSIVE POLYMERS

PEG is the polymer of choice for development of pharmaceutical protein-polymer bioconjugates due to its excellent low toxicity and the wide array of conjugation methods that have been developed throughout the years. Nevertheless, new polymers with similar or even better properties are currently under investigation to obtain derivatives with enhanced biological functionalities or physicochemical and biopharmaceutical features.

Stimuli-sensitive polymers have recently emerged as innovative materials for the production of tuneable biopharmaceuticals. Typically, the structure and solubility of these materials change in response to slight alterations in the environmental conditions, offering novel opportunities to produce therapeutic compounds that can be structurally modified by internal or external stimuli, such as pH and temperature.

Temperature-sensitive materials can be obtained by isopropylacrylamide polymerization (poly-N-isopropylacrylamide, pNIPAAm) and by isopropylacrylamide and acrylamide copolymerization (poly-N-isopropylacrylamide-co-acrylamide, pNIPAAm-coAm) by atom transfer radical polymerization

(ATRP) or reversible addition fragmentation transfer (RAFT) techniques. *In vitro*, these materials display higher toxicity profiles than PEG. However, no cytotoxicity was observed *in vivo* at low doses and after oral administration [158].

In water, these polymers undergo a temperature sol/gel transition at the *lower critical solution temperature* (LCST). The polymers are soluble below the LCST, whereas they dehydrate, collapse and become insoluble above the LCST. Polymers with different LCSTs can be designed by changing the composition, architecture and length of the final polymer chain [159].

pNIPAAm and pNIPAAm-coAm can be conjugated to proteins by either "grafting-to" or "growing-from" approaches. pNIPAAm conjugation to BSA by RAFT polymerization produced a derivative that underwent a phase transition by temperature-induced aggregation. As the polymer length increased, the transition temperature of the conjugate decreased and the size of the aggregate increased [154].

The combination of N-isopropylacrylamide (NIPAAm) and acrylamide (Am) monomers can produce polymers (pNIPAAm-coAm) with sharp responsiveness at a temperature that is close to human body temperature. These materials have been investigated for their potential to endow proteins with triggerable properties in disease sites, such as tumors or inflamed tissues, at temperatures that are higher than that of normal tissues.

Conjugation of pNIPAAm-co-Am (5 kDa, 37°C LCST) with avidin yielded a derivative that underwent thermal reversible aggregation [160] at a slightly higher LCST than that of the polymer itself, as reported in the literature [154]. The pNIPAAm-co-Am hindrance decreased the recognition of biotinylated antibodies by 15%. A similar PEGylation reduced the affinity by about 64%. After intravenous administration to rats, the pNIPAAm-co-Am-avidin AUC was twice as high as that obtained with the native protein; the distribution and elimination half-lives were respectively 8 and 35 times shorter than that of the PEGylated derivative. These results indicate that the pNIPAm-co-Am bioconjugate distributes more rapidly in the peripheral compartment than PEG-avidin. Thus, it also suggests that pNIPAm-co-Am can be exploited to promote the passive diffusion of the conjugate into disease tissues.

11.7. PEGylated PHARMACEUTICALS ON THE MARKET

In the last two decades, PEGylation has been used in the industrial world, as evidenced by the presence of several PEGylated products on the market. Several successful examples of commercialized PEGylated proteins and peptides are reported in Table 11.2.

PEG bovine adenosine deaminase (marketed as Adagen® from Enzon/NPS Pharma) was the first PEGylated protein to enter the market in March 1990 [76]. Adenosine deaminase is an enzyme that is required for the metabolism of adenosine from food and for the turnover of nucleic acids in tissues. Its genetic

TABLE 11.2 Commercial PEGylated proteins approved by the US Food and Drug Administration for human treatment

Commercial name	Active substance	Indication	Approval year
Adagen® (pegademase bovine)	Adenosine deaminase	Severe combined immunodeficiency disease (SCID)	1990
Oncaspar® (Pegaspargase)	L-asparaginase	Leukaemia	1994
PEGINTRON® (Peginterferon alfa-2b)	Interferon α-2b	Hepatitis C	2000
PEGASYS® (Peginterferon alfa-2a)	Interferon α-2a	Hepatitis C	2002
Neulasta® (Pegfilgrastim)	Granulocyte-colony stimulating factor	Neutropenia associated with cancer chemotherapy	2002
Somavert® (Pegvisomat)	Human growth hormone antagonist	Acromegaly	2003
Mucagen® (Pegaptanib sodium)	anti-VEGF aptamer	Age-related macular degeneration	2004
Mircera® (mPEG-epoetin beta)	Erytropoietin-β	Anaemia associated with chronic kidney disease	2007
Cimzia® (PEG-anti-TNF-α antibody fragment)	Anti-TNF-α Fab′	Rheumatoid arthritis Crohn's disease	2008

deficiency causes severe combined immune deficiency (SCID disease). Adagen® was approved for SCID disease treatment in children as an alternative to bone marrow transplantation. The PEGylated bovine adenosine deaminase is obtained by random conjugation of lysines with 11−17 chains of 5 kDa PEG molecules. Such polymer conjugation increased the enzyme activity by a factor of 1800 and resulted in a long-lasting permanence in the bloodstream [161].

PEGylated L-asparaginase (ONCASPAR® from Enzon/Rhone-Poulenc Rorer) is an enzyme derivative approved by the FDA in 1994 for the treatment of acute lymphoblastic leukemia in children. ONCOSPAR® has been obtained by random and multiple conjugations with 5 kDa linear PEG chains. PEGylation led to a product with a 357-hour half-life, compared to the 24-hour half-life of the native protein, and with a strongly reduced immunogenicity [162,163].

In 2000 and 2002, mono-PEGylated interferon-α2b (PEGINTRON® from Enzon/Schering-Plough) and mono-PEGylated interferon-α2a (PEGASYS® from Nektar/Roche) were approved for the treatment of the hepatitis C virus. While the two interferons have similar amino acid sequences, different conjugation protocols were used. PEGINTRON® was obtained using linear 12 kDa succinimidyl-carbonate-activated PEG molecules. This derivative is a mixture of 15 positional polymer-conjugated isomers [164]. Bioconjugation under conditions of unusually low pH preferentially led to the modification of His^{34}, an amino acid that is located far from the receptor recognition site. The modification of His^{34} produces a derivative that regenerates the native cytokine *in vivo* by PEG release. These features, combined with the low PEG molecular weight, resulted in a high residual activity that was found to be about 37% of that of the native interferon for the His^{34} isomer and about 28% for the isomer mixture. After subcutaneous administration, the bioconjugate showed a 7-fold lower clearance rate and a 5-fold greater *in vivo* half-life than interferon-α2b [165]. PEGASYS® was obtained by the conjugation of one branched 40 kDa PEG to the protein amino groups. The random conjugation of the Lysines and terminal amino groups resulted in a mixture of isomers that maintained 7% of the native cytokine activity. Nevertheless, this bioconjugate displayed a very high systemic bioavailability after subcutaneous administration compared to that of the 12 kDa PEG conjugated interferon-α2b. The circulation half-life shifted from 3–8 hours to 65 hours after PEGylation, the distribution volume decreased from 31–73 L to 8–12 L and the systemic clearance decreased from 6.6–29.2 to 0.06–0.10 L/h [166,167]. In all cases, clinical studies showed that both PEGylated products significantly increased the virological responses, compared to the unPEGylated interferons. The dosage schedule could thus be reduced from three times a week to once a week. PEGylation also resulted in an increase in antiviral activity (from 12 to 135 times as compared to the unPEGylated cytokine) [168]. A large clinical survey showed that, despite differences in residual activity and the pharmacokinetic profiles, the efficacy and side-effect profiles of the two PEGylated products were similar.

In 2002, a 20 kDa PEG N-terminal conjugate of rh-G-CSF (NEULASTA®, from Amgen) received FDA approval for the human treatment of neutropenia associated with cancer chemotherapy. rh-G-CSF stimulates the production of the infection-fighting white blood cells (neutrophils) that are depleted by cancer chemotherapy. NEULASTA® is given once for each cycle of high-dose chemotherapy and no sooner than 14 days before the next chemo infusion, while NEUPOGEN® (Amgen/Roche), the unPEGylated form of rh-G-CSF, requires daily injections for about 14 days until neutrophil counts reach the physiologic level [169].

Pegvisomant (SOMAVERT® from Nektar/Pfizer) is the PEGylated form of an rh-GH muteine. The covalent conjugation of 4–6 5 kDa PEG chains endows a derivative with human growth hormone antagonist properties. The bioconjugate selectively binds to the growth hormone receptor on cell surfaces, thus antagonizing the effect of endogenous growth hormone, and inhibits the release of many cytokines, such as insulin-like growth factor-I (IGF-I) and IGF binding protein-3 (IGFBP-3) [170]. SOMAVERT® was approved for acromegaly treatment by the FDA in 2002 [171].

The linear 30 kDa PEG derivative of Epoetin β (MIRCERA® from Roche) was released in 2007 as an erythropoiesis-stimulating agent for the treatment of symptomatic anemia associated with chronic kidney disease (CKD). Exogenous replacement of erythropoietin by the recombinant hormone epoetin (r-HuErythropoietin) is a well-tolerated therapy for anemia in patients with CKD. Currently, two erythropoietins (epoetin-α and epoetin-β) are available for therapeutic treatment according to a once-a-week administration protocol. A carbohydrate-modified recombinant erythropoietin, darbepoetin-α, with a higher half-life and lower clearance rate is available in the market. Compared to unPEGylated Epoetin-β, MIRCERA® shows a slower association with and a faster dissociation from the receptor. These features reduce its specific activity *in vitro*. Nevertheless, the loss of biological activity is compensated by improved pharmacokinetic properties, such as a longer half-life and an increased overall activity *in vivo* [172,173].

Certolizumab Pegol (CIMZIA® from UCB Pharma) is a PEGylated humanized tumor necrosis factor alpha (TNF-α) inhibitor approved by the FDA in 2008 for the therapeutic treatment of Crohn's disease in patients that are refractory to conventional therapy. CIMZIA® has been obtained by site-specific conjugation of a 20 kDa PEG and an unpaired cysteine. The recommended initial dose of the drug is 400 mg, and the dosing treatment consists of an administration every two weeks. In patients who respond clinically to the treatment, the recommended maintenance regimen is 400 mg every four weeks [174].

A number of new PEGylated proteins, including PEG-Hemoglobin SB1 that is proposed as a blood substitute and infusion fluid for the treatment of stroke, are currently under development or in advanced clinical trials [175,176].

PEG-hirudin is in advanced clinical experimentation as an anticoagulant drug in patients undergoing chronic hemodialysis [177].

PEG-uricase, also known as Pegloticase (PURICASE® by Savient Pharmaceuticals) is in Phase II of a clinical trial to treat hyperuricemia and gout in patients with non-Hodgkin's lymphoma who cannot tolerate, or have not responded adequately to, conventional therapy [178].

A PEGylated derivative of arginine deiminase is investigated in Phase I/II clinical trials in patients with unresectable hepatocellular carcinoma [179] and metastatic melanoma [180]. Melanoma is auxotrophic for arginine because it lacks argininosuccinate synthetase (ASS), a key enzyme required for the synthesis of arginine from citrulline via the urea cycle.

PEGylated glutaminase is also under investigation in an anti-cancer therapy combined with a glutamine analog to remove circulating glutamine that competes with the drug for tumor uptake, thus increasing the drug efficacy *in vivo*. The results of the Phase II clinical trial with patients in late stage colorectal and lung cancers have been recently reported [181].

REFERENCES

[1] F.M. Veronese, Peptide and protein PEGylation: A review of problems and solutions, Biomaterials 22 (2001) 405–417.

[2] F.M. Veronese, J.M. Harris, Introduction and overview of peptide and protein pegylation, Adv. Drug. Del. Rev. 54 (2002) 453–456.

[3] P. Caliceti, F.M. Veronese, Pharmacokinetic and biodistribution properties of poly(ethylene glycol)-protein conjugates, Adv. Drug. Deliv. Rev. 55 (2003) 1261–1277.

[4] M.S. Hershfield, Biochemistry and immunology of poly(ethylene glycol)-modified adenosine deaminase (PEG-ADA), ACS Symposium Series 680 (1997) 145–154.

[5] R. Rabkin, D.C. Dahl, Renal uptake and disposal of proteins and peptides, in: K.L. Audus, T.J. Raub (Eds.), Biological Barriers in Protein Delivery, Plenum Press, New York, 1993, p. 299.

[6] K. Kartre, The conjugation of proteins with poly(ethylene glycol) and other polymers, Adv. Drug. Deliv. Rev. 10 (1993) 91–114.

[7] B.M. Brenner, T.H. Hostetter, H.D. Humes, Glomerular permeselectivity: Barrier function based on discrimination of molecular size and charge, Am. J. Physiol. 234 (1978) F455–F460.

[8] W.M. Deen, M.P. Bohrer, B.M. Brenner, Macromolecule transport across glomerular capillaries: Application of pore theory, Kidney Int. 16 (1979) 353–365.

[9] U.B. Kompella, V.H.L. Lee, Pharmacokinetics of peptide and protein drugs, in: V.H.L. Lee (Ed.), Peptide and Protein Drug Delivery, Marcel Dekker, New York, 1991, p. 391.

[10] R. Rabkin, D.C. Dahl, Renal uptake and disposal of proteins and peptides, in: K.L. Audus, T.J. Raub (Eds.), Biological Barriers in Protein Delivery, Plenum Press, New York, 1993, p. 299.

[11] Y. Takakura, T. Fujita, M. Hashida, H. Sezaki, Disposition of macromolecules in tumor bearing mice, Pharm. Res. 7 (1990) 339–346.

[12] T. Yamaoka, Y. Tabata, Y. Ikada, Fate of water-soluble polymers administered via different routes, J. Pharm. Sci. 84 (1995) 349–354.

[13] B. Chen, K. Jerger, J.M.J. Fréchet, F.C. Szoka, The influence of polymer topology on pharmacokinetics: Differences between cyclic and linear PEGylated poly(acrylic acid) comb polymers, J. Control. Release 140 (2009) 203–209.

[14] J.M. Harris, Poly(ethylene glycol) Chemistry: Biotechnical and Biomedical Applications, Plenum Press, New York, 1991.

[15] M.R. Sherman, L.D. Williams, M.C.P. Saifer, J.A. French, L.W. Kwak, J.J. Oppenheim, Conjugation of high, molecular weight poly(ethylene glycol) to cytokines: Granulocyte-macrophage colony stimulating factors as model substrates, in: J.M. Harrris, S. Zalipsky (Eds.), Poly(ethylene glycol): Chemistry and Biological Applications, ACS, Washington, DC, 1997, p. 155.

[16] R. Nakaoka, Y. Tabata, T. Yamaoka, Y. Ikada, Prolongation of the serum half11-life period of superoxide dismutase by poly(ethylene glycol) modification, J. Control. Release 46 (1997) 253–262.

[17] W.J. Arendshorst, L.G. Navar, Renal circulation and glomerular hemodynamics, in: R.W. Schrier, C.W. Gottshalk (Eds.), Diseases of the Kidney, Little, Brown, Boston, MA, 1988, p. 65.
[18] G.A. Chernysheva, V.I. Smol'yakova, M.B. Plotnikov, I.V. Cherkashina, Pharmacokinetics of high-molecular-weight poly(ethylene oxide), Pharm. Chem. J. 40 (2006) 467–469.
[19] M. Hamidi, A. Azadi, P. Rafiei, Pharmacokinetic consequences of pegylation, Drug. Delivery 13 (2006) 399–409.
[20] M.J. Knauff, D.P. Bell, P. Hirtzer, Z.P. Luo, J.D. Young, N.V. Kartre, Relationship of effective molecular size to systemic clearance in rats of recombinant interleukin-2 chemically modified with water soluble polymers, J. Biol. Chem. 263 (1998) 15064–15070.
[21] K. Yang, A. Basu, M. Wang, R. Chintala, M.C. Hsieh, S. Liu, et al., Tailoring structure-function and pharmacokinetic properties of single-chain Fv proteins by site-specific PEGylation, Protein Engineering 16 (2003) 761–770.
[22] Y. Zhai, Y. Zhao, J. Lei, Z. Su, G. Ma, Enhanced circulation half11-life of site-specific PEGylated rhG-CSF: Optimization of PEG molecular weight, J. Biotechnol. 142 (2009) 259–266.
[23] A. Basu, K. Yang, M. Wang, S. Liu, R. Chintala, T. Palm, et al., Structure-function engineering of interferon-beta-1b for improving stability, solubility, potency, immunogenicity, and pharmacokinetic properties by site-selective mono-PEGylation, Bioconjugate Chem. 17 (2006) 618–630.
[24] Y.W. Jo, Y.S. Youn, S.H. Lee, B.M. Kim, S.H. Kang, M. Yoo, et al., Long-acting interferon-alfa2a modified with a trimer-structured polyethylene glycol: Preparation, *in vitro* bioactivity, *in vivo* stability and pharmacokinetics, Int. J. Pharmaceut. 309 (2006) 87–93.
[25] G.R. Foster, Pegylated interferons for the treatment of chronic hepatitis C: Pharmacological and clinical differences between peginterferon-alpha-2a and peginterferon-alpha-2b, Drugs 70 (2010) 147–165.
[26] Y. Nojima, Y. Suzuki, K. Yoshida, F. Abe, T. Shiga, T. Takeuchi, et al., Lactoferrin conjugated with 40-kDa branched poly(ethylene glycol) has an improved circulating half-life, Pharm. Res. 26 (2009) 2125–2132.
[27] I.L. Koumenis, Z. Shahrokh, S. Leong, V. Hsei, L. Deforge, G. Zapata, Modulating pharmacokinetics of an anti-interleukin-8 F(ab′)(2) by amine-specific PEGylation with preserved bioactivity, Int. J. Pharm. 198 (2000) 83–95.
[28] E.M. Renkin, J.P. Gilmore, in: J. Orloff, R.W. Berliner (Eds.), Renal Physiology, American Physiological Society, Bethesda, MD, 1973, pp. 185–248.
[29] A.P. Chapman, P. Antoniw, M. Spitali, S. West, S. Stephens, D.J. King, Therapeutic antibody fragments with prolonged *in vivo* half11-lives, Nat. Biotechnol. 17 (1999) 780–783.
[30] R. Clark, K. Olson, G. Fuh, M. Marian, D. Mortensen, G. Teshima, et al., Long-acting growth hormones produced by conjugation with polyethylene glycol, J. Biol. Chem. 271 (1996) 21969–21977.
[31] F.M. Veronese, P. Caliceti, A. Pastorino, O. Schiavon, L. Sartore, L. Banci, et al., Preparation, physicochemical and pharmacokinetic characterization of monomethoxypoly(ethylene glycol)-derivatized superoxide dismutase, J. Control. Release 10 (1989) 145–154.
[32] P. Bailon, A. Palleroni, C.A. Schaffer, C.L. Spence, W.J. Fung, J.E. Porter, et al., Rational design of a potent, long lasting form of interferon: A 40 kDa branched polyethylene glycol-conjugated interferon-2a for the treatment of hepatitis C, Bioconjugate Chem. 12 (2001) 195–202.
[33] P. van Der Auwera, E. Platzer, Z.X. Xu, R. Schulz, O. Feugeas, R. Capdeville, et al., Pharmacodynamics and pharmacokinetics of single doses of subcutaneous pegylated human

G-CSF mutant (Ro 25-8315) in healthy volunteers: Comparison with single and multiple daily doses of filgrastim, Am. J. Hematol. 66 (2001) 245–251.
[34] T. Yamaoka, Y. Tabata, Y. Ikada, Comparison of body distribution of poly(vinyl alcohol) with other water-soluble polymers after intravenous injection, J. Pharm. Pharmacol. 47 (1995) 479–486.
[35] Y. Yabe, M. Nishikawa, A. Tamada, Y. Takakura, M. Hashida, Targeted delivery and improved therapeutic potential of catalase by chemical modification: Combination with superoxide dismutase, J. Pharmacol. Exp. Ther. 289 (1999) 1176–1184.
[36] C.D. Conover, C.W. Gilbert, K.L. Shum, R.L. Shorr, The impact of polyethylene glycol conjugation on bovine hemoglobin's circulatory half1 1-life and renal effects in a rabbit top-loaded transfusion model, Artif. Organs 21 (1997) 907–915.
[37] J. Lamka, O. Schiavon, M. Laznicek, P. Caliceti, F.M. Veronese, Distribution of catalase, ribonuclease, superoxide dismutase modified by monomethoxy(polyethylene glycol) into rat cerebral lymph and lymphatic nodes, Physiol. Res. 44 (1995) 307–313.
[38] S.A. Chen, R.J. Sawchuk, R.C. Brundage, C. Horvath, H.V. Mendenhall, R.A. Gunther, et al., Plasma and lymph pharmacokinetics of recombinant human interleukin-2 and polyethylene glycol-modified interleukin-2 in pigs, J. Pharmacol. Exp. Ther. 293 (2000) 248–259.
[39] Y. Matsumura, H. Maeda, A new concept for macromolecular therapeutics in cancer chemotherapy: Mechanism of tumoritropic accumulation of proteins and the antitumor agent smancs, Cancer Res. 46 (1986) 6387–6392.
[40] T. Sawa, J. Wu, T. Akaike, H. Maeda, Tumor-targeting chemotherapy by a xanthine oxidase polymer conjugate that generates oxygen-free radicals in tumor tissue, Cancer Res. 60 (2000) 666–671.
[41] T.L. Cheng, B.M. Chen, L.Y. Chan, P.Y. Wu, J.W. Chern, S.R. Roffler, Poly(ethylene glycol) modification of b-glucuronidase-antibody conjugates for solid-tumor therapy by targeted activation of glucuronide prodrugs, Cancer Immunol. Immun. 44 (1997) 305–315.
[42] P.M. Deckert, A. Jungbluth, N. Montalto, M.A. Clark, R.D. Finn, C. Williams, et al., Pharmacokinetics and microdistribution of polyethylene glycol-modified humanized A33 antibody targeting colon cancer xenografts, Int. J. Cancer 87 (2000) 832–839.
[43] C. Delgado, R.B. Pedley, A. Herraz, R. Boden, J.A. Boden, P.A. Keep, et al., Enhanced tumor specificity of an anti-carcinoembrionic antigen Fab′ fragment by poly(ethylene glycol) (PEG) modification, Brit. J. Cancer 73 (1996) 175–182.
[44] R.K. Jain, Physiological barriers to delivery of monoclonal antibodies and other macromolecules to tumors, Cancer Res. 50 (1990) 814–819.
[45] J.S. Patton, C.S. Fishburn, J.G. Weers, The lungs as a portal of entry for systemic drug delivery, Proc. Am. Thorac. Soc. 1 (2004) 338–344.
[46] N. Seersholm, M. Wencker, N. Banik, K. Viskum, A. Dirksen, A. Kok-Jensen, et al., Does a1-antitrypsin augmentation therapy slow the annual decline in FEV1 in patients with severe hereditary a1-antitrypsin deficiency? Wissenschaftliche Arbeitsgemeinschaft zur Therapie von Lungenerkrankungen (WATL) a1-AT study group, Eur. Respir. J. 10 (1997) 2260–2263.
[47] A.M. Cantin, D.E. Woods, D. Cloutier, E.K. Dufour, R. Leduc, Polyethylene glycol conjugation at Cys232 prolongs the half1 1-life of a1 proteinase inhibitor, Am. J. Respir. Cell Mol. Biol. 27 (2002) 659–665.
[48] Y.S. Youn, M.J. Kwon, D.H. Na, S.Y. Chae, S. Lee, K.C. Lee, Improved intrapulmonary delivery of site-specific PEGylated salmon calcitonin: Optimization by PEG size selection, J. Control. Release 125 (2008) 68–75.

[49] Y.S. Youn, D.H. Na, K.C. Lee, High-yield production of biologically active mono-PEGylated salmon calcitonin by site-specific PEGylation, J. Control. Release 117 (2007) 371–379.

[50] K.C. Lee, S.Y. Chae, T.H. Kim, S. Lee, E.S. Lee, Y.S. Youn, Intrapulmonary potential of polyethylene glycol-modified glucagon-like peptide-1s as a type 2 anti-diabetic agent, Regul. Peptides 152 (2009) 101–107.

[51] Y.S. Youn, J.E. Jeon, S.Y. Chae, S. Lee, K.C. Lee, PEGylation improves the hypoglycaemic efficacy of intranasally administered glucagon-like peptide-1 in type 2 diabetic *db/ db* mice, Diabetes Obes. Metab. 10 (2008) 343–346.

[52] D.H. Na, Y.S. Youn, E.J. Park, J.M. Lee, O.R. Cho, K.R. Lee, et al., Stability of PEGylated salmon calcitonin in nasal mucosa, J. Pharm. Sci. 93 (2004) 256–261.

[53] K.C. Lee, M.O. Park, D.H. Na, Y.S. Youn, S.D. Lee, S.D. Yoo, et al., Intranasal delivery of PEGylated salmon calcitonins: Hypocalcemic effects in rats, Calcified Tissue Int. 73 (2003) 545–549.

[54] Y. Nojima, Y. Suzuki, K. Iguchi, T. Shiga, A. Iwata, T. Fujimoto, et al., Development of poly (ethylene glycol) conjugated lactoferrin for oral administration, Bioconjugate Chem. 19 (2008) 2253–2259.

[55] P. Caliceti, F.M. Veronese, Improvement of the physico-chemical and biopharmaceutical properties of insulin by poly(ethylene glycol) conjugation, STP Pharma. Sci. 9 (1999) 107–113.

[56] A. Bernkop-Schnürch, S. Thaler, Polycarbophil–cysteine conjugates as platforms for oral (poly)peptide delivery systems, J. Pharm. Sci. 89 (2000) 901–909.

[57] P. Calceti, S. Salmaso, G. Walker, A. Bernkop-Schnürch, Development and *in vivo* evaluation of an oral insulin-PEG delivery system, Eur. J. Pharm. Sci. 22 (2004) 315–323.

[58] T. Matsuda, T. Matsubara, S. Hino, Immunogenic and allergenic potentials of natural and recombinant innocuous proteins, J. Biosci. Bioeng. 101 (2006) 203–211.

[59] A.S. De Groot, D.W. Scott, Immunogenicity of protein therapeutics, Trends Immunol. 28 (2007) 482–490.

[60] F.F. Davis, G.M. Kazo, M.L. Nucci, A. Abuchowski, Reduction of immunogenicity and extension of circulating half11-life of peptides and proteins, in: V.H.L. Lee (Ed.), Peptide and Protein Drug Delivery, Mercel Dekker, New York, 1991, p. 831.

[61] F.M. Veronese, P. Caliceti, O. Schiavon, Branched and linear poly(ethylene glycol): Influence of the polymer structure on enzymological, pharmacokinetic, and immunological properties of protein conjugates, J. Bioact. Comp. Polym. 12 (1997) 196–207.

[62] P. Caliceti, O. Schiavon, F.M. Veronese, Immunological properties of uricase conjugated to neutral soluble polymers, Bioconjugate Chem. 12 (2001) 515–522.

[63] A.P. Chapman, PEGylated antibodies and antibody fragments for improved therapy: A review, Adv. Drug Deliv. Rev. 54 (2002) 531–545.

[64] S. Wanwetswinkel, S. Plaisance, Z. Zhi-yong, I. Vanlinthout, K. Brepoeles, I. Lasters, et al., Pharmacokinetic and thromolytic properties of cysteine-linked polyethylene glycol derivatives of staphylokinase, Blood 95 (2000) 936–942.

[65] A.W. Richter, E. Akerblom, Antibodies against polyethylene glycol produced in animals by immunization with monomethoxy polyethylene glycol modified proteins, Int. Arch. Allergy Appl. Immunol. 70 (1983) 124–131.

[66] J. Shiea, K.W. Liao, Y.M. Wang, Y.C. Chang, B.J. Huang, C.J. Wu, et al., Measurement of poly(ethylene glycol) by cell-based anti-poly(ethylene glycol) ELISA, Anal. Chem. 82 (2010) 2355–2362.

[67] Y.C. Su, B.M. Chen, K.H. Chuang, T.L. Cheng, S.R. Roffler, Sensitive quantification of PEGylated compounds by second-generation anti-Poly(ethylene glycol) monoclonal antibodies, Bioconjugate Chem. 21 (2010) 1264–1270.
[68] J.C. Tsuji, K. Hirose, Studies on antigenicity of polyethylene glycol (PEG)-modified uricase, Int. J. Immunopharmacol. 7 (1985) 725–730.
[69] T.L. Cheng, P.Y. Wu, M.F. Wu, J.W. Chern, S.R. Roffler, Accelerated clearance of polyethylene glycol-modified proteins by anti-polyethylene glycol IgM, Bioconjugate Chem. 10 (1999) 520–528.
[70] M.J. Roberts, M.D. Bentley, J.M. Harris, Chemistry for peptide and protein PEGylation, Adv. Drug. Deliv. Rev. 54 (2002) 459–476.
[71] A. Bendele, J. Seely, C. Richey, G. Sennello, G. Shopp, Short communication: Renal tubular vacuolation in animals treated with polyethylene-glycol-conjugated proteins, Toxicol. Sci. 42 (1998) 152–157.
[72] I. Hamad, A.C. Hunter, J. Szebeni, S.M. Moghimi, Poly(ethylene glycol)s generate complement activation products in human serum through increased alternative pathway turnover and a MASP-2-dependent process, Mol. Immunol. 46 (2008) 225–232.
[73] G.T. Hermanson, Bioconjugate Techniques, Academic Press, San Diego, CA, 1996.
[74] F.F. Davis, PEG-adenosine deaminase and PEG-asparaginase, Adv. Exp. Med. Biol. 519 (2004) 51–58.
[75] L.M. Graham, PEGASPARAGINASE: A review of clinical studies, Adv. Drug. Deliv. Rev. 55 (2003) 1293–1302.
[76] Y. Levy, M.S. Hershfield, C. Fernandez-Mejia, S.H. Polmar, D. Scudiery, M. Berger, et al., Adenosine deaminase deficiency with late onset or recurrent infections: Response to treatment with polyethylene glycol modified adenosine deaminase, J. Pediatr. 113 (1988) 312–317.
[77] J.M. Harris, R.B. Chess, Effect of pegylation on pharmaceuticals, Nat Rev Drug Discov. 2 (2003) 214–221.
[78] S. Hermeling, D.J. Crommelin, H. Schellekens, W. Jskoot, Structure–immunogenicity relationships of therapeutic proteins, Pharm. Res. 21 (2004) 897–903.
[79] G.E. Francis, D. Fisher, C. Delgado, F. Malik, A. Gardiner, D. Neale, PEGylation of cytokines and other therapeutic proteins and peptides: The importance of biological optimisation of coupling techniques, Int. J. Hematol. 68 (1998) 1–18.
[80] H.A. Klok, Biological-synthetic hybrid block copolymers: Combining the best from two worlds, J. Polym. Sci. A Polym. Chem. 43 (2005) 1–17.
[81] O.B. Kinstler, D.N. Brems, S.L. Lauren, A.G. Paige, J.B. Hamburger, M.J. Treuheit, Characterization and stability of N-terminal PEGylated rhG-CSF, Pharm. Res. 13 (1996) 996–1002.
[82] T.D. Bartley, J.M. Bogenberger, R.A. Bosselman, P. Hunt, O.B. Kinstler, B.B. Samal, Monopegylated proteins that stimulate megakaryocyte growth and differentiation, US Patent US 5795569, Amgen, 1998.
[83] O. Kinstler, G. Molineux, M. Treuheit, D. Ladd, C. Gegg, Mono-N-terminal poly(ethylene glycol)–protein conjugates, Adv. Drug. Deliver. Rev. 54 (2002) 477–485.
[84] P. Anderlini, D. Przepiorka, R. Champlin, M. Korbling, Biologic and clinical effects of granulocyte colony stimulating factor in normal individuals, Blood 88 (1996) 2819–2825.
[85] K. Welte, J. Gabrilove, M. Bronchud, E. Platzer, G. Morstyn, Filgrastim (r-metHuG-CSF): The first 10 years, Blood 88 (1996) 1907–1929.
[86] M. Fukuda, M. Oka, Y. Ishida, H. Kinoshita, K. Terashi, M. Fukuda, et al., Effects of renal function on pharmacokinetics of recombinant human granulocyte colony-stimulating factor in lung cancer patients, Antimicrob. Agents Chemother. 45 (2001) 1947–1951.

[87] E. Johnston, J. Crawford, S. Blackwell, T. Bjurstrom, P. Lockbaum, L. Roskos, et al., Randomized, dose-escalation study of SD/01 compared with daily filgras-tim in patients receiving chemotherapy, J. Clin. Oncol. 18 (2000) 2522–2528.

[88] P.I. Guerra, C. Acklin, A.A. Kosky, J.M. Davis, M.J. Treuheit, D.N. Brems, PEGylation prevents the N-terminal degradation of megakaryocyte growth and development factor, Pharm. Res. 15 (1998) 1822–1827.

[89] T.R. Ulich, J. del Castillo, G. Senaldi, O. Kinstler, S. Yin, S. Kaufman, et al., Systemic hematologic effects of PEG-rHuMGDF11-induced megakaryocyte hyperplasia in mice, Blood 87 (1996) 5006–5015.

[90] J.M. McFarland, M.B. Francis, Reductive alkylation of proteins using Iridium catalyzed transfer hydrogenation, J. Am. Chem. Soc. 127 (2005) 13490–13491.

[91] M. Morpurgo, F.M. Veronese, D. Kachensky, J.M. Harris, Preparation of characterization of poly(ethylene glycol) vinyl sulfone, Bioconjugate Chem. 7 (1996) 363–368.

[92] G.P. Adams, J.E. McCartney, M.S. Tai, H. Oppermann, J.S. Huston, W.F. Stafford III, et al., Highly specific *in vivo* tumor targeting by monovalent and divalent forms of 741F8 anti-c-erbB-2 single-chain Fv, Cancer Res. 53 (1993) 4026–4034.

[93] A. Natarajan, C.Y. Xiong, H. Albrecht, G.L. DeNardo, S.J. DeNardo, Characterization of site-specific ScFv PEGylation for tumor-targeting pharmaceuticals, Bioconjugate Chem. 16 (2005) 113–121.

[94] M.X. Li, H.W. Yeung, L.P. Pan, S.I. Chan, Trichosanthin, a potent HIV-1 inhibitor, can cleave supercoiled DNA *in vitro*, Nucleic Acids Res. 19 (1991) 6309–6312.

[95] S.W. Tsao, K.T. Yan, H.W. Yeung, Selective killing of choriocarcinoma cells *in vitro* by trichosanthin, a plant protein purified from root tubers of the Chinese medicinal herb *Trichosanthes kirilowii*, Toxicon 24 (1986) 831–840.

[96] M.S. McGrath, K.M. Hwang, S.E. Caldwell, I. Gaston, K.C. Luk, P. Wu, et al., GLQ223: An inhibitor of human immunodeficiency virus replication in acutely and chronically infected cells of lymphocyte and mononuclear phagocyte lineage, Proc. Natl. Acad. Sci. U.S.A. 86 (1989) 2844–2848.

[97] Q. An, Y. Lei, N. Jia, X. Zhang, Y. Bai, J. Yi, et al., Effect of site-directed PEGylation of trichosanthin on its biological activity, immunogenicity, and pharmacokinetics, Biomol. Eng. 24 (2007) 643–649.

[98] V. Briz, E. Poveda, V. Soriano, HIV entry inhibitors: Mechanisms of action and resistance pathways, J Antimicrob. Chemoth. 57 (2006) 619–627.

[99] M.R. Boyd, K.R. Gustafson, J.B. McMahon, R.H. Shoemaker, B.R. O'Keefe, T. Mori, et al., Discovery of cyanovirin-N, a novel human immunodeficiency virus-inactivating protein that binds viral surface envelope glycoprotein gp120: Potential applications to microbicide development, Antimicrob. Agents Chemother. 41 (1997) 1521–1530.

[100] H. Zappe, M.E. Snell, M.J. Bossard, PEGylation of cyanovirin-N, an entry inhibitor of HIV, Adv. Drug Deliv. Rev. 60 (2008) 79–87.

[101] C.A. Bewley, S. Otero-Quintero, The potent anti-HIV protein cyanovirin-N contains two novel carbohydrate binding sites that selectively bind to Man(8) D1D3 and Man(9) with nanomolar affinity: Implications for binding to the HIVenvelope protein gp120, J. Am. Chem. Soc. 123 (2001) 3892–3902.

[102] S. Salmaso, S. Bersani, A. Scomparin, F. Mastrotto, R. Scherpfer, G. Tonon, et al., Tailored PEG for rh-G-CSF analogue site-specific conjugation, Bioconjugate Chem. 20 (2009) 1179–1185.

[103] M. Petersen, P. Jonson, S. Petersen, Amino acid neighbours and detailed conformational analysis of cysteines in proteins, Protein Eng. 12 (1999) 535–548.

[104] T. Creighton, Disulphide-coupled protein folding pathways, Philos. Trans. R. Soc. Lond. B. Biol. Sci. 348 (1995) 5–10.

[105] S. Brocchini, A. Godwin, S. Balan, J.W. Choi, M. Zloh, S. Shaunak, Disulfide bridge based PEGylation of proteins, Adv. Drug. Deliv. Rev. 60 (2008) 3–12.

[106] M. Zloh, S. Shaunak, S. Balan, S. Brocchini, Identification and insertion of 3-carbon bridges in protein disulfide bonds: A computational approach, Nat. Protocols 2 (2007) 1070–1083.

[107] M. Denton, H. Scheraga, Spectroscopic, immunochemical, and thermodynamic properties of carboxymethyl (cys6, cys127)-hen egg white lysozyme, J. Protein Chem. 10 (1991) 213–232.

[108] S. Balan, J. Choi, A. Godwin, I. Teo, C. Laborde, S. Heidelberger, et al., Site-specific PEGylation of protein disulfide bonds using a three-carbon bridge, Bioconjugate Chem. 18 (2007) 61–76.

[109] S. Shaunak, A. Godwin, J.W. Choi, S. Balan, E. Pedone, D. Vijayarangam, et al., Site-specific PEGylation of native disulfide bonds in therapeutic proteins, Nat. Chem. Biol. 2 (2006) 312–313.

[110] T. Hohsaka, N. Muranaka, C. Komiyama, K. Matsui, S. Takaura, R. Abe, et al., Position-specific incorporation of dansylated non-natural amino acids intostreptavidin by using a four-base codon, FEBS Lett. 560 (2004) 173–177.

[111] D. Kajihara, R. Abe, I. Iijima, C. Komiyama, M. Sisido, T. Hohsaka, FRET analysis of protein conformational change through position-specific incorporation of fluorescent amino acids, Nat. Methods 3 (2006) 923–929.

[112] T. Watanabe, N. Muranaka, I. Iijima, T. Hohsaka, Position-specific incorporation of biotinylated non-natural amino acids into a protein in a cell-free translation system, Biochem. Biophys. Res. Commun. 361 (2007) 794–799.

[113] L. Wang, A. Brock, B. Herberich, P.G. Schultz, Expanding the genetic code of *Escherichia coli*, Science 292 (2001) 498–500.

[114] J.W. Chin, S.W. Santoro, A.B. Martin, D.S. King, L. Wang, P.G. Schultz, Addition of *p*-Azido-l-phenylalanine to the genetic code of *Escherichia coli*, J. Am. Chem. Soc. 124 (2002) 9026–9027.

[115] C.W. Tornoe, C. Christensen, M. Meldal, Peptidotriazoles on solid phase: [1,2,3]-Triazoles by regiospecific copper(I)-catalyzed 1,3-dipolar cycloadditions of terminal alkynes to azides, J. Org. Chem. 67 (2002) 3057–3064.

[116] A. Deiters, T.A. Cropp, D. Summerer, M. Mukherji, P.G. Schultz, Site-specific PEGylation of proteins containing unnatural amino acids, Bioorg. Med. Chem. Lett. 14 (2004) 5743–5745.

[117] B. Le Droumaguet, K. Velonia, Click chemistry: A powerful tool to create polymer-based macromolecular chimeras, Macromol. Rapid Comm. 29 (2008) 1073–1089.

[118] N. Shozen, I. Iijima, T. Hohsaka, Site-specific incorporation of PEGylated amino acids into proteins using nonnatural amino acid mutagenesis, Bioorg. Med. Chem. Lett. 19 (2009) 4909–4911.

[119] S.A. Robertson, J.A. Ellman, P.G.J. Schultz, A general and efficient route for chemical aminoacylation of transfer RNAs, J. Am. Chem. Soc. 113 (1991) 2722–2729.

[120] H. Sato, K. Yamamoto, E. Hayashi, Y. Takahara, Transglutaminase-mediated dual and site-specific incorporation of poly(ethylene glycol) derivatives into a chimeric interleukin-2, Bioconjugate Chem. 11 (2000) 502–509.

[121] R. Griffin, R. Casadio, C.M. Bergamini, Transglutaminases: Nature's biological glues, Biochem. J. 368 (2002) 377–396.

[122] Y. Zhu, A. Rinzema, J. Tramper, J. Bol, Microbial transglutaminase: A review of its production and application in food processing, Appl. Microbiol. Biotechnol. 44 (1995) 277–282.

[123] J.E. Folk, Mechanism and basis for specificity of transglutaminase catalyzed ε-(γ-glutamyl) lysine bond formation, Adv. Enzymol. Relat. Areas Mol. Biol. 54 (1983) 1–56.

[124] P.J. Coussons, N.C. Price, S.M. Kelly, B. Smith, L. Sawyer, Factors that govern the specificity of transglutaminase-catalysed modification of proteins and peptides, Biochem. J. 282 (1992) 929–930.

[125] A. Fontana, B. Spolaore, A. Mero, F.M. Veronese, Site-specific modification and PEGylation of pharmaceutical proteins mediated by transglutaminase, Adv. Drug. Deliver. Rev. 60 (2008) 13–28.

[126] H. Sato, E. Hayashi, N. Yamada, M. Yatagai, Y. Takahara, Further studies on the site-specific protein modification by microbial transglutaminase, Bioconjugate Chem. 12 (2001) 701–710.

[127] A. Mero, B. Spolaore, F.M. Veronese, A. Fontana, Transglutaminase-mediated PEGylation of proteins: Direct identification of the sites of protein modification by mass spectrometry using a novel monodisperse PEG, Bioconjugate Chem. 20 (2009) 384–389.

[128] C. Maullu, D. Raimondo, F. Caboi, A. Giorgetti, M. Sergi, M. Valentini, et al., Site-directed enzymatic PEGylation of the human granulocyte colony-stimulating factor, FEBS J. 276 (2009) 6741–6750.

[129] S. DeFrees, Z.G. Wang, R. Xing, A.E. Scott, J. Wang, D. Zopf, et al., GlycoPEGylation of recombinant therapeutic proteins produced in *Escherichia coli*, Glycobiology 16 (2006) 833–843.

[130] B.A. Konkle, L.S. Ebbesen, E. Erhardtsen, R.P. Bianco, T. Lissitchkov, L. Rusen, et al., Randomized, prospective clinical trial of recombinant factor VIIa for secondary prophylaxis in hemophilia patients with inhibitors, J. Thromb. Haemost. 5 (2007) 1904–1913.

[131] C.M. Lindley, W.T. Sawyer, B.G. Macik, J. Lusher, J.F. Harrison, K. Baird-Cox, et al., Pharmacokinetics and pharmacodynamics of recombinant factor VIIa, Clin. Pharmacol. Ther. 55 (1994) 638–648.

[132] H.R. Stennicke, H. Ostergaard, R.J. Bayer, M.S. Kalo, K. Kinealy, P.K. Holm, et al., Generation and biochemical characterization of glycoPEGylated factor VIIa derivatives, Thromb. Haemost. 100 (2008) 920–928.

[133] P. Sen, S. Ghosh, M. Ezban, U.R. Pendurthi, L. Vijaya Mohan Rao, Effect of glycoPEGylation on factor VIIa binding and internalization, Haemophilia 16 (2010) 339–348.

[134] R.B. Greenwald, PEG drugs: An overview, J. Control. Release 74 (2001) 159–171.

[135] R. Webster, E. Didier, P. Harris, N. Siegel, J. Stadler, L. Tilbury, et al., PEGylated proteins: Evaluation of their safety in the absence of definitive metabolism studies, Drug Metab. Dispos. 35 (2007) 9–16.

[136] D. Filpula, H. Zhao, Releasable PEGylation of proteins with customized linkers, Adv. Drug Deliv. Rev. 60 (2008) 29–49.

[137] S. Lee, R.B. Greenwald, J. McGuire, K. Yang, C. Shi, Drug delivery systems employing 1,6-elimination: Releasable poly(ethylene glycol) conjugates of proteins, Bioconjugate Chem. 12 (2001) 163–169.

[138] R.B. Greenwald, H. Zhao, K. Yang, P.A. Reddy, A. Martinez, A new aliphatic amino prodrug system for the delivery of small molecules and proteins utilizing novel PEG derivatives, J. Med. Chem. 47 (2004) 726–734.

[139] H. Zhao, K. Yang, A. Martinez, A. Basu, R. Chintala, H.C. Liu, et al., Linear and branched bicin linkers for releasable PEGylation of macromolecules: Controlled release

in vivo and *in vitro* from mono- and multi-PEGylated proteins, Bioconjugate Chem. 17 (2006) 341–351.

[140] D.R. Filpula, K. Yang, A. Basu, I. Pastan, Polymer-linked *Pseudomonas* exotoxin immunotoxin, P.C.T. WO 2006/065867 (2006).

[141] H. Tsubery, M. Mironchik, M. Fridkin, Y. Shechter, Prolonging the action of protein and peptide drugs by a novel approach of reversible polyethylene glycol modification, J. Biol. Chem. 279 (2004) 38118–38124.

[142] A.S. Maisel, Nesiritide: A new therapy for the treatment of heart failure, CardioVasc. Toxico. 3 (2003) 37–42.

[143] M. Nesher, Y. Vachutinsky, G. Fridkin, Y. Schwarz, K.M. SassonFridkin, Y. Shechter, et al., Reversible Pegylation prolongs the hypotensive effect of atrial natriuretic peptide, Bioconjugate Chem. 19 (2008) 342–348.

[144] Y.S. Youn, D.H. Na, S.D. Yoo, S.C. Song, K.C. Lee, Carbohydrate-specifically polyethylene glycol-modified ricin A-chain with improved therapeutic potential, Int. J. Biochem. Cell. Biol. 37 (2005) 1525–1533.

[145] S. Salmaso, A. Semenzato, S. Bersani, F. Mastrotto, A. Scomparin, P. Caliceti, Site-selective protein glycation and PEGylation, Eur. Polym. J. 44 (2008) 1378–1389.

[146] J. Nicolas, G. Mantovani, D.M. Haddleton, Living radical polymerization as a tool for the synthesis of polymer-protein/peptide bioconjugates, Macromol. Rapid. Commun. 28 (2007) 1083–1111.

[147] M. Kato, M. Kamigaito, M. Sawamoto, T. Higashimura, Polymerization of methyl methacrylate with the carbon tetrachloride/dichlorotris-(triphenylphosphine)ruthenium(II)/ methylaluminum bis(2,6-di-tert-butylphenoxide) initiating system: Possibility of living radical polymerization, Macromolecules 28 (1995) 1721–1723.

[148] J.S. Wang, K. Matyjaszewski, Controlled/"living" radical polymerization. Atom transfer radical polymerization in the presence of transition-metal complexes, J. Am. Chem. Soc. 117 (1995) 5614.

[149] D.H. Solomon, E. Rizzardo, P. Cacioli, US Patent US 4581429, Commonwealth Scientific and Industrial Research Organization, 1986.

[150] P. Corpart, D. Charmot, T. Biadatti, S. Z. Zard, D. Michelet, WO 9858974, 1998.

[151] B.S. Lele, H. Murata, K. Matyjaszewski, A.J. Russell, Synthesis of uniform protein-polymer conjugates, Biomacromolecules 6 (2005) 3380–3387.

[152] V. Depp, A. Alikhani, V. Grammer, B.S. Lele, Native protein-initiated ATRP: A viable and potentially superior alternative to PEGylation for stabilizing biologics, Acta Biomater. 5 (2009) 560–569.

[153] D. Bontempo, H.D. Maynard, Streptavidin as a macroinitiator for polymerization: *In situ* protein–polymer conjugate formation, J. Am. Chem. Soc. 127 (2005) 6508–6509.

[154] C. Boyer, V. Bulmus, J. Liu, T.P. Davis, M.H. Stenzel, C. Barner-Kowollik, Well-defined protein–polymer conjugates via *in situ* RAFT polymerization, J. Am. Chem. Soc. 129 (2007) 7145–7154.

[155] W. Gao, W. Liu, J.A. Mackay, M.R. Zalutsky, E.J. Toone, A. Chilkoti, *In situ* growth of a stoichiometric PEG-like conjugate at a protein's N-terminus with significantly improved pharmacokinetics, Proc. Natl. Acad. Sci. U.S.A. 106 (2009) 15231–15236.

[156] J.P. Magnusson, S. Bersani, S. Salmaso, C. Alexander, P. Caliceti, *In situ* growth of side-chain peg polymers from functionalized human growth hormonesa new technique for preparation of enhanced protein–polymer conjugates, Bioconjugate Chem. 21 (2010) 671–678.

[157] J. Nicolas, V. San Miguel, G. Mantovani, D.M. Haddleton, Fluorescently tagged polymer bioconjugates from protein derived macroinitiators, Chem. Commun. (2006) 4697–4699.

[158] H. Malonne, F. Eeckman, D. Fontaine, A. Otto, L. DeVos, A. Moes, et al., Preparation of poly(N-isopropylacrylamide) copolymers and preliminary assessment of their acute and subacute toxicity in mice, Eur. J. Pharm. Biopharm. 61 (2005) 188–194.
[159] C. de Las Heras Alarcon, S. Pennadam, C. Alexander, Responsive polymers for biomedical applications, Chem. Soc. Rev. 34 (2005) 276–285.
[160] S. Salmaso, S. Bersani, S.S. Pennadam, C. Alexander, P. Caliceti, Avidin bioconjugate with a thermoresponsive polymer for biological and pharmaceutical applications, Int. J. Pharm. 340 (2007) 20–28.
[161] M.S. Hershfield, in: J.M. Harris (Ed.), Polyethylene Glycol Chemistry: Biotechnical and Biomedical Applications, Plenum Press, NewYork, 1992, pp. 145–154.
[162] L.M. Holle, Pegasparagase: An alternative, Ann. Pharmacother. 3 (1997) 616–624.
[163] U. Narta, S. Kanwar, W. Azmi, Pharmacological and clinical evaluation of l-asparaginase in the treatment of leukemia, Crit. Rev. Oncol. Hematol. 61 (2007) 208–221.
[164] Y.S. Wang, S. Youngster, M. Grace, J. Bausch, R. Bordensc, D.F. Wyssa, Structural and biological characterization of pegylated recombinant interferon alpha-2b and its therapeutic implications, Adv. Drug Deliv. Rev. 54 (2002) 547–570.
[165] P. Glue, J.W.S. Fang, R. Sabo, R. Rouzier-Panis, C. Raffanel, S.K. Gupta, et al., PEG-interferon-α2b: Pharmacokinetics, pharmacodynamics, safety and preliminary efficacy data, Clin. Pharmacol. Ther. 68 (2000) 556–567.
[166] M. Silva, J. Poo, F. Wagner, A randomised trial to compare the pharmacokinetic, pharmacodynamic, and antiviral effects of peginterferon alfa-2b and peginterferon alfa-2a in patients with chronic hepatitis C (COMPARE), J. Hepatol. 45 (2006) 204–213.
[167] F. Sierra, Peginterferon alfa-2b vs. peginterferon alfa-2a in the treatment of hepatitis C, is there any difference? J. Hepatol. 46 (2007) 349–350.
[168] M.W. Modi, M.W. Lamb, M. Shiomi, Peginterferon alfa-2a (40KD): A potent long-acting form of interferon alfa-2a for the treatment of hepatitis C, Adv. Exp. Med. Biol. 519 (2003) 59–67.
[169] M. Green, H. Koelbl, J. Baselga, A randomized double-blind multicenter phase III study of fixed-dose single-administration PEGfilgrastim *versus* daily filgrastim in patients receiving myelosuppressive chemotherapy, Ann. Oncol. 14 (2003) 29–35.
[170] K. Olson, R. Gehant, V. Mukku, K. O'Connell, B. Tomlinson, K. Totpal, et al., Preparation and characterization of poly(ethylene glycol)ylated human growth hormone antagonist, in: J.M. Harris, S. Zalipsky (Eds.), Poly(ethylene glycol): Chemistry and Biological Applications, ACS, American Chemical Society, Washington, DC, 1997, pp. 170–181.
[171] A.J. van der Lely, R.K. Hutson, P.J. Trainer, G.M. Besser, A.L. Barkan, L. Katznelson, et al., Long-term treatment of acromegaly with pegvisomant, a growth hormone receptor antagonist, Lancet 358 (2001) 1754–1759.
[172] M. Klinger, M. Arias, V. Vargemezis, A. Besarab, W. Sulowicz, T. Gerntholtz, et al., Efficacy of intravenous methoxy polyethylene glycol-epoetin beta administered every 2 weeks compared with epoetin administered 3 times weekly in patients treated by haemodialysis or peritoneal dialysis: A randomized trial, Am. J. Kidney Dis. 50 (2007) 989–1000.
[173] Committee for medicinal products for human use, European Medicines Agency, Mircera: Summary for the public, in: Mircera: European public assessment report, EMEA. EMEA/H/C/739, London, 2007. Available: http://www.emea.europa.eu/humandocs/PDFs/EPAR/mircera/H-739-en1.pdf.
[174] P. Rutgeerts, S. Schreiber, B. Feagan, D.L. Keininger, L. O'Neil, R.N. Fedorak, CDP870 Crohn's Disease Study Group Certolizumab pegol, a monthly subcutaneously administered

Fc-free anti-TNFalpha, improves health-related quality of life in patients with moderate to severe Crohn's disease, Int. J. Colorectal Dis. 23 (2008) 289–296.

[175] O.S. Kwon, U.T. Chung, Y.B. Chung, Pharmacokinetics of PEG-Hemoglobin SB1, a hemoglobin-based oxygen carrier, after its intravenous administration in beagle dogs, Arch. Pharm. Res. 27 (2004) 259–264.

[176] J. Lee, J. Lee, S. Yoon, K. Nho, Pharmacokinetics of 125I-radiolabelled PEG-hemoglobin SB1, Artif. Cells Blood Substit. Immobil. Biotechnol. 34 (2006) 277–292.

[177] K. Ulbricht, E. Bucha, K.A. Pöschel, G. Stein, G. Wolf, G. Nowak, The use of PEG-Hirudin in chronic hemodialysis monitored by the ecarin clotting time: Influence on clotting of the extracorporeal system and hemostatic parameters, Clin. Nephrol. 65 (2006) 180–190.

[178] M.R. Sherman, M.G. Saifer, F. Perez-Ruiz, PEG-uricase in the management of treatment-resistant gout and hyperuricemia, Adv. Drug. Deliv. Rev. 60 (2008) 59–68.

[179] F. Izzo, P. Marra, G. Beneduce, G. Castello, P. Vallone, V. De Rosa, et al., PEGylated arginine deiminase treatment of patients with unresectable hepatocellular carcinoma: Results from phase I/II studies, J. Clin. Oncol. 22 (2004) 1815–1822.

[180] P.A. Ascierto, S. Scala, G. Castello, A. Daponte, E. Simeone, A. Ottaiano, et al., Pegylated arginine deiminase treatment of patients with metastatic melanoma: Results from phase I and II studies, J. Clin. Oncol. 23 (2005) 7660–7668.

[181] C. Mueller, S. Al-Batran, E. Jaeger, B. Schmidt, M. Bausch, C. Unger, et al., A phase IIa study of PEGylated glutaminase (PEG-PGA) plus 6-diazo-5-oxo-Lnorleucine (DON) in patients with advanced refractory solid tumors, J. Clin. Oncol. 26 (Suppl. 20) (2008). abstr. 2533.

Chapter 12

Peptide and Protein Application in Tissue Repair and Regeneration

Patrícia Pereira
Research Institute for Medicines and Pharmaceutical Sciences (iMed.UL), Faculty of Pharmacy, University of Lisbon, Av. Prof. Gama Pinto, 1649-003 Lisbon, Portugal

Chapter Outline

12.1. INTRODUCTION: MEDICAL DEVICES

According to the US Food and Drug Administration (FDA), there are three regulatory classes for medical devices, depending on the risk that they present to the patient. The regulatory level is determined in order to legally commercialize the device: Class I – General Controls; Class II – Special Controls and Class III – Premarket Approval. Class III is the most stringent regulatory category, and devices falling within this category require a premarket approval

Peptide and Protein Delivery. DOI: 10.1016/B978-0-12-384935-9.10012-4

(PMA) to ensure their effectiveness and safety prior to entering the market. In this chapter, Class III devices will be reviewed.

12.2. PERIODONTAL DEFECTS

Periodontal disease is characterized as a chronic bacterial infection which affects the gums and the bone supporting the teeth. Very often, surgical regenerative procedures are needed to treat the condition. According to the American Academy of Periodontology, periodontal regeneration includes the regeneration of cementum, functionally aligned periodontal ligament, alveolar bone and gingiva [1]. In 1996, the FDA approved Emdogain®, an embryonic enamel matrix derivative (EMD) device containing amelogenins extracted from six-month-old piglets [2]. This EMD is indicated for treatment of intrabony defects without furcations (resulting from moderate or severe periodontitis), by topical application onto surgically exposed root surfaces. Amelogenins are a family of hydrophobic proteins (due to their high proline, histidine, glutamine and leucine content) that represent approximately 90% of the organic constituent of the enamel matrix [3]. In Emdogain®, amelogenins which have been stabilized by heat treatment are dissolved in an acidic solution of polyethylene glycol alginate (PGA) and supplied in a pre-filled syringe (30 mg protein/ml PGA). The thixotropic rheology (the ability of a fluid to suffer a decrease in viscosity with time when subjected to constant shear) of PGA allows the application of the proteins as a viscous formulation (2.5 Pa). Moreover, upon application of a shear force (by means of the syringe in which the device is supplied), the viscosity of the formulation decreases, thus facilitating the coating of the surgically exposed root surfaces [4]. During application, physiological conditions will cause a decrease in viscosity and an increase in pH, which in turn causes the amelogenins to precipitate, forming a matrix on the root surface.

In vitro, EMD has been shown to enhance the proliferation, total protein synthesis and mineralized nodule formation of periodontal ligament (PDL) cells, while having no significant effect on cell migration, attachment and spreading [5]. In another study, the application of high doses of EMD had a detrimental effect on the viability of human periodontal ligament fibroblasts (PDLF); however, cellular proliferation appeared to be improved following exposure, as well as cellular attachment to diseased root surfaces [6]. It has also been suggested that EMD has the potential to direct the differentiation of a mesenchymal pluripotential cell line (C2C12) to osteoblasts or chondroblasts [7], and its regulatory role on cementoblast and osteoblast activities has also been documented [8]. Moreover, the ability of EMD to promote periodontal regeneration may also be related to its ability to reduce dental plaque: EMD has an inhibitory effect on the growth of periodontal gram-negative pathogens (which are associated with reduced wound healing), whilst having no significant effect on gram-positive bacteria [9].

In preclinical animal models, EMD has been found to promote the regeneration of all periodontal tissues, namely acellular cementum, periodontal ligaments and alveolar bone [10]. In particular, EMD has the potential to increase the osteoinductive properties of the graft material [11]. For example, when EMD was administered together with demineralized freeze-dried bone allograft (DFDBA) above a certain threshold concentration in a nude mouse muscle implantation assay, osteoinduction was enhanced. In a rat femur drill-hole injury model, locally applied EMD significantly increased the volume of newly formed bone trabeculae when compared to the blank vehicle [12]. These studies demonstrate that, in addition to stimulatory effects on bone growth, EMD has both osteoconductive and cementoconductive properties. In clinical trials, the following inferences can be drawn:

(i) the periodontal regeneration promoted by EMD is only observed when surgical treatment of deep intrabony defects is performed;
(ii) the combination of EMD and periodontal surgery of deep intrabony defects leads to significantly higher improvements in clinical parameters when compared to open flap debridement alone;
(iii) the combination of EMD and guided tissue regeneration (GTR) does not improve the outcomes when compare to EMD or GTR alone;
(iv) the combination of EMD with some types of bone graft/substitute can result in the improvement of certain parameters, compared to EMD treatment alone;
(v) the combination of EMD and coronally repositioned flaps on the treatment of recession-type defects may promote formation of cementum and bone;
(vi) the application of EMD may improve periodontal regeneration in mandibular class II bifurcations [13].

In June 2009, Emdogain® received FDA approval, for use in conjunction with bone substitute materials and also for the treatment of gingival recession defects.

In 2005, the FDA approved a bone grafting material containing a biological therapeutic agent, named GEM21S™ (Growth Factor Enhanced Matrix), to treat periodontal-related defects, such as intrabony, furcation and gingival recession [14]. The GEM21S device comprises two components:

(i) a synthetic, highly porous (pore diameter 1 to 500 μm) and resorbable beta-tricalcium phosphate (β-TCP) scaffold, to provide a matrix for bone growth, prevent the collapse of soft tissues and promote the stabilization of the blood clot;
(ii) purified recombinant human platelet-derived growth factor-BB (rhPDGF-BB), a growth factor that is released at the site of injury during blood clotting, to act as a chemoattractant and mitogen for mesenchymal cells (including osteogenic cells) and promote angiogenesis [15].

The β-TCP particles (0.25 to 1 mm in diameter) are supplied sterile in a "cup", with the rhPDGF-BB supplied as a sterile 0.5 ml solution (0.3 mg/ml, in 20 mM sodium acetate buffer, pH 6). Both components are stored under refrigeration and have a shelf life of 36 months. Prior to application, the β-TCP particles are transferred to the cup, in an aseptic environment, before addition of the rhPDGF-BB solution. Upon saturation of the β-TCP particles with the rhPDGF-BB solution, the mixture is rested for 10 minutes to completely hydrate the β-TCP particles and entrap the rhPDGF-BB within the pores; this in turn facilitates a more uniform distribution of the growth factor at the implant location. It has been shown both *in vitro* and *in vivo* that the growth factor is rapidly released from the particles, and that the non-specific adsorption has no effect on its structural integrity and biological activity [16]. In a large, multi-center, randomized, controlled trial, the following treatments were used for advanced periodontal osseous defects:

(i) β-TCP + 0.3 mg/ml rhPDGF-BB in buffer;
(ii) β-TCP + 1.0 mg/ml rhPDGF-BB in buffer; and
(iii) b-TCP + buffer (active control).

Clinical attachment levels (CAL), gingival recession (GR), linear bone growth (LBG) and percentage bone fill (% BF) were assessed [17]. After six months of healing, an improvement in the clinical outcomes was observed for both test and active control treatments. The rate of gain in CAL at 3 months appeared to be more rapid in the β-TCP + 0.3 mg/ml rhPDGF-BB group than in the control group, although no significant differences were found. The improvement in LB and % BF was significantly higher for the β-TCP + 0.3 mg/ml rhPDGF-BB group when compared to the other groups, supporting the rationale for the clinical use of this concentration of growth factor. The classification of this type of peptide formulation as a medical device has been the subject of discussion, since the same peptide is also commercialized in a different formulation as a human medicine.

12.3. BONE REGENERATION

The healing of bone fractures, as well as the reconstruction of critical-sized bone defects continuously challenges orthopedists, traumatologists and maxillofacial surgeons. Despite the fact that autologous bone grafts are often used in clinical practice, increased donor morbidity, limited graft accessibility and the financial costs involved have led to the ongoing search for alternative approaches to bone repair and regeneration [18]. The application of osteoinductive peptide/protein growth factors (*de novo* bone formation inducers) is promising, particularly the bone morphogenetic proteins (BMPs) – a class of proteins present in bone matrix, whose role in bone formation and repair has been investigated extensively. BMPs subtypes 1 through 7 and 9 have osteoinductive properties; that is, they have the ability to provide the appropriate signal for the differentiation of mesenchymal stem cells into osteoblasts [19]. However, as BMPs are rapidly cleared from the repair site [20] and are non-specific for bony tissue [21], appropriate delivery systems are required to simultaneously ensure efficient local retention of the growth factors, enabling the osteoinductive response, and prevent undesired ectopic bone formation in non-bony tissues of the body. So far, only three bovine-collagen based delivery devices have been approved by the FDA for rhBMPs 2 and 7, with limited clinical applications. Collagen-based carriers, namely type I collagen, are biocompatible and biodegradable, and provide an osteoconductive matrix for the ingrowth of newly formed bone, since they promote the deposition of minerals and they also bind non-collagenous matrix proteins which initiate mineralization [19]. In 2001, the FDA approved OP-1 Implant™, under a Humanitarian Device Exemption (HDE), which is composed of 3.3 mg of rhBMP-7 (also known as osteogenic protein 1, rhOP-1) and 1 g purified of bovine collagen type I. It is indicated for the treatment of nonunion of tibia of at least nine months duration, in cases where previous procedures have failed and the use of autografts is not possible [22]. The medical device is supplied in one vial that is stable over 18 months when refrigerated, and is reconstituted with 2 to 3 ml of sterile saline (0.9%), gently swirled to allow an adequate mixing and rested for 2 min to allow expansion to a maximum of 4 ml. OP-1 Implant™ can then be applied at the site of the bone defect with the aid of a sterile spatula. A prospective, randomized clinical trial of tibial non-unions that had persisted for nine months compared the clinical and radiographic results of the efficacy of OP-1 Implant™ versus autograft [23]: 124 tibial non-unions in 122 patients were randomized to receive intramedullary nails with either autograft or OP-1 implant. The follow-up of patients monitored:

(i) the severity of pain at the fracture site;
(ii) the ability to walk bearing full weight;
(iii) the need for repeat of surgical treatment;
(iv) radiographic evaluation of healing; and
(v) physician satisfaction.

It was found that, nine months after the procedure, clinical union had been achieved in 81% of the patients in the OP-1 implant group against 85% in the autograft group. Radiographic analysis showed that the fractures had healed completely in 75% of the patients in the OP-1 implant group and 84% in the autograft group, but this difference was not statistically significant. It was also found that more than 20% of the patients in the autograft group complained of persistent pain at the donor site. It was thus concluded that OP-1 Implant™ was an effective and safe alternative to bone grafts in the treatment of tibial non-unions.

In 2004, the FDA approved the device OP-1® Putty under a HDE. This comprises rhOP-1, type I bovine bone collagen matrix and carboxymethyl cellulose (CMC) [24]. It is indicated for posterolateral lumbar spinal fusion, in patients in which previous surgery has failed or autologous bone and bone marrow harvest are not feasible. The device is provided in two components: (1) one sterile vial containing 3.5 mg of lyophilized rhOP-1 and 1 g of purified bovine collagen type I and, (2) one sterile vial containing 230 mg of CMC. Both components have a shelf life of 18 months when refrigerated. To produce the product, the CMC is added to the rhOP-1 and the mixture reconstituted with 250 ml of sterile saline (0.9%) solution, with stirring. This procedure allows the improvement of handling characteristics due to the putty-like consistency of the product. Once reconstituted, the product is applied to each side of the spine. Several clinical studies have assessed the effect of OP-1® Putty [25–27]. Despite the limited number of subjects, the efficacy and safety profiles have led to the conclusion that the device represents a viable bone substitute for spinal fusion applications. It is worth noting that in the devices mentioned above, the protein and carrier polymer interact through non-covalent ionic interactions.

A rhBMP-2 containing device was approved by the FDA in 2002, under the name of INFUSE® Bone Graft, as an autograft replacement for spinal fusion procedures in patients with degenerative disc disease, in combination with LT-CAGE® Lumbar Tapered Fusion Device [27a]. In 2004, INFUSE® Bone Graft was approved for the treatment of acute, open tibial shaft fractures, stabilized by intramedullary (IM) nail fixation [28], and in 2007 was approved as an alternative to autologous bone graft for sinus augmentations [29]. The device is supplied containing 4.2 or 12.7 mg of lyophilized rhBMP-2, 5 or 10 ml of sterile water and adsorbable bovine type I collagen sponge (ACS). At the time of surgery, the rhBMP-2 is reconstituted and uniformly applied to the ACS prior implantation. ACS is manufactured by lyophilization of a dispersion of bovine type I collagen (obtained through an alkaline treatment step), which is cross-linked with formaldehyde and sterilized by ethylene oxide [30]. The alkaline treatment causes cleavage of the glutamine and asparagine side chains on the collagen molecule, giving it an isoelectric point (pI) of 5.1. In contrast, rhBMP-2 has a pI of 8.5, and upon reconstitution the solution has a pH of 4.5. When this solution is added to the ACS, a protein–matrix complex is generated through electrostatic interaction; rhBMP-2 and collagen being positively and negatively charged, respectively [31].

The efficacy of the INFUSE® Bone Graft has been extensively studied in all of its indications. In a prospective, randomized, non-blinded, two-year multicenter study, 279 patients with degenerative disc disease (DDD) were randomly assigned to receive either rhBMP-2/ACS or autogenous bone from the iliac crest [32]. Results showed that the mean operative time and blood loss were less in the rhBMP-treated group. Based on radiographs and computed tomography scans, at 24 months after surgery the fusion rate was higher for the same group (94.5%) when compared to the control group (88.7%). The Oswestry back and leg pain scores and neurological status improved in both groups in a similar way. Other studies using different surgical approaches have also proved the superiority of the use of INFUSE® Bone Graft in spinal fusion applications [33–36]. INFUSE® Bone Graft in the treatment of open tibial fractures was also investigated in a large international clinical trial named the BESTT (BMP-2 Evaluation in Surgery for Tibial Trauma) Study Group [37]. A total of 450 patients, from 11 different countries, with open tibia fracture were enrolled in this prospective, randomized, controlled study and were assigned to one of the following groups:

(i) standard care – IM nailing with routine soft tissue management;
(ii) standard care + 0.75 mg/ml rhBMP-2/ACS;
(iii) standard care + 1.5 mg/ml rhBMP-2/ACS (INFUSE® Bone Graft).

The implant was placed over the fracture at the point of wound closure. The primary outcome measurement was the proportion of patients who required a second intervention due to delayed or nonunion within 12 months post-operation. The results revealed that the patient group which had received INFUSE® Bone Graft showed a 44% decrease in the need for secondary interventions, relative to the control group, and also showed improved healing of the fracture (74% vs. 54%), with reduced infection and faster wound healing (83% vs. 65% at six weeks post-operation).

For the examination of the safety and efficacy of INFUSE® Bone Graft in sinus floor augmentations, a randomized, multicenter pivotal study was performed [28]. 160 patients were treated with either INFUSE® Bone Graft ($n = 82$) or autograft ($n = 78$). The procedure involved the insertion of INFUSE® Bone Graft, followed by a period of time for osteoinduction, after which the dental implant was placed. Following a timeframe that allowed osseointegration, prostheses were placed (functional loading) and successful implant restoration assessed at 6 (primary endpoint), 12, 18 and 24 months (secondary endpoints) of functional loading. At the primary endpoint, the implant was successfully restored on 79% of the patients in the study group, against 90.8% of autograft patients; the same trend was observed at the secondary endpoints. However, patients in the control group had statistically significant higher adverse reactions, and it was thus concluded that the benefits (although less than the observed for the control patients) outweighed the risks. Another large clinical study investigated the efficacy of two doses of rhBMP-2/ACS (0.75 and 1.5 mg/ml) against an empty control (ACS alone) in 80 patients requiring extraction socket augmentation [38]. The results showed that the INFUSE® Bone Graft treated sites had twice the amount of new bone compared to the control. In addition, analysis of bone density and histology showed no differences between rhBMP-2-induced bone and native bone.

12.4. SURGICAL PROCEDURES

In certain types of surgery, there is a need to use adjuncts to promote better closure of wounds or attain the desired levels of hemostasis. The range of both types of adjuncts is vast, so in this subsection only formulations containing proteins or peptides will be discussed.

12.4.1. Closure Adjuncts

In 2010, the FDA approved ProGel™ Pleural Air Leak Sealant, which was intended to be applied to the visceral pleura, after closure by sutures or staples, to seal visible air leaks ($\geq$ 2 mm) that had been incurred during open resection of lung parenchyma [39]. The device includes: (1) a "chemistry kit" of two pre-loaded cartridges containing 2 ml of Human Serum Albumin (HSA) and polyethylene glycol disuccinimidyl succinate (PEG-(SS)2) as a dried white powder and (2) an "applicator kit" with sterile water. The shelf life of the device is 12 months under refrigeration, and application involves reconstitution of the PEG-(SS)2 and immediate mixing with the HAS for sealing of the tissue. On mixing, the liquid components quickly cure *in situ* to form a flexible hydrogel by chemical cross-linking of the PEG-(SS)2 and free amine groups of the HSA (formulated at basic pH) [40]. The succinate ester linkages of the hydrogel facilitate its hydrolytic degradation and clearance, primarily through the kidneys. The hydrogel has been proven to have better adhesion, and to withstand higher pressures than a fibrin glue in the rat air lung leak model [40]. In a prospective, randomized, multicenter clinical trial, 161 patients were randomized in a 2:1 ratio to receive ProGel™ Pleural Air Leak Sealant or control (no sealant) if they had at least one significant air leak ($\geq$ 2 mm) after pulmonary resection [41]. One month post-operation, 35% of the ProGel™-treated patients remained leak free, compared to 14% of the control group. The length of hospitalization was also lower for the treatment group (six days) compared to the control group (seven days), and the incidence of adverse effects in both groups was similar.

In 2005, the FDA approved DuraSeal™ Sealant System, another type of surgical sealant for application in sutured dural membranes during cranial surgery, to provide a watertight closure [42]. DuraSeal™ Spine Sealant System was granted approval by the FDA for the same applications but during spine surgery [43]. It is composed of two solutions: one of PEG-(SS)2 and one of trilysine amine (a synthetic peptide) with FD&C Blue no. 1 dye. Mixing of the two solutions generates the hydrogel as described above for ProGel™. Because the gel may swell by up to 50%, the dye is incorporated to monitor the thickness of the layer that is delivered at the site of application. A prospective, non-randomized, single-arm, multicenter clinical trial was conducted to evaluate the effectiveness and safety of the DuraSeal™ Sealant System as an adjunct to sutured dural repair during cranial surgery to provide watertight closure [42]. Of the 111 patients included in the trial, 98.2% showed no intra-operative cerebrospinal fluid (CSF) leakage, with an overall rate of surgical wound infection of 8.1%. It was concluded that the benefits of the use of DuraSeal™

Sealant System outweighed the risks of illness or injury. The effectiveness and safety of this system as an adjunct to sutured dural repair to provide watertight closure in patients undergoing an intentional durotomy during spinal surgery was evaluated by a prospective, randomized, two-arm, single-blind, multicenter study. The study involved a total of 158 patients, 102 of which received DuraSeal™ Spine Sealant System and 56 of which received Standard Care (control) methods [43]. All patients in the treatment group achieved intra-operative sealing, whereas the control group only achieved 62.5% sealing. The presence of CSF leakage and the incidence of surgical site infection in the 90 days following the procedure was not statistically different between the two groups. Overall, the adverse events that occurred were also comparable between both groups.

12.4.2. Hemostasis Adjuncts

Bioglue® Surgical Adhesive was approved in 2001 by FDA as an adjunct to standard methods of achieving hemostasis (sutures or staples) in open surgical repair of large vessels such as aorta, carotid and femoral arteries [44]. The device has two components: (1) purified bovine serum albumin (BSA) and (2) glutaraldehyde. These are dispensed by a controlled delivery system that comprises a reusable delivery device and applicator tips. It has a shelf life of three years when stored at up to 25°C. On dispensing, the solutions are mixed in a pre-defined ratio in the applicator tip to initiate cross-linking. This results from covalent bonds being formed between the aldehyde groups of glutaraldehyde and the free amine groups (e.g. lysine side chains) of BSA. Upon application, cross-linking continues with the tissue proteins at the repair site, and a flexible mechanical seal is created, independent of the host's clotting mechanism. The efficacy and safety of Bioglue® Surgical Adhesive in cardiac and vascular surgical repairs were assessed in a prospective, randomized, multicenter controlled trial [45]. Of a total of 151 patients, 76 were assigned to standard repair plus Bioglue® Surgical Adhesive and 75 to standard repair alone (control group). Patients underwent cardiac, aortic or peripheral vascular procedures. Hemostasis was achieved in 81.1% of Bioglue® Surgical Adhesive vascular anastomoses and in 51% of control standard vascular repair anastomoses. The occurrence of adverse effects was equivalent between both groups, with the exception of a lower incidence of neurological effects in the Bioglue®-treated group.

12.5. HUMAN MEDICINES APPROVED OR UNDER DEVELOPMENT

In this subsection, peptides and proteins that are now classified as human medicines, or those that have demonstrated the therapeutic potential to proceed to the clinic will be discussed.

12.5.1. Wound Healing

Wound healing is a complex, cellular and biochemical process involving many growth factors working in concert to restore anatomical and functional integrity. The principal (overlapping) stages of wound repair include [46]:

(i) blood vessel constriction and blood coagulation, which occurs within seconds;
(ii) an inflammatory phase, over *ca.* 24 h, involving neutrophil and macrophage infiltration, bacterial phagocytosis and release of growth factors for cell migration and proliferation;
(iii) cell migration and proliferation over 72 h, involving collagen synthesis and deposition, re-epithelialization, granulation tissue formation, angiogenesis and wound closure;
(iv) cell matrix deposition and remodeling, occurring over 15–20 days – cells no longer undergo apoptosis and the collagen fibers are laid into a well-defined extracellular matrix.

This complex, sequential, reparative process is easily disrupted by infection, disease or simply by old age, leading to chronic, non-healing wounds (e.g. pressure sores, venous and diabetic ulcers). As the molecular basis of wound healing has been clarified, it has been realized that growth factors (soluble signaling proteins) play a preponderant role, so the potential for these molecules to act as therapeutic agents in the treatment of chronic wounds has been investigated extensively. However, to date, the only human medicine containing a recombinant human growth factor (rhPDGF-BB, also known as becaplermin) to be approved for topical application is Regranex®, approved by the FDA in 1997 and by the European Medicines Evaluation Agency (EMEA, now European Medicines Agency) in 1999. Becaplermin is a homodimeric protein of approximately 23.4 kDa composed of two B chains linked together by a disulphide bond, mainly released from the platelets' α-granules, and also by fibroblasts, macrophages, endothelial cells and keratinocytes [47]. The growth factor acts by binding and activating specific cell surface tyrosine kinase receptors, PDGF receptor-α and PDGF receptor-β [48]. In Regranex®, becaplermin (100 μg/g) is formulated as a gel supplied in multi-use tubes. The excipients include sodium CMC (gelling agent), sterile water, methyl parahydroxybenzoate (methylparaben), propyl parahydroxybenzoate (propylparaben) and *m*-cresol as preservatives, sodium chloride, sodium acetate and glacial acetic acid (to provide an adequate pH for the maximum stability of the protein) and L-lysine hydrochloride at a concentration that stabilizes the protein in the gel formulation. The gel has a shelf life of one year when refrigerated, and once opened is stable for up to six weeks. Regranex® is indicated, in association with other wound care treatments, for healing by induced granulation of full-thickness, neuropathic, chronic diabetic ulcers below 5 cm^2.

The efficacy of becaplermin in neuropathic diabetic ulcers has been assessed on a number of controlled, randomized, multicenter clinical trials (Table 12.1) [49–52]. Inclusion criteria for all studies were:

(i) a chronic foot ulcer with a duration of eight weeks or more;
(ii) an adequate blood supply documented; and
(iii) an absence of infection based on clinical examination.

A combined analysis based on the studies presented above (922 patients in total) showed that becaplermin gel (100 μg/g) significantly increased complete healing, when compared to placebo gel (50% vs. 36%, $p = 0.007$). The time required for complete healing was also decreased in the becaplermin gel-treated group (14.1 weeks vs. 20.1 weeks for placebo, $p = 0.01$). Moreover, the adverse effects were similar in nature and occurrence for all treatment groups [53]. It was thus concluded that treatment with Regranex® once a day in conjunction with good wound care was an effective and safe treatment for patients with full thickness ulcers. More recently, in a clinical trial and in postmarketing use, it has been found that in patients treated with three or more tubes of Regranex®, the rate of death from systemic malignancies increased 5-fold compared to those who had received none (however, the risk of getting new cancers was not increased). Due to this, the FDA has required the manufacturers of Regranex® to place a black box warning on the label, clearly stating these findings and advising caution when the medicine has to be used in patients with known malignancy [54]. Such side-effects are due to the intrinsic physiological action of PDGF-BB in the promotion of cellular proliferation and angiogenesis.

Another growth factor formulation for the treatment of chronic, venous, leg ulcers currently being investigated in a multicenter Phase I/II clinical trial in Norway and Sweden is ChronSeal® [55]. ChronSeal® is a topical application therapy (gel), containing recombinant human Hepatocyte Growth Factor (rhHGF) as the main active ingredient. HGF is a growth factor primarily produced by stromal fibroblasts, amongst other cell types. It regulates a variety of biological processes, such as vascular permeability and angiogenesis, cell migration and matrix deposition and degradation [56]. An impaired regulation of HGF has been shown to be related to the pathophysiology of chronic wounds [57]. Due to the normal pathway of development of human medicines, the components of ChronSeal® remain undisclosed.

12.5.2. Scar Treatments

The formation of scars occurs following trauma, injury or surgery to any tissue or organ in the human body. They consist of a failure in tissue regeneration, in that the absent normal tissue is replaced by an extracellular matrix composed mainly of collagen types I and III and fibronectin [58]. Scarring can result in major clinical problems, such as loss of function and restriction of tissue

TABLE 12.1 Clinical trials of becaplermin I neuropathic diabetic ulcers. Reprinted from reference [47] with permission from Sage Publications

Author	Year	Study design	Comparison	No. patients enrolled	Findings
Steed	1995	Double-blind placebo controlled	Becaplermin gel 30 μg/g vs. placebo gel	61 vs. 57	Complete ulcer healing at 20 weeks: 48% vs. 25%, $p = 0.01$
Wieman	1998	Double-blind placebo controlled	Becaplermin gel 100 μg/g vs. 30 μg/g vs. placebo gel	123 vs. 132 vs. 127	Complete ulcer healing at 20 weeks: 50% vs. 36% vs. 35%, $p = 0.007$ Time for complete healing: 86 days (100 μg/g) vs. 127 (placebo) $p = 0.013$
Wieman	1998	Evaluator-blind placebo controlled	Becaplermin gel 100 μg/g vs. good wound care	128 vs. 122	Complete ulcer healing at 20 weeks: 36% vs. 32%.
d'Hemecourt	1998	Randomized double-blind placebo controlled	Becaplermin gel 100 μg/g vs. CMC gel vs. good wound care	34 vs. 70 vs. 68	Complete ulcer healing at 20 weeks: 44.1% vs. 35.7% vs. 22%.

movement and/or growth, besides the aesthetic appearance and derived psychological disturbances.

The skin is the most frequently injured tissue, and thus the most prone to the occurrence of scarring. Investigation of the mechanisms by which scarring occurs has led to the identification of the importance of transforming growth factor β3 (TGFβ3) in a scar-free healing response, and thus its promising potential in antiscarring therapy [58]. TGFβ3 is a morphogenetic factor synthesized primarily by fibroblasts and keratinocytes, which has been shown to inhibit scarring and to promote better collagen organization *in vivo* [59]. The recombinant form of TGFβ3, avotermin (Juvista), applied as an intradermal injection, has already entered clinical trials. A Phase I clinical trial confirmed that the application of avotermin, at a dosage range between 50 and 10000 ng per 3 mm punch-biopsy wound, is well tolerated and showed no impaired wound healing or enhanced systemic exposure to TGFβ3 at 2–4 hours after application [60]. The effect of prophylactic administration of avotermin on skin scarring was investigated in seven randomized, double-blinded clinical trials in healthy volunteers [61]. In all studies, the administration of avotermin to full thickness incisions at the time of surgery resulted in a significant improvement of scar appearance when compared to placebo and/or standard care. Moreover, significant improvements have been seen for avotermin at concentrations ranging from 50 to 500 ng/μl per linear cm of wound margin. The scars resulting from the treatment had a close resemblance to the surrounding normal skin, with the beneficial effect of being poorly visible 12 months after surgery. The local intradermal application of avotermin has also been shown to be safe, with no clinical differences seen for all treatment groups [61].

A formulation of recombinant human interleukin 10 (rhIL-10 or ilodekain) named Prevascar® is also undergoing clinical investigation. IL-10 is produced by epidermal cells and infiltrating mononuclear cells [62], and plays a major role in the suppression of the inflammatory response and activity of macrophages [63]. Based on pre-clinical studies, rhIL-10 is thought to reduce scarring by the following mechanisms:

(i) modulation of recruitment and differentiation of inflammatory cells and reduction of secretion of pro-inflammatory cytokines by these cells;
(ii) decreased extracellular matrix production (ECM) and increased ECM breakdown by upregulation of proteolytic enzymes;
(iii) downregulation of TGFβ1 (another isoform of the TGF β family, known to induce scar formation) [59] expression and activity [61].

In one prospective, randomized, double-blind, placebo controlled (within subject) Phase II clinical trial, ilodekain administered intradermally in full thickness incisions (healthy volunteers) has proved to achieve significant improvement in scar appearance at concentrations of 5 and 25 μg/100 μl per linear cm of wound margin at 12 months after administration, compared to

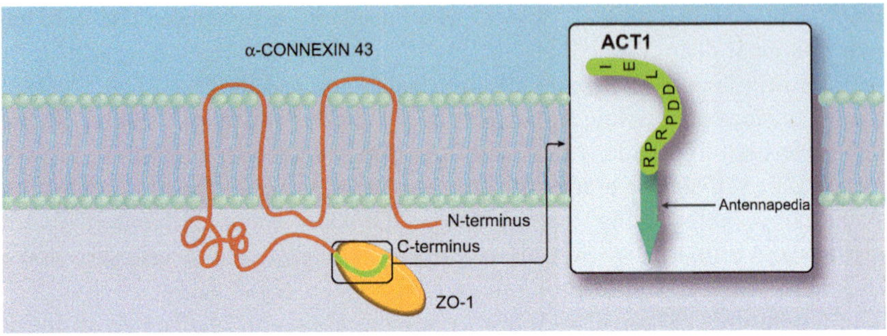

FIGURE 12.1 Model of the gap-junctional protein α-connexin 43 (red) with its C–terminus (green) bound to ZO-1 (orange); the ACT1 fusion peptide is shown in the inset. *Reprinted from reference [66] with permission from Future Medicine.*

placebo or standard care. The occurrence of side-effects was comparable in all groups [61]. The precise components of Prevascar® formulation remain undisclosed.

A novel, synthetic, 24 amino acid peptide analog of heat shock protein 20, termed AZX-100, is already in Phase II clinical trials for administration following excision of keloid scars [64]. *In vitro*, AZX-100 has been shown to decrease stress fiber formation, and to alter the morphology of human dermal keloid fibroblasts, whereas *in vivo* it significantly improved collagen organization in a Siberian hamster scarring model [65]. The formulation in which AZX-100 is administered is not yet in the public domain. Another peptide about to enter clinical phase testing is ACT1, a polypeptide based on the carboxy terminal amino acid sequence of α-connexin 43, linked to an antennapedia internalization sequence (RQPKIWFPNRRKPWKKRPRPDDLEI) (Figure 12.1). By competitively inhibiting the interaction of endogenous α-connexin 43 molecules with *zonula occludens* (ZO)-1, ACT-1 reduces scar progenitor tissue and promotes the regeneration of skin structure and function *in vivo* [66].

12.5.3. Nerve Regeneration

The repair of post-traumatic lesions in nerves poses a major challenge for restorative medicine and microsurgery. Despite progress in surgery, functional recovery is often incomplete and research has turned to the investigation of new approaches to treat such conditions. So far, only two recombinant proteins have been translated into the clinical setting, both of which are intended for use in spinal cord injuries. It should be noted that the precise components of the formulations are unknown, due to patent protection. The first protein is a recombinant human monoclonal antibody to human Nogo-A protein of the IgG4/k class. This was granted Orphan Designation by the European Medicines Agency in 2009 [67]. The protein inactivates Nogo-A, a membrane protein

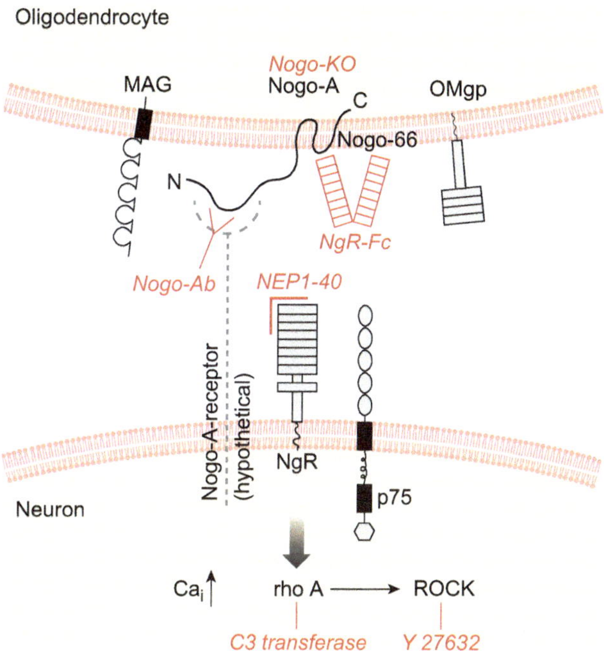

FIGURE 12.2 Nogo-A, MAG and OMgp, the principal inhibitors of neurite growth in CNS myelin, interact with a receptor complex comprising NgR, p75 and additional components. Methods of blocking Nogo and its actions are shown in red. As well as gene deletions (Nogo-KO), Nogo-A, which is shown with its two main active sites facing the extracellular space, can be neutralized by specific antibodies or by a soluble Nogo-66-binding fusion protein comprising domains of the receptor subunit NgR. The NgR subunit itself can be blocked by the NEP1-40 peptide derived from the first 40 amino acids of the Nogo-66 region of Nogo-A. As MAG and OMgp also bind to NgR, NEP1-40 may be a particularly potent reagent. Nogo-A and Nogo-66 activate Rho-A and its downstream target ROCK, the activity of which can be blocked by C3 transferase and the inhibitor Y27632, respectively. *Reprinted from reference [68] with permission from Elsevier.*

expressed in the central nervous system by oligodendrocytes which inhibit neural growth (Figure 12.2) [68]. At the time that the orphan designation was awarded, clinical trials in patients with acute spinal cord injury were ongoing. A human Phase I clinical trial was initiated in May 2006 in patients with complete spinal cord injury between C-5 and T-12, within a time interval of 4 to 14 days post-injury. The human anti-Nogo-A IgG4 was administered in a continuous intrathecal infusion in four increasing dose regimens and is under clinical review [69].

The second recombinant protein in clinical use is a recombinant derivative of C3 transferase, which was granted Orphan Designation in 2005 by the FDA, and in 2008 by the European Medicines Agency [70]. C3 transferase is a toxin produced by *Clostridium botulinum* and is a specific inhibitor of Rho, a guanosine triphosphatase, activated by myelin and ECM inhibitors (Figure 12.2). Upon activation, Rho binds to Rho kinase (ROCK), an important regulator of axonal growth and cellular apoptosis. The recombinant version of C3 is extended with a peptide transport signal, to enhance cellular permeability [71]. By blocking Rho, the intention is that nerve cells will be able to repair themselves and regrow. The protein will be available as a kit, in combination with a fibrin sealant, under the commercial name of Cethrin® [72,73]. A Phase I/IIa multicenter clinical trial was initiated in 2005 [73], for a single dose application of Cethrin® to the dura mater at the time of spinal decompression (within seven days of injury) in complete spinal cord injuries from T2 to T12 and C4 to T1. The results of this trial in regard to the improvement of neurological function and safety were promising and a Phase II study is currently taking place [69].

12.5.4. Bone Regeneration

Discussion of applications for bone repair and regeneration have been made above. In Europe some of the formulations classified by the FDA as medical devices are considered as human medicines and regulated on that basis. For example, the European equivalent of OP-1® Putty is Opgenra™ (approved 2009 by the European Medicines Agency) and the European counterpart of OP-1 Implant™ is Osigraft® (approved 2001). Similarly, INFUSE® Bone Graft is commercialized in Europe under the trade name of InductOs™ (approved in 2002), as an alternative to autologous bone graft in patients with degenerative disc disease, and for the treatment of tibia fractures stabilized by intramedullary (IM) nail fixation.

12.6. FUTURE OUTLOOK

In this chapter several case studies have been described which present the application of peptide and protein drugs in various surgical techniques and medicines. The formulation of these drugs is relatively complex and draws on existing polymeric carriers for their delivery to a local site. This is interesting, since we see a move away from more familiar protein delivery systems which are generally based on subcutaneous or intravenous injection (e.g. of monoclonal antibodies) or polymer encapsulation technologies for controlled delivery (e.g. Zoladex® Implant). It is therefore important to be aware of this class of medical devices, since it offers great potential for emerging biomaterials research to be translated into medical practice. The design of the formulation is clearly central to the method of application and will continue to draw on multidisciplinary research themes such as protein engineering, surface adsorption, polymer synthesis, protein–polymer interactions, etc. The outcomes of the human clinical trials have been vital to determining the scope and utility of these medical devices and play a clear role in directing the field.

The clinical need for this class of medical devices will continue to grow as the field of regenerative medicine becomes ever more integrated into healthcare.

REFERENCES

[1] American Academy of Periodontology, Glossary of Periodontal Terms, fourth ed., American Academy of Periodontology, Chicago, IL, USA, 2001.

[2] United States Food and Drug Administration, Premarket approval application No. P930021. http://www.accessdata.fda.gov/cdrh_docs/pdf/p930021.pdf (1996) (accessed 29.10.10).

[3] S.J. Brookes, C. Robinson, J. Kirkham, W.A. Bonass, Biochemistry and molecular biology of amelogenin proteins of developing dental enamel, Arch. Oral Biol. 40 (1995) 1–14.

[4] S. Gestrelius, C. Andersson, A.C. Johansson, E. Persson, A. Brodin, L. Rydhag, et al., Formulation of enamel matrix derivative for surface coating. Kinetics and cell colonization, J. Clin. Periodontol. 24 (1997) 678–684.

[5] S. Gestrelius, C. Andersson, D. Lidstrom, L. Hammarstrom, M. Somerman, *In vitro* studies on periodontal ligament cells and enamel matrix derivative, J. Clin. Periodontol. 24 (1997) 685–692.

[6] D.R. Davenport, J.M. Mailhot, J.C. Wataha, M.A. Billman, M.M. Sharawy, M.K. Shrout, Effects of enamel matrix protein application on the viability, proliferation, and attachment of human periodontal ligament fibroblasts to diseased root surfaces *in vitro*, J. Clin. Periodontol. 30 (2003) 125–131.

[7] M. Ohyama, N. Suzuki, Y. Yamaguchi, M. Maeno, K. Otsuka, K. Ito, Effect of enamel matrix derivative on the differentiation of C2C12 cells, J. Clin. Periodontol. 73 (2002) 543–550.

[8] Y. Tokiyasu, T. Takata, E. Saygin, M. Somerman, Enamel factors regulate expression of genes associated with cementoblasts, J. Periodont. 71 (2000) 1829–1839.

[9] A. Spahr, S.P. Lyngstadaas, C. Boeckh, C. Andersson, A. Podbielski, B. Haller, Effect of the enamel matrix derivative Emdogain on the growth of periodontal pathogens *in vitro*, J. Clin. Periodontol. 29 (2002) 62–72.

[10] L. Hammarstrom, L. Heijl, S. Gestrelius, Periodontal regeneration in a buccal dehiscence model in monkeys after application of enamel matrix proteins, J. Clin. Periodontol. 24 (1997) 669–677.

[11] B.D. Boyan, T.C. Weesner, C.H. Lohmann, D. Andreacchio, D.L. Carnes, D.D. Dean, et al., Porcine fetal enamel matrix derivative enhances bone formation induced by demineralized freeze dried bone allograft *in vivo*, J. Periodont. 71 (2000) 1278–1286.

[12] F. Kawana, Y. Sawae, T. Sahara, S. Tanaka, K. Debari, M. Shimizu, et al., Porcine enamel matrix derivative enhances trabecular bone regeneration during wound healing of injured rat femur, Anat. Rec. 264 (2001) 438–446.

[13] A. Sculean, F. Schwarz, J. Becker, M. Brecx, The application of an enamel matrix protein derivative (Emdogain) in regenerative periodontal therapy: A review, Med. Princ. Pract. 16 (2007) 167–180.

[14] United States Food and Drug Administration, Premarket Approval Application No. P040013 <http://www.accessdata.fda.gov/cdrh_docs/pdf4/P040013b.pdf (2005) (accessed 29.10.10).

[15] J.O. Hollinger, C.E. Hart, S.N. Hirsch, S. Lynch, G.E. Friedlaender, Recombinant human platelet-derived growth factor: Biology and clinical applications, J. Bone Jt. Surg. Am. 90 (Suppl. 1) (2008) 48–54.

[16] C.S. Young, P.A. Ladd, C.F. Browning, A. Thompson, J. Bonomo, K. Shockley, et al., Release, biological potency, and biochemical integrity of recombinant human platelet-derived

growth factor-BB (rhPDGF-BB) combined with Augment(TM) Bone Graft or GEM 21S beta-tricalcium phosphate (beta-TCP), J. Control. Release 140 (2009) 250–255.
[17] M. Nevins, W.V. Giannobile, M.K. McGuire, R.T. Kao, J.T. Mellonig, J.E. Hinrichs, et al., Platelet-derived growth factor stimulates bone fill and rate of attachment level gain: Results of a large multicenter randomized controlled trial, J. Periodont. 76 (2005) 2205–2215.
[18] G.B. Bishop, T.A. Einhorn, Current and future clinical applications of bone morphogenetic proteins in orthopaedic trauma surgery, Int. Orthop. 31 (2007) 721–727.
[19] M.F. Termaat, F.C. Den Boer, F.C. Bakker, P. Patka, H.J. Haarman, Bone morphogenetic proteins. Development and clinical efficacy in the treatment of fractures and bone defects, J. Bone Jt. Surg. Am. 87 (2005) 1367–1378.
[20] H. Seeherman, J. Wozney, R. Li, Bone morphogenetic protein delivery systems, Spine 27 (2002) S16–S23.
[21] Y. Okubo, K. Bessho, K. Fujimura, Y. Konishi, K. Kusumoto, Y. Ogawa, et al., Osteoinduction by recombinant human bone morphogenetic protein-2 at intramuscular, intermuscular, subcutaneous and intrafatty sites, Int. J. Oral Maxillofac. Surg. 29 (2000) 62–66.
[22] United States Food and Drug Administration, Premarket approval application No. H010002. http://www.accessdata.fda.gov/cdrh_docs/pdf/H010002b.pdf (2001) (accessed 29.10.10).
[23] G.E. Friedlaender, C.R. Perry, J.D. Cole, S.D. Cook, G. Cierny, G.F. Muschler, et al., Osteogenic protein-1 (bone morphogenetic protein-7) in the treatment of tibial nonunions, J. Bone Jt. Surg. Am. 83-A (Suppl. 1) (2001) S151–S158.
[24] United States Food and Drug Administration, Premarket Approval Application No. H020008. http://www.accessdata.fda.gov/cdrh_docs/pdf2/H020008b.pdf (2004) (accessed 29.10.10).
[25] A.R. Vaccaro, T. Patel, J. Fischgrund, D.G. Anderson, E. Truumees, H. Herkowitz, et al., A pilot safety and efficacy study of OP-1 putty (rhBMP-7) as an adjunct to iliac crest autograft in posterolateral lumbar fusions, Eur. Spine J. 12 (2003) 495–500.
[26] A.R. Vaccaro, T. Patel, J. Fischgrund, D.G. Anderson, E. Truumees, H. Herkowitz, et al., A 2-year follow-up pilot study evaluating the safety and efficacy of op-1 putty (rhbmp-7) as an adjunct to iliac crest autograft in posterolateral lumbar fusions, Eur. Spine J. 14 (2005) 623–629.
[27] A.R. Vaccaro, T. Patel, J. Fischgrund, D.G. Anderson, E. Truumees, H.N. Herkowitz, et al., A pilot study evaluating the safety and efficacy of OP-1 Putty (rhBMP-7) as a replacement for iliac crest autograft in posterolateral lumbar arthrodesis for degenerative spondylolisthesis, Spine 29 (2004) 1885–1892.
[27a] United States Food and Drug Administration, Premarket Approval Application No. P000058. http://www.accessdata.fda.gov/cdrh_docs/pdf/P000058b.pdf (2002) (accessed 29.10.10).
[28] United States Food and Drug Administration, Premarket Approval Application No. P000054. http://www.accessdata.fda.gov/cdrh_docs/pdf/P000054b.pdf (2004) (accessed 29.10.10).
[29] United States Food and Drug Administration, Premarket Approval Application No. P050053. http://www.accessdata.fda.gov/cdrh_docs/pdf5/P050053b.pdf (2007) (accessed 29.10.10).
[30] W. Friess, H. Uludag, S. Foskett, R. Biron, Bone regeneration with recombinant human bone morphogenetic protein-2 (rhBMP-2) using absorbable collagen sponges (ACS): Influence of processing on ACS characteristics and formulation, Pharm. Dev. Technol. 4 (1999) 387–396.
[31] W. Friess, H. Uludag, S. Foskett, R. Biron, C. Sargeant, Characterization of absorbable collagen sponges as rhBMP-2 carriers, Int. J. Pharm. 187 (1999) 91–99.
[32] J.K. Burkus, E.E. Transfeldt, S.H. Kitchel, R.G. Watkins, R.A. Balderston, Clinical and radiographic outcomes of anterior lumbar interbody fusion using recombinant human bone morphogenetic protein-2, Spine. 27 (2002) 2396–2408.

[33] J.K. Burkus, H.S. Sandhu, M.F. Gornet, M.C. Longley, Use of rhBMP-2 in combination with structural cortical allografts: Clinical and radiographic outcomes in anterior lumbar spinal surgery, J. Bone Jt. Surg. Am. 87 (2005) 1205–1212.

[34] R.W. Haid Jr., C.L. Branch Jr., J.T. Alexander, J.K. Burkus, Posterior lumbar interbody fusion using recombinant human bone morphogenetic protein type 2 with cylindrical interbody cages, Spine J. 4 (2004) 527–538. [discussion: 538–9].

[35] S.D. Glassman, L. Carreon, M. Djurasovic, M.J. Campbell, R.M. Puno, J.R. Johnson, et al., Posterolateral lumbar spine fusion with INFUSE bone graft, Spine J. 7 (2007) 44–49.

[36] K. Singh, J.D. Smucker, S. Gill, S.D. Boden, Use of recombinant human bone morphogenetic protein-2 as an adjunct in posterolateral lumbar spine fusion: A prospective CT-scan analysis at one and two years, J. Spinal Disord. Tech. 19 (2006) 416–423.

[37] S. Govender, C. Csimma, H.K. Genant, A. Valentin-Opran, Y. Amit, R. Arbel, et al., Recombinant human bone morphogenetic protein-2 for treatment of open tibial fractures: A prospective, controlled, randomized study of four hundred and fifty patients, J. Bone Jt. Surg. Am. 84-A (2002) 2123–2134.

[38] J.P. Fiorellini, T.H. Howell, D. Cochran, J. Malmquist, L.C. Lilly, D. Spagnoli, et al., Randomized study evaluating recombinant human bone morphogenetic protein-2 for extraction socket augmentation, J. Periodont. 76 (2005) 605–613.

[39] United States Food and Drug Administration, Premarket Approval Application No. P010047. http://www.accessdata.fda.gov/cdrh_docs/pdf/P010047b.pdf (2010) (accessed 29.10.10).

[40] H. Kobayashi, T. Sekine, T. Nakamura, Y. Shimizu, *In vivo* evaluation of a new sealant material on a rat lung air leak model, J. Biomed. Mater. Res. 58 (2001) 658–665.

[41] M.S. Allen, D.E. Wood, R.W. Hawkinson, D.H. Harpole, R.J. McKenna, G.L. Walsh, et al., Prospective randomized study evaluating a biodegradable polymeric sealant for sealing intraoperative air leaks that occur during pulmonary resection, Ann. Thorac. Surg. 77 (2004) 1792–1801.

[42] United States Food and Drug Administration, Premarket Approval Application No. P040034. http://www.accessdata.fda.gov/cdrh_docs/pdf4/P040034b.pdf (2005) (accessed 29.10.10).

[43] United States Food and Drug Administration, Premarket Approval Application No. P080013. http://www.accessdata.fda.gov/cdrh_docs/pdf8/P080013b.pdf (2009) (accessed 29.10.10).

[44] United States Food and Drug Administration, Premarket Approval Application No. P010013. http://www.accessdata.fda.gov/cdrh_docs/pdf/P010003b.pdf (2001) (accessed 29.10.10).

[45] H.H. Chao, D.F. Torchiana, Bioglue: Albumin/glutaraldehyde sealant in cardiac surgery, J. Card. Surg. 18 (2003) 500–503.

[46] V. Falanga, Wound healing and its impairment in the diabetic foot, Lancet 366 (2005) 1736–1743.

[47] N. Papanas, E. Maltezos, Growth factors in the treatment of diabetic foot ulcers: New technologies, any promises? Int. J. Low. Extrem. Wounds 6 (2007) 37–53.

[48] C.H. Heldin, B. Westermark, Mechanism of action and *in vivo* role of platelet-derived growth factor, Physiol. Rev. 79 (1999) 1283–1316.

[49] D.L. Steed, Clinical evaluation of recombinant human platelet-derived growth factor for the treatment of lower extremity diabetic ulcers. Diabetic Ulcer Study Group, J. Vasc. Surg. 21 (1995) 71–78. [discussion: 79–81].

[50] T.J. Wieman, J.M. Smiell, Y. Su, Efficacy and safety of a topical gel formulation of recombinant human platelet-derived growth factor-BB (becaplermin) in patients with chronic neuropathic diabetic ulcers. A phase III randomized placebo-controlled double-blind study, Diabetes Care 21 (1998) 822–827.

[51] T.J. Wieman, Clinical efficacy of becaplermin (rhPDGF-BB) gel. Becaplermin Gel Studies Group, Am. J. Surg. 176 (1998) 74S–79S.
[52] P.A. d'Hemecourt, J.M. Smiell, M.R. Karim, Sodium carboxymethylcellulose aqueous-based gel vs. becaplermin gel in patients with nonhealing lower extremity diabetic ulcers. Wounds-Compend, Clin. Res. Pract. 10 (1998) 69–75.
[53] J.M. Smiell, T.J. Wieman, D.L. Steed, B.H. Perry, A.R. Sampson, B.H. Schwab, Efficacy and safety of becaplermin (recombinant human platelet-derived growth factor-BB) in patients with nonhealing, lower extremity diabetic ulcers: A combined analysis of four randomized studies, Wound Repair Regen. 7 (1999) 335–346.
[54] United States Food and Drug Administration, Product Description Regranex. http://www.accessdata.fda.gov/drugsatfda_docs/label/2008/103691s5074lbl.pdf (2008) (accessed 29.10.10).
[55] AB. Kringle Pharma-Europe. Phase I/II Dose Ranging CHRONSEAL® Study in Venous Leg Ulcers. ClinicalTrials.gov Identifier: NCT00797706, ClinicalTrials.gov® (2010).
[56] K. Conway, P. Price, K.G. Harding, W.G. Jiang, The molecular and clinical impact of hepatocyte growth factor, its receptor, activators, and inhibitors in wound healing, Wound Repair Regen. 14 (2006) 2–10.
[57] K. Conway, F. Ruge, P. Price, K.G. Harding, W.G. Jiang, Hepatocyte growth factor regulation: An integral part of why wounds become chronic, Wound Repair Regen. 15 (2007) 683–692.
[58] M.W. Ferguson, S. O'Kane, Scar-free healing: From embryonic mechanisms to adult therapeutic intervention, Philos. Trans. R. Soc. B-Biol. Sci. 359 (2004) 839–850.
[59] M. Shah, D.M. Foreman, M.W. Ferguson, Neutralisation of TGF-beta 1 and TGF-beta 2 or exogenous addition of TGF-beta 3 to cutaneous rat wounds reduces scarring, J. Cell Sci. 108 (Pt 3) (1995) 985–1002.
[60] M.W. Ferguson, J. Duncan, J. Bond, J. Bush, P. Durani, K. So, et al., Prophylactic administration of avotermin for improvement of skin scarring: Three double-blind, placebo-controlled, phase I/II studies, Lancet 373 (2009) 1264–1274.
[61] N.L. Occleston, S. O'Kane, N. Goldspink, M.W. Ferguson, New therapeutics for the prevention and reduction of scarring, Drug Discov. Today 13 (2008) 973–981.
[62] K.W. Moore, A. O'Garra, R. de Waal Malefyt, P. Vieira, T.R. Mosmann, Interleukin-10, Annu. Rev. Immunol. 11 (1993) 165–190.
[63] A.J. Singer, R.A. Clark, Cutaneous wound healing, N. Engl. J. Med. 341 (1999) 738–746.
[64] Capstone Therapeutics, A study to evaluate the safety and efficacy of AZX100 drug product following excision of keloid scars, ClinicalTrials.gov Identifier: NCT00825916, ClinicalTrials.gov (2010).
[65] L.B. Lopes, E.J. Furnish, P. Komalavilas, C.R. Flynn, P. Ashby, A. Hansen, et al., Cell permeant peptide analogues of the small heat shock protein, HSP20, reduce TGF-beta1-induced CTGF expression in keloid fibroblasts, J. Invest. Dermatol. 129 (2009) 590–598.
[66] G.S. Ghatnekar, M.P. O'Quinn, L.J. Jourdan, A.A. Gurjarpadhye, R.L. Draughn, R.G. Gourdie, Connexin43 carboxyl-terminal peptides reduce scar progenitor and promote regenerative healing following skin wounding, Regen. Med. 4 (2009) 205–223.
[67] Committee for Orphan Medicinal Products, European Medicines Agency, Public summary of positive opinion for orphan designation of recombinant human monoclonal antibody to human Nogo-A protein of the IgG4/k class for the treatment of spinal cord injury, Document reference: EMEA/COMP/623774/2008, European Medicines Agency (2009).
[68] M.E. Schwab, Nogo and axon regeneration, Curr. Opin. Neurobiol. 14 (2004) 118–124.
[69] G.W. Hawryluk, J. Rowland, B.K. Kwon, M.G. Fehlings, Protection and repair of the injured spinal cord: A review of completed, ongoing, and planned clinical trials for acute spinal cord injury, Neurosurg. Focus 25 (2008) E14.

[70] Committee for Orphan Medicinal Products, European Medicines Agency, Public summary of positive opinion for orphan designation of recombinant derivative of C3 transferase for the treatment of traumatic spinal cord injury, Document reference: EMEA/COMP/449959/2008, European Medicines Agency (2009).

[71] L. McKerracher, H. Higuchi, Targeting Rho to stimulate repair after spinal cord injury, J. Neurotrauma 23 (2006) 309–317.

[72] M.G. Fehlings, D.C. Baptiste, Current status of clinical trials for acute spinal cord injury, Injury 36 (Suppl. 2) (2005) B113–B122.

[73] A. Jha, D.P. Lammertse, S.W. Charlifue, G.H. Clayton, A review of current clinical trials for improving neurological outcomes, Top. Spinal Cord Inj. Rehabil. 11 (2005) 86–100.

Chapter 13

Product Quality During Manufacture and Supply

Alex Berrill[1], **Jamie Biddlecombe**[2] **and Daniel Bracewell**[3]

[1] *Department of Biochemical Engineering, University College London, Torrington Place, London, WC1E 7JE, UK, a.berrill@ucl.ac.uk. T: +44 (0) 2076793794,* [2] *MedImmune, Milstein Building, Granta Park, Cambridge, CB21 6GH, UK, biddlecombej@medimmune.com. T: + 44 (0) 1223898232,* [3] *Department of Biochemical Engineering, University College London, Torrington Place, London, WC1E 7JE, UK*

Chapter Outline

Peptide and Protein Delivery. DOI: 10.1016/B978-0-12-384935-9.10013-6

13.1. ADVERSE EVENTS CAUSED BY THE LOSS OF PRODUCT QUALITY

Therapeutic protein products are effective against a wide variety of diseases. These proteins can be used to reduce morbidity, slow disease progression, and/or replace essential proteins that are not produced endogenously by the patient [1]. However, if the product quality is lacking and foreign species are present – such as protein aggregates of active ingredients or other dangerous contaminants – endogenous antibodies may be produced. These have the potential to reduce or eliminate the efficacy of the drug, or to cross-react and neutralize other endogenous proteins. The drug's route of administration, product quality, dosing frequency and duration, as well as the individual patient's immune system, can all impact such an immune response. This can have devastating results, compounding patient suffering and potentially leading to mortality [2].

Currently, little is known about the range of parameters that influence aggregated product immunogenicity. However, the most potent foreign particles for eliciting an immune response tend to be large protein aggregates in their native conformation with repetitive arrays of antigens [2]. This has an evolutionary significance, as the immune system has adapted to being particularly sensitive to invasions by microbial organisms which also possess repetitive arrays of antigens.

Harmful species other than those deriving from cell culture can also enter the drug formulation through routes such as leaching or extraction from materials in contact with the drug product. Contaminants potentially include: tungsten residues from glass molding for needles; hydroxyl ions from glass syringes; vultac agents from rubber stoppers; plasticizing agents from plastic syringes and silicone oil used as a syringe lubricant. As an example, in 1998, contaminated Epoietin-α in pre-filled (Eprex™) syringes caused production of a neutralizing antibody for endogenous Epoietin-α. This caused patients to cease producing red blood cells (pure red cell aplasia) [3–6]. The contamination was due to polysorbate 80 leaching vulcanizing agents from the uncoated rubber stopper. These agents are thought to enhance the immune response [7]. This was addressed by switching to polytetrafluoroethylene (PTFE) coated rubber stoppers.

13.2. TYPES OF PRODUCT QUALITY LOSS

Product can be lost by a wide array of mechanisms. The damage usually involves the loss of the secondary or tertiary protein structure. These mechanisms are categorized as chemical or physical damage.

13.2.1. Chemical Damage

Chemical damage is when the structure of the protein is altered due to a chemical modification of the protein.

13.2.1.1. Oxidation

Oxidation is a major protein degradation pathway which can result in the covalent modification of amino acid residues in the protein chain. Oxidizing agents such as peroxides, dissolved oxygen, metal ions, light and free radicals can catalyze the reaction. In processing and formulation, metal catalysts may come from metal contaminated buffers. A number of amino acids are susceptible to oxidation, in particular methionine and cysteine, but also histidine, tyrosine, tryptophan and phenylalanine. Oxidation can result in reduced antigenicity and enhanced susceptibility to proteases [8,9]. The clinical impact of antibody oxidation is not yet fully understood [10].

The metal-catalyzed oxidation of histidine and methionine residues has been demonstrated to cause the aggregation and precipitation of human Relaxin [8]. The location of these amino acids within the protein structure affected their susceptibility to oxidation; when buried in the core of the protein they were less susceptible to such oxidation than when situated on the surface of the molecule. Reactive oxygen species are generated at these specific sites which, rather than diffusing into the bulk medium, attack labile functional groups nearby thus causing protein damage and aggregation [11].

Immunoglobulin G (IgG) is also susceptible to oxidation of methionine and cysteine residues. Methionine residues in the Fc region may be particularly susceptible to oxidation, and this region has in fact been shown to be susceptible to metal-catalyzed oxidation, affecting its ability to bind macrophages. Unpaired cysteine residues residing in the variable region of the molecule can also be targets for oxidation [10,12].

13.2.1.2. Deamidation

Deamidation can result in changes to protein hydrophobicity, charge, mass and to the formation of ammonia [13]. Isoaspartate and aspartate are the major and minor products, respectively, of the partial hydrolysis of a succinimide intermediate. This intermediate arises through nucleophilic attack on the neighboring nitrogen at the C-terminal side of an asparagine residue. Particular groups of amino acids in the protein chain are thought to be targets for deamidation, namely asparagine-glycine and asparagine-serine. Asparagine-threonine has also been identified as being prone to attack, while glutamine also undergoes a similar deamidation reaction, although to a lesser extent [13,14]. Deamidation of asparagine to aspartate, or isoaspartate, will increase the negative charge on the protein. A study of the charged isoforms of antibodies found that they contained varying levels of isoAsp which had formed as a result of Asn deamidation, and explained their charge heterogeneity [15]. The formation of isoAsp can also arise from nucleophilic attack. A study found that deamidation to isoAsp could be detected in accelerated degradation of a therapeutic antibody at 30°C. Asparagine residues were found to be susceptible to succinimide-mediated isomerization to aspartic acid at pH 5 [14], and it has also been reported that deamidation in antibodies could occur at high pH [16]. Deamidation in antibodies has been identified in both the heavy and light chains [17]. When the molecule is structurally intact deamidation is slow, but this can become accelerated upon unfolding of the molecule. Certain groups of amino acid sequences have been identified as being more prone to deamidation than others, aiding the development of more stable antibodies in the future [18].

13.2.1.3. Hydrolysis

Hydrolysis of the peptide bonds between amino acids in the protein primary structure occurs under acidic and basic conditions. Sites in the amino acid sequence which are particularly susceptible to hydrolysis are those in the X-asparagine-Y sequence in which the asparagine-Y bond has been recorded as being a hundred times more labile under hydrolyzing conditions than other peptide bonds. Hydrolysis is a severe degradation and would not be expected under conditions normally encountered during formulation and product storage [9,19].

13.2.1.4. Other

Other common protein degradations include the Maillard reaction, disulphide bond breakage and reshuffling, non-disulphide cross-linking and isomerization [19].

13.2.2. Physical Damage

Physical damage is when the structure of the protein is altered due to a physical modification.

13.2.2.1. Aggregation

Protein aggregation is one of the most common routes for protein instability, and in therapeutic protein production can render the product unfit for release [19]. In disease states, protein aggregation can occur as amyloid fibrillogenesis, known to be responsible for neurodegeneration in Parkinson's and Alzheimer's diseases [20,21]. Protein aggregation can be reversible or irreversible and the aggregates formed can be soluble/insoluble, covalent/non-covalent and native/non-native (Figure 13.1) [22–27]. Non-native protein aggregation is the typical mechanism encountered in the biopharmaceutical industry and it will be termed "protein aggregation" in this chapter [23]. The large number of different theories and models that have been proposed to describe protein aggregation demonstrate that the phenomenon varies significantly depending on the aggregation conditions and the protein being studied. For a multi-domain protein such as a monoclonal antibody, different domains may denature independently and irreversibly via different routes depending on the denaturing conditions [28,29]. The denaturation method can therefore affect the structure of the aggregates

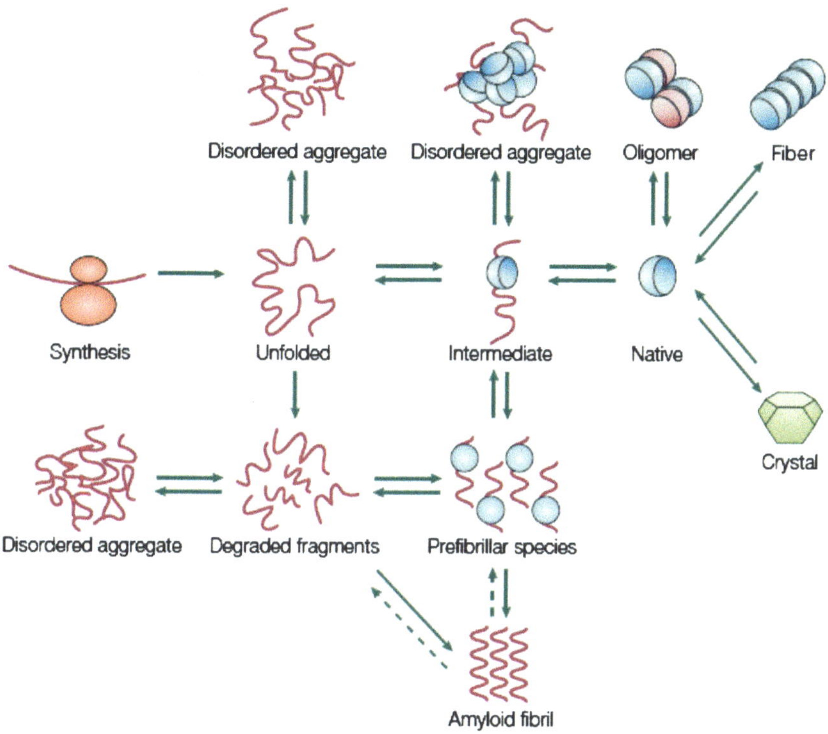

FIGURE 13.1 An example of the many mechanisms for polypeptide aggregation. *Reprinted from reference [25] with permission from Frokjaer and Otzen (2005), Nature Publishing Group.*

formed. The Fab domain has been found to be more sensitive to heat and the Fc region more sensitive to low pH. The aggregation mechanism therefore needs to be assessed for each new protein in order to minimize its occurrence.

Protein aggregation can occur through chemical or physical degradation and is dependent on the thermodynamic stability of the protein's native state. The driving force behind protein aggregation is the reduction in free surface energy by the removal of hydrophobic residues from contact with the solvent.

During the process, there can be a lag phase where the loss of native structure is undetectable and the protein solution remains clear [23]. Protein aggregation is a nucleation/growth phenomenon and this lag phase arises since there is an energy barrier to nucleation which exists because of the free energy necessary to create a new solid–liquid interface. The energy barrier is highest when reaching a critical size for the new phase; this is when nucleus growth can proceed. Insoluble aggregates form when the aggregates are large enough to exceed their solubility. The growth of these aggregates will occur in the direction that creates an orientation with the lowest free energy to assemble, which can result in an ordered morphology e.g. fibrils.

During the long-term storage of therapeutic proteins, it is possible that soluble aggregates can remain present at undetectable levels for several months. However, when they reach the critical level for nucleation, large aggregates can rapidly assemble (taking days to weeks). Seeding with preformed aggregates has proved that they cause a reduction in the lag phase, as was observed during the refolding process of Interleukin 1β in which a critical concentration of protein had to be exceeded before aggregation would occur [30]. Nucleation is considered to be the initiating condition for most protein aggregation phenomena in solution. It is therefore beneficial to reduce soluble aggregates in the bulk drug product to aid long-term stability.

Protein aggregation can be initiated by a number of factors, including temperature (e.g. freeze-thaw), ionic strength and interfacial exposure (solid–liquid, liquid–liquid, gas–liquid). These factors interact with the engineering environment of the production bioprocess which dictates the micro-environment the protein product encounters, e.g. good mixing during a change in pH or temperature will prevent local extremes in these conditions and increase interfacial renewal and interfacial exposure.

Chemical degradation of proteins includes oxidation (tyrosine), deamidation and disulfide bond shuffling [24]. The most common chemically induced aggregation pathway is disulphide bond formation or exchange [31]. Such modifications may result in exposure of hydrophobic residues and cause subsequent aggregation [9,19]. A study of the aggregation of a lyophilized monoclonal antibody found that aggregation coincided with reduced free thiol, and therefore disulphide bond formation [32]. Physical degradation includes adsorption to interfaces, protein unfolding and any resultant aggregation. Chi *et al.* looked at factors that affect the physical degradation of proteins in aqueous solutions [23]. They showed that aggregation maybe controlled by conformational and colloidal stability. Therefore, solution conditions should be selected to enable conformation stability and prevent attractive intermolecular forces.

13.2.2.2. Fragmentation

Fragmentation is a common degradation route for antibodies. The papain cleavage site in the hinge region is known to be particularly susceptible to fragmentation, though the mechanism is kinetic rather than enzymatic [33]. Cleaved antibodies are generally found in very low quantities and can be limited by appropriate formulation of the drug [10].

13.3. PARAMETERS CRITICAL TO PRODUCT QUALITY

13.3.1. Molecular Properties

A protein's structure governs its susceptibility to damage and even subtle differences can be highly significant. For example, human and bovine serum albumins show different mechanisms of aggregation under conditions of shear at air–liquid interfaces [34]. The amino acid sequences of currently approved therapeutic antibodies (IgGs) are very similar, nevertheless there is no "one-size-fits-all" stable formulation, as each antibody has distinct stability characteristics. The minor differences that exist between the IgGs are to be found in the antigen binding regions of the molecule, the complementary determining regions (CDRs). Being on the surface of the protein, these regions are in direct contact with the solution environment and contribute to the individuality of each IgG with regard to its chemical and physical stability. The relative number and position of hydrophobic residues in a protein has a significant influence. A small reduction in hydrophobic residues can have considerable effect on increasing protein stability [31].

13.3.2. Temperature

Exposure to high temperatures (typically > 50°C) causes proteins to undergo thermal unfolding, resulting in an increased rate of degradation and loss of activity. Low temperatures can also reduce the stability of proteins by reducing the strength of hydrogen bonding [9,31,35]. Forced degradation studies have shown that elevated temperatures (30°C) cause deamidation in recombinant antibodies [14]. A study into the stability of monoclonal antibodies during storage demonstrated that the degree of fragmentation increased with storage temperature in samples stored in acetate buffer and NaCl [33]. Samples stored at −20, 5, 30 and 40°C showed increased Fab and Fab + Fc fragmentation at the higher temperatures. The study concluded that fragmentation during storage was a kinetic process and not caused by the presence of proteases from the host cell.

13.3.3. Solution Conditions

The pH of a protein solution affects the charge on the protein's amino acid groups, and hence its overall solubility and stability. A study of the thermal stability of IgG at different pHs concluded that the antibody was sensitive to pH change, and that the Fc region was the most sensitive domain to low pH conditions [29].

An increase in ionic strength has been shown to unravel and extend the structure of β-lactoglobulin, causing a greater degree of flexibility in the morphology of the aggregates [36]. In a related fashion, the addition of sodium chloride in freeze-dried formulations of bovine IgG has been shown to contribute to the formation of insoluble aggregation upon rehydration [37].

13.3.4. Interfacial Effects

Shear occurs throughout bioprocessing, being associated with centrifuging, pumping, mixing, filtration, fill-finish operations and protein drug delivery systems (Figure 13.2) [38–41]. Shear forces in fluids describe the velocity gradients that occur due to the difference in velocity between the bulk fluid and

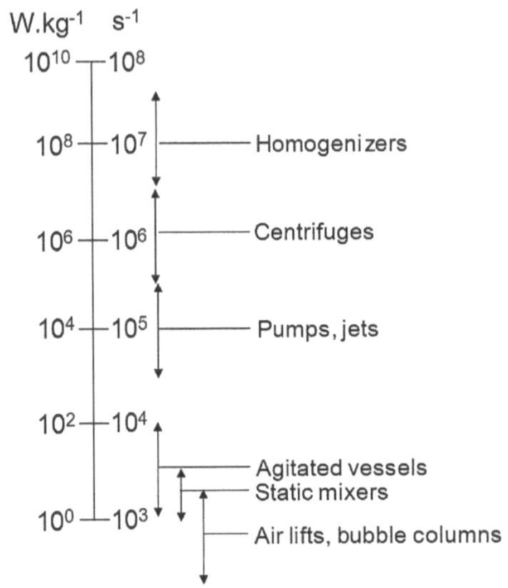

FIGURE 13.2 Estimates of fluid energy dissipation rates (W kg^{-1}) and shear rates (s^{-1}) in bioprocess equipment. *Adapted from reference [41] with permission from Yim and Shamlou (2000), Springer.*

a stationary surface. A surface may be the source of the shear if it is in motion through the fluid, as in the case of a moving pump head, or due to fluid moving at high velocity past a stationary surface, as in the case of fluid flowing through a pipe. It may also be caused by gas bubbles moving through a liquid. In the case of an object moving through a fluid, the liquid "layer" directly at the surface of the moving object will have the same velocity as the surface. Movement is imparted to the fluid directly above this layer by viscous drag, yet this and each subsequent layer moves more slowly than the surface. The result is a velocity gradient, called the shear rate, is proportional to the shear force applied to the fluid. Shear stress, τ, is equal to the force per unit area applied, and is proportional to shear rate where the constant is viscosity, μ. To better represent the real flow intensities that biological materials may encounter in processing situations, a more complex solution for shear flow accounts for the shear strain rate, which is defined as the sum of the elongation rate and the shear rate. The solution of these hydrodynamic flows in the complex geometries that exist throughout the bioprocess requires the use of computational fluid dynamics (CFD).

Wang *et al.* reviewed how proteins can be denatured by shaking or shearing at interfaces [19]. A suggested mechanism for aggregation during mechanical stress is constant interface renewal (Figure 13.3) [42]. This causes reversible or irreversible adsorption and/or partial unfolding of the protein at an interface. These interfaces are usually formed between a gas–liquid, liquid–liquid or solid–liquid. Hydrodynamic flow may then transport the damaged proteins into solution [9,31,42].

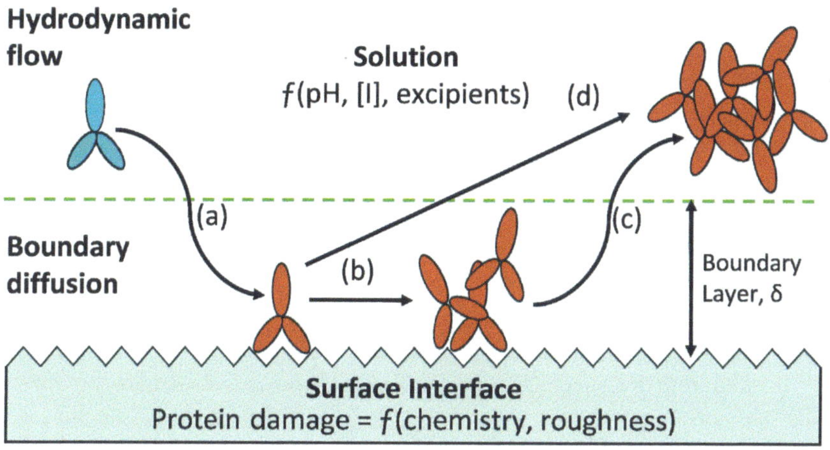

FIGURE 13.3 Proposed mechanism for the aggregation of proteins at interfaces under shear conditions. The protein in solution is brought to the surface by hydrodynamic flow where upon it adsorbs and unfolds (a). Once adsorbed to the surface it may aggregate with other protein molecules forming small aggregates that can act as nucleation sites for further aggregation (b). The damaged molecule turnover will be affected by the velocity gradient (shear) of fluid passing the surface, where high shear encourages transport back into the liquid where further aggregation may occur (c) or (d).

Protein aggregation resulting from adsorption to various interfaces is protein specific [43,44]. Proteins that are structurally compact and stable are often termed "hard" proteins (e.g. lysozyme and α-chymotrypsin), and they will tend to bind only to hydrophobic interfaces. Proteins that are larger and less stable in structure are termed "soft" (e.g. bovine serum albumin and α-lactalbumin). Antibodies, like other "soft" proteins, will bind to a variety of interfaces, both hydrophobic and hydrophilic, via non-specific adsorption.

For a protein (e.g. β-lactoglobulin) adsorbed at the interface, an applied stress such as expansion or compression may lead to exposure of hydrophobic patches and protein aggregation [45]. Upon extensive compression (leaving less than half of the initial area available) the rotational mobility of β-lactoglobulin reduces, coinciding with the formation of anti-parallel strands; this is often associated with protein aggregate formation. The same group that documented this discovery later used chicken egg ovalbumin to discover that protein films, generated by equilibrium adsorption from the bulk, behaved as a densely packed colloidal repulsive particle system, where the protein had significant rotational mobility, predominantly retaining globular folds and exhibiting distinct (lateral) diffusion. Applied stresses on the surface film (by compression of the interface) may result in protein denaturation and aggregation. This process changes a surface film from a colloidal particle into a gelled system.

Work by Kiese *et al.* in 2008 looked at comparing aggregate formation caused by stirring and shaking, with and without the presence of polysorbate 20, and the effect of fill volumes (varying headspace) on protein stability during shaking, also with and without polysorbate 20 [42]. It was found that the two different types of stress probably resulted in different mechanisms of aggregation. In the absence of polysorbate 20, stirring was shown to be a very high stress method, producing highly turbid samples with large, visible protein aggregates. Intermediate, smaller, soluble aggregates did not increase – presumably as they were short lived in solution [42]. Stirring stress could have been a result of:

(i) shear;
(ii) interfacial effects between the antibody and stirrer bar/glass bottom or between the antibody and the hydrophobic surface of the teflon (polytetrafluorethylene) coated stirrer bar;
(iii) cavitation;
(iv) local thermal effects of stirring due to the friction between stirrer bar on glass bottom; and
(v) rapid transportation of either aggregated or adsorbed species into solution, thus facilitating further aggregation within the bulk.

Shaking produced smaller visible and sub-visible particles (~2 μm). The mechanism for damage during shaking was due to renewal of the air–liquid interface (Figure 13.3). This was supported when a reduction in damage was observed with air headspace removal. The hypothesized mechanisms for shaking and stirring were supported when polysorbate 20 was added. The surfactant protected the shaken antibody at much lower quantities because it competed for the air–liquid interface. In the stirred sample there were mechanisms other than interfacial effects so surfactant had less of a protective effect.

13.3.4.1. Gas–liquid Interfaces

Shear effects in the presence of air–liquid interfaces have been studied by agitation for several biopharmaceutical proteins, including insulin [46,47], rhGH [48], rFXIII [49], albutropin [50], β-lactoglobulin, HSA, BSA [34] and IgG [40]. Antibodies that were subjected to stirring and shaking stresses at low levels of shear in the presence of air resulted in precipitates of up to 25 μm in diameter, as measured by dynamic light scattering [40]. These recordings have been supported as increases in turbidity measurements when antibody solutions have been mechanically agitated with air [51–54]. In general, the damaging effects of interfacial shear on proteins may take the form of changes in secondary, tertiary and quaternary structure, altered activity and aggregation due to exposure of hydrophobic amino acid residues [55].

Molecular functionalities, such as electrostatic repulsion, molecular size, exposed hydrophobicity, chemical reactivity and bulkiness all contribute to a protein's behavior in proximity to air–liquid interfaces [56]. Maa *et al.* set up experiments to investigate the effect of the air–liquid interface on recombinant human growth hormone (rhGH) (22.1 kDa) and recombinant human DNase (rhDNase) (32.7 kDa) dissolved in low salt concentration buffers at pH 7.4 and 6.3 respectively [57]. Aggregation was measured with SE-HPLC, scanning microcalorimetry and SDS-PAGE. In the presence of shear without air, rhGH and rhDNase remained practically intact, with air addition rhDNase remained relatively stable but rhGH denatured and was found to have a lower surface tension and higher foaming tendency. Increasing protein concentration, air–liquid interfacial areas and shear all resulted in increased air–liquid interface renewal and protein damage.

13.3.4.2. Liquid–liquid Interfaces

A hydrophobic liquid such as oil behaves in a similar fashion to air when it forms an interface with an aqueous solution [58]. In 2005, Nichols *et al.* investigated how the interaction between an amyloid-beta peptide and a defined aqueous–chloroform interface influences the rate of aggregation and the structure of these aggregates [59]. They showed that the aggregation rate was significantly enhanced and that the structure of the aggregates and stability were notably distinct from proto-fibrils prepared solely in buffer.

An aggregation study carried out by Jones *et al.* in 2005 looked at the short-term (5 hours) effects of silicone oil on protein solutions [60]. Silicone oil (0.5% w/v) was added to 1mg/mL solutions of Ribonuclease A (RNase A), Lysozyme, Bovine Serum Albumin (BSA), Concanavalin A (ConA). An increase in turbidity (OD_{360}) in the presence of silicone oil over time was found as a result of protein aggregation. The more hydrophobic proteins such as BSA and ConA had an increased likelihood of aggregation relative to the more hydrophilic proteins examined. Tertiary structure changes showed that phenylalanine residues were pushed towards polar environments when exposed to silicone oil.

13.3.4.3. Solid–liquid Interfaces

During bioprocess purification, protein adsorption to solids such as chromatography beads is essential but in the case of drug delivery, it is advantageous to avoid it. The stability of a protein in its adsorbed state may differ from that in its native soluble state. There is a wide variety of literature describing the adsorption of proteins to solid surfaces [61–64]. The process is influenced by the electrostatic charge and hydrophobicity of the substrate and protein, and structural stability of the protein. The main forces involved in protein adsorption to solid surfaces are protein structure rearrangement, dehydration of sorbent surface and redistribution of charged groups. After adsorption, a protein's secondary structure may change significantly, potentially resulting in destabilization of the protein; for example antibodies have been shown to undergo a conformational change, highlighting the effect of surface interactions on the three-dimensional structure of the protein (Figure 13.3) [19,65].

The level of antibody adsorption can be enough to reduce their concentration in solution [66]. This was observed when mouse monoclonal antibodies adsorbed onto the surface of glass shake flasks. Coating the glass or adding Pluronic F127 minimized the loss. In 1995, Buijs *et al.* examined the adsorption of monoclonal antibodies and $F(ab')_2$ fragments onto polymeric surfaces [61]. By analyzing with transmission electroscopic microscopy (TEM) and single particle optical sizing (SPOS) they found that they could observe how antibodies and $F(ab)_2$ fragments orientate themselves on charged and uncharged, hydrophilic and hydrophobic surfaces. They found that electrostatic interactions had minimal impact on IgG and $F(ab')_2$ adsorption affinity for hydrophobic polystyrene surfaces. However, on hydrophilic surfaces, electrostatic interaction played a big role in adsorption.

Investigation of insulin adsorption to solid surfaces in the presence of agitation found that aggregation occurred when protein monomers adsorbed to the surface, and then underwent a conformational change that resulted in aggregation with other altered state monomers [46]. It was proposed that insulin dimers and trimers would reversibly adsorb to a hydrophobic surface, whereas monomers would adsorb and form unfolded molecules. Insulin aggregates were found to be joined non-covalently, and interaction of hydrophobic residues exposed by the partial unfolding of protein molecules at the interface was proposed as the likely aggregation mechanism [47]. In 1984, Feingold *et al.* investigated how insulin aggregation was affected by vibration and contact material [67]. Hydrophilic contact materials like polyamide and cellulose butyrate (2% of total insulin aggregated after 96 hours vibration) seemed more compatible with insulin stability than hydrophobic ones such as polypropylene (16% aggregated) and polyvinylchloride (37% aggregated). Addition of polyethylenepolypropyleneglycol increased stability by 3–5 times in most cases.

In 2007, Biddlecombe *et al.* used a device capable of producing quantifiable shear rates at a solid–liquid interface in the absence of air [68,69]. IgG_4 antibody concentration and its aggregation were monitored by SE-HPLC and turbidity (350 nm). The rate of damage was found to increase rapidly at the higher shear rate corresponding to the more aggressive processing operation, indicating the importance of minimizing such environments during processing. Different solid interface properties caused different monomer reduction rates. This, along with observations in other systems, suggested that the mechanism was due to adsorption and renewal at the surface proportional to the mixing/shear in the immediate microenvironment (Figure 13.3).

13.3.5. Leachables and Extractables

Leachables are contaminants that have migrated from packaging or processing components, and extractables are species released from materials of construction after exposure to product over time under specific solvent and temperature conditions [70]. Contaminants can be leached or extracted from many of the surfaces that the dosage form comes into contact with, including: tungsten residues left over from molding glass syringes; hydroxyl ions from glass syringes (raising pH); rubber leachates from stoppers; plasticizing agents from plastic syringes; silicone oil lubricant from syringes; polymer extractables; aldehydes [71,72]. Harmful agents leached from elastomers contained in syringes have been well documented [6,73–76].

Metal ions and peroxides can typically leach from manufacturing equipment and final packaging containers [70]. Metal ions (e.g. tungsten residuals) and peroxides can cause oxidation and hence protein degradation by the formation of peroxyl radicals. Tungsten oxide vapor deposits and tungsten particles can result from aging tungsten wires that have been used to create the needle mounting hole in prefilled syringes. Bee *et al.* investigated the effect of tungsten metal, tungsten trioxide (WO_3) and soluble sodium tungstate salt (Na_2WO_4) on the stability of IgG_1 [2]. The aggregation of antibodies caused by tungsten was measured by size exclusion chromatography (SEC) and found to be dependent upon: low pH (< 6) to form soluble tungsten polyanions; a minimum threshold level of tungsten polyanions (3 ppm at pH 5); the ratio of tungsten to antibody and the mode of mixing the tungsten with the antibodies.

13.4. REGULATING PRODUCT QUALITY

The manufacture and sale of biopharmaceuticals is highly regulated. In this section we consider two areas of the highest importance:

(i) the existing regulations pertaining to protein aggregation and their known limitations;
(ii) the consideration in selecting the final product package.

13.4.1. Regulation of Aggregates/Particles in the Product

Biopharmaceuticals such as monoclonal antibodies have the tendency to form subvisible particles, a term applied to particles over a range of sizes from 0.1 to 100 μm. The pharmacopeia requirements state that particles larger than 10 μm and 25 μm are to be maintained at 6000 and 600 particles/container respectively. These requirements were originally set with extraneous, process related particles in mind, to prevent blood vessel occlusion during intravenous administration of the drug product. With regard to protein particles, specifically, there is considerable discussion in the industry concerning their potential to elicit an immune response, though there is no direct clinical evidence of immunogenicity caused by these particles [76a].

Interpreting the Q6B statement from the International Conference on Harmonization (ICH) suggests that the level to control particulates at is dependent on the potential risks of the drug product under its intended route of administration. The main concern for drug development is to demonstrate that the level of subvisible, or visible, particles in a drug product are under control during formulation, fill-finish and subsequent storage.

The levels of subvisible particles in a drug product are most commonly measured by the light obscuration (LO) method described in the United States Pharmacopeia (USP<788>) and European Pharmacopeia (EP 2.9.19). Although the method was not originally intended for measuring protein particles, has a limited ability to distinguish between different types of particles and can be affected by high viscosity formulations [76b], it remains the standard for QC lot-release of product.

13.4.2. Considerations for Parenteral Based Products

Parenteral drug administration involves piercing the skin or mucous membranes and is the usual route for protein and peptide therapeutics. Dosage forms for injection-based products are generally solutions, emulsions, or suspensions, and are all required to be sterile. These dosage forms represent one of the highest risk drug products due to the possibility of the rapid introduction of harmful species into the patient's bloodstream (Table 13.1). The main

TABLE 13.1 Examples of packaging concerns for common classes of drug products [80]

Degree of concern associated with the route of administration	Likelihood of packaging component-dosage form interaction		
	High	Medium	Low
Highest	Inhalation aerosols and solutions; injections and injectable suspensions	Sterile powders and powders for injection; inhalation powders	
High	Ophthalmic solutions and suspensions; transdermal ointments and patches; nasal aerosols and sprays		
Lowest	Topical solutions and suspensions; topical and lingual aerosols; oral solutions and suspensions	Topical powders; oral powders	Oral tablets and oral (hard and soft gelatine) capsules

concerns include protein aggregates formed from the interaction of protein with silicon oil (see Section 13.4.2.4), and contaminants from container closure systems such as leachates from glass, rubber or plastic containers, or the delaminating of glass particles from glass vials [78,79].

13.4.2.1. Selection of Final Dosage Packaging

A container is suitable if it is: safe; compatible; functions properly and protects the dosage from degradation over its entire shelf life. Protection is necessary from factors such as temperature, light, loss of solvent, exposure to reactive gases, absorption of water vapor and microbial contamination. For reasons mentioned previously, the long-term effects of protein exposure to container materials can cause instability which may lead to immunogenicity of the product. Therefore, selection of a suitable packaging process and container is critical, and the selection process involves assessing initial suitability for the intended use, followed by continued assessment against accepted quality control (QC) tests. Elastomeric components such as rubber stoppers typically need to meet "USP <381> Elastomeric closures for Injections Requirements" to prove safe, while plastic components typically need to meet "USP <87> Biological Reactivity Tests" to prove safe [81].

Glass used for vials and syringes has historical evidence of safety and compatibility, but is also known to adsorb protein. In some cases, glass packaging components need to meet additional criteria to prevent significant interactions between the packaging component and the dosage form.

Pre-filled syringes have more components than vials. This is a major concern for the Food and Drug Administration (FDA), particularly when materials that might facilitate protein aggregation are used. Manufacturers have therefore had to develop methods to reduce the use of these materials. For instance, manufacturers have introduced alternative materials to replace tungsten as the heat resistant material used in glass forming. Such technology is now standard and available to stabilize sensitive proteins.

13.4.2.2. Quality Control of Packaging Components

The quality control that ensures container consistency aims to limit unintended variations in the materials of construction and/or container manufacturing process and aims to prevent adverse effects on drug quality. The assessment involves two parts; (1) determining the concentrations of extracted chemical species from container components that may migrate into the dosage form; and (2) their toxicological assessment. The use of stability studies for monitoring container consistency in terms of compatibility and protection is accepted. However, there is currently no general policy concerning the monitoring of a packaging system and components with regard to dose quality.

Ideally, extraction studies should involve the drug product. For example, it may be advisable to identify extractables and obtain a quantitative extraction profile of a packaging component and to compare this periodically to the profile from a new batch of the packaging component. A composition change may occur due to slight changes in the raw materials used and how they are processed e.g. using a new raw material supplier or using a different mold release agent.

Upon manufacture, the package containing the final dosage form is sealed and inspected for particulate matter arising from foreign contaminants or aggregation. Inspection may be a visual assessment by an operator or an automated procedure.

13.4.2.3. Air–liquid Interfaces in Pre-filled Syringes and Vials

In practice it is difficult to eliminate air during the filling and packaging of pre-filled syringes and vials. Headspace air in a vial or syringe can accelerate oxidation, denaturation and aggregation of antibodies. To reduce protein damage, air can be replaced with an inert gas to minimize oxidation, or air can be completely removed from the final container [66]. This can be achieved with maximum filling of vials or vacuum placement of pre-filled syringe stoppers, although this is more technically challenging and therefore involves greater expense. A further preventative measure is to increase protein concentration. It has been reported that the gas–liquid interface formed during agitation is less damaging to proteins as protein concentration is increased [82]. This is thought to be due to the critical nature of the gas–liquid interface to protein ratio which is greatest at lower concentrations. An undesirable consequence of formulating at high protein concentration is that the protein is more prone to self-interact and aggregate [55]. Formulating at these high concentrations can also result in very high viscosities, which can affect the ease with which the protein can be delivered, especially for subcutaneous injection [83].

13.4.2.4. Syringe Lubricants

Syringes are coated with lubricants (e.g. silicone oil) to limit protein adsorption and facilitate stopper movement [70]. Silicone oil is the most common lubricant but it can cause undesirable liquid–liquid interactions with the protein solution and lead to protein aggregation. Methods to reduce the use of silicone oil include baking the silicone onto the syringe barrel. This requires heating the siliconized syringe at a specific temperature for an appropriate time. An alternative is to use a reactive silicone system applied as liquid which is then polymerized. One option to avoid the use of silicone completely is to use a chemical vapor deposition or plasma technology to generate non-silicone lubricant films on the barrel or piston, or on both. This lubricant film could be a fluorocarbon to improve drug/stopper compatibility and stopper movement.

13.5. BIOPROCESS DESIGN CONSIDERATIONS

During manufacture, shipping and distribution, proteins are subjected to harsh conditions that would not be present in their natural environment. These conditions include those described in early parts of this chapter: temperature; pH and ionic strength; oxidizing and reducing conditions; high concentrations; microbial contamination; light exposure; interfacial effects (gas–liquid, liquid–liquid and solid–liquid) and the presence of proteases [9,70,84]. These conditions can lead to degradation products and hence reduce efficacy of the drug, while the formation of aggregates could induce adverse immune responses in the patient. The instabilities of proteins *in vitro* provide significant challenges to their successful manufacture and delivery. The large amount of resources invested to develop a viable drug candidate will be wasted if the product cannot be delivered to the patient in a stable and efficacious form.

13.5.1. Materials of Construction

Exposure to a variety of materials used for constructing process equipment and product containers can lead to denaturation of the protein. In 2008, Tyagi *et al.* investigated whether solid nano–particles that had been shed from a positive displacement piston pump nucleated IgG aggregation and particle formation from an IgG formulation [77]. They showed, using SE-HPLC and infrared (IR) spectroscopy, that nanoparticles of foreign materials shed by pumps can serve as heterogeneous nuclei for formation of protein microparticles.

Filtration is used at many different stages of a bioprocess. Sources of protein denaturation and degradation during filtration can be due to: high trans-membrane pressures; protein adsorption to membranes; and the effect of pumping and high flow rates [9,55,70]. High shear during ultrafiltration operations has been observed to cause protein aggregation [85,86]. Other reported causes of denaturation during ultrafiltration are fouling of the filter membranes caused by protein adsorption [87] and changes to the ionic strength of the buffer during diafiltration of IgG [88]. The loss of surfactants due to adsorption on filtration membranes has also been reported [89], resulting in the loss of protein stability during storage.

13.5.2. The Interfacial Environment During Processing, Product Shipping and Storage

The filling process involves transferring the formulated protein into the final drug product container, usually a vial or a syringe. Ensuring the compatibility of the protein formulation with the filling equipment prior to manufacturing is highly recommended to ensure that the manufacturing process does not induce unforeseen degradation in the final drug product. Filling must be carried out in a highly sterile environment (class 100) to avoid any contamination. Vial and stopper materials need to be carefully considered and validated to ensure that no product denaturation occurs due to incompatibility with the contact surfaces. Leachables, extractables and introduction of foreign particles must all be avoided. The filling speed should be low enough to avoid aggregation by interfacial effects such as foaming, but high enough to maximize throughput [9,70,90]. It has been demonstrated that the type of pump selected for vial filling can also affect the level of protein aggregates in the final dosage product [24]. For example, one study showed that with a piston pump the protein solution acted as a lubricant and the antibody used was particularly prone to damage in this environment [24]. The type of pump used, pumping rates, nozzle sizes and materials involved can all affect product quality. Mixing and pumping must be optimized, as inadequate mixing will lead to heterogeneity, while very intense mixing can cause shear stress and exposure to air–liquid interfaces. High pressures and turbulence can likewise lead to damage [9,70]. It is also possible that mixing and shear can occur from the storage and transport of the dosage form (such as vibrations from air transportation) or from syringe stopper movement.

13.5.3. Freeze-thaw

Freeze-thaw of the purified bulk drug product is often the first operation in the formulation process. Storing the purified drug in a frozen form allows the creation of a large product inventory, allowing increased manufacturing flexibility. Holding the drug in a frozen form can increase the stability of the product by reducing the likelihood of microbial growth and slowing degradation processes, particularly those mediated by the presence of water – such as hydrolysis. Also, a protein held in a frozen matrix is less likely to interact with other proteins to form aggregates [9,70].

Despite its advantages, subjecting the protein to freezing and thawing can itself be the cause of protein damage. Cold denaturation can result in aggregation due to the weakening of hydrogen bonds at low temperatures [9,31,35]. Cryoconcentration can also occur, due to variations in the rates at which solution components freeze, which can lead to the precipitation of buffer components and protein stabilizers causing dramatic changes to solution pH upon freezing (this is a particular issue for sodium phosphate buffers), leading to aggregation of the protein [91]. During large-scale operation, this can lead to requirements for equipment to control the rate of freeze-thaw. Proteins may also unfold at ice–water interfaces [92] and air–liquid interfaces formed from air trapped during rapid freezing [49]. Optimization of buffer conditions and the excipients used can mitigate these effects. For example, the addition of polysorbate 80 has been shown to minimize the aggregation of recombinant hemoglobin under short-term storage conditions, and the addition of sucrose was found to stabilize this protein during long-term storage at –20°C [93].

13.5.4. Formulation and Final Dosage Form

The dosage forms predominantly favored for therapeutic monoclonal antibodies are liquid and freeze-dried formulations. The required shelf life for a therapeutic protein is typically 18–24 months [55,94]. The challenge for liquid formulations is to achieve long-term stability and successful administration of the drug at very high concentrations (1–125 mg/mL). Lyophilized formulations are generally more stable than liquid formulations, although the cost of development and manufacture is significantly greater [10,95].

To achieve long-term stability, the formulation needs to be tailored for each new drug candidate [10]. The protein should be formulated at the concentration required for dosing, with optimum buffer pH, conductivity and excipients for maximum stability during transport, storage and final delivery to the patient. The chemical stability of an IgG_1, known to be prone to deamidation, was significantly increased through optimization of pH, buffer species and storage temperatures [96]. Slight dimerization was observed in an IgG devoid of preservatives during storage for three years at 4°C. Inclusion of thimerosal and phenol caused a significant increase in aggregation, highlighting the need for careful selection of formulation excipients [97].

13.5.4.1. Surfactants

To protect proteins from interfacial damage during processing, storage and shipping, surfactants can be used [31]. They work by competing with proteins at hydrophobic interfaces or by binding directly to the protein. Due to the wide variety of additives, a group of researchers have come up with methods to speed up the surfactant scouting and selection process [98].

Poloxamer surfactants may increase the viscosity of protein solutions; this limits protein backbone motion and inhibits protein aggregation. Studies of ovalbumin with pectin have shown that their interaction in a bulk solution causes a significantly slower adsorption of ovalbumin to the air–water interface [99,100].

In 2009, Joshi *et al.* looked at the mechanism of how polysorbate 80 minimizes lysozyme loss through adsorption, aggregation and preserved native structure and activity [63]. Hydrophilic silanized and hydrophobic silica surfaces were used as substrates for protein and surfactant adsorption, which was monitored *in situ* with ellipsometry. At the hydrophobic surface, polysorbate reduced the amount of lysozyme adsorbed, but had little effect on adsorption to a hydrophilic surface. Similar observations were noticed when polysorbate was used as a pre-coat. This emphasized the importance of the direct interaction between polysorbate and the solid surface compared to polysorbate protein association.

13.5.4.2. Lyophilization and Spray-drying

If the drug shelf life in liquid form is not acceptable then freeze-drying, also known as lyophilization or spray-drying, can be considered. The removal of water stabilizes the protein by preventing water-mediated degradation pathways such as hydrolysis. In the case of a monoclonal antibody, high residual moisture levels (up to 8%) were found to reduce the chemical stability, although the physical stability was not affected and may even have been increased [101].

Lyophilization immobilizes the protein in a glassy phase, eliminating diffusion-induced protein–protein interactions, thus preventing chemical and physical degradation [9,70]. The liquid formulation is first frozen to make it suitable for drying, after which primary drying removes the ice by sublimation. Secondary drying then removes protein (or "bound") water to reach a target residual moisture level which is typically $< 1\%$ [55]. The extent of the secondary drying step, and hence residual moisture levels, has been found to influence the aggregation of bovine IgG, demonstrating the importance of optimizing each step of the process [37]. Denaturation and aggregation during freeze-drying may only become apparent once the protein is rehydrated. Stresses that occur during drying can depend on the method used. Stresses found in freeze-thaw operations can also occur during bulk freeze-drying such as denaturation at ice–water interfaces [92,102]. Lyophilization of IgG can result in excessive dehydration of the antibody when a carbohydrate excipient is not present, and can lead to increased levels of aggregation. When a carbohydrate excipient such as sucrose or trehalose is used in the formulation, the stability is improved by maintaining the native structure of the antibody through a water replacement mechanism [103,104]. Due to the specific interaction of the excipient with the protein, the amount of excipient required to form a stable lyophilized antibody formulation can be determined by a specific molar ratio of stabilizing molecule to antibody [105].

Spray-drying is an alternative to freeze-drying. The formulation is sprayed through an atomizer or spray nozzle which produces tiny droplets of drug product that are then heated by a hot gas to evaporate the liquid. Spray drying is cheaper and quicker than freeze-drying and turns the formulation solution into a powder in a single step. If products degrade using this method then freeze-drying is the usual alternative. Spray-drying can also cause aggregation due to shear [106], and air–liquid interface denaturation [107,108]. Both disulfide-linked and non-covalent soluble aggregates of a monoclonal antibody have been observed during storage after spray-drying [32]. Excipients and surfactants are used to aid refolding of the protein molecule during the rehydration process when aggregation may otherwise be preferred [55]. It has been shown that spray-drying of bovine IgG in the absence of protective excipients resulted in the formation of insoluble aggregates. Aggregation was then reduced, though not eliminated, by the addition of a surfactant [37]. A recent study demonstrated a significant improvement in the stability of an IgG_1 during storage after spray-drying through successful development of an effective formulation using excipients [109].

13.6. CONCLUSION

Progress is continuously being made in retaining product quality during manufacture and supply. There are many examples of how to stabilize proteins in aqueous solutions and dried solids based on previous experience and well designed experiments. This selection of appropriate conditions allows such products to remain stable for supply as effective medicines. We are, however, just beginning to appreciate how various excipients can modulate protein stability and behavior at interfaces. As more research is carried out in these areas, the selection and design of appropriate molecules and conditions to maintain stability during manufacture and supply based on fundamental understanding becomes possible. Combining this information with quality by design (QbD) approaches will better enable robust manufacturing and supply of biopharmaceutical products.

REFERENCES

[1] P. Goddard, Therapeutic proteins – a pharmaceutical perspective, Adv. Drug Deliv. Rev. 6 (1991) 103–131.

[2] A.S. Rosenberg, Effects of protein aggregates: An immunologic perspective, AAPS J. 8 (2006) E501–E507.

[3] K. Boven, S. Stryker, J. Knight, A. Thomas, M. van Regenmortel, D.M. Kemeny, et al., The increased incidence of pure red cell aplasia with an Eprex formulation in uncoated rubber stopper syringes, Kidney Int. 67 (2005) 2346–2353.

[4] F. Locatelli, J. Olivares, R. Walker, M. Wilkie, B. Jenkins, C. Dewey, et al., Novel erythropoiesis stimulating protein for treatment of anemia in chronic renal insufficiency, Kidney Int. 60 (2001) 741–747.

[5] F. Locatelli, Once weekly treatment with epoetin-beta, Nephrol. Dial. Transplant. 20 (Suppl. 6) (2005) vi26–vi30.

[6] A.P. Villalobos, S.R. Gunturi, G.A. Heavner, Interaction of polysorbate 80 with erythropoietin: A case study in protein-surfactant interactions, Pharm. Res. 22 (2005) 1186–1194.

[7] M.H. Ryan, G.A. Heavner, M. Brigham-Burke, F. McMahon, M.F. Shanahan, S.R. Gunturi, B. Sharma, F.X. Farrell, An *in vivo* model to assess factors that may stimulate the generation of an immune reaction to erythropoietin, Int. Immunopharmacol 6 (2006) 647–655.

[8] S. Li, T.H. Nguyen, C. Schoneich, R.T. Borchardt, Aggregation and precipitation of human relaxin induced by metal-catalyzed oxidation, Biochemistry 34 (1995) 5762–5772.

[9] S.Y. Patro, E. Freund, B.S. Chang, Protein formulation and fill-finish operations, Biotechnol. Annu. Rev. 8 (2002) 55–84.

[10] A.L. Daugherty, R.J. Mrsny, Formulation and delivery issues for monoclonal antibody therapeutics, Adv. Drug Deliv. Rev. 58 (2006) 686–706.

[11] F. Zhao, E. Ghezzo-Schoneich, G.I. Aced, J. Hong, T. Milby, C. Schoneich, Metal-catalyzed oxidation of histidine in human growth hormone. Mechanism, isotope effects, and inhibition by a mild denaturing alcohol, J. Biol. Chem. 272 (1997) 9019–9029.

[12] L. Margiloff, L. Chaplia, A. Chow, P.C. Singhal, J. Mattana, Metal-catalyzed oxidation of immunoglobulin G impairs Fc receptor-mediated binding to macrophages, Free Radic. Biol. Med. 25 (1998) 780–785.

[13] J.L. Reubsaet, J.H. Beijnen, A. Bult, R.J. van Maanen, J.A. Marchal, W.J. Underberg, Analytical techniques used to study the degradation of proteins and peptides: Chemical instability, J. Pharm. Biomed. Anal. 17 (1998) 955–978.

[14] R.J. Harris, B. Kabakoff, F.D. Macchi, F.J. Shen, M. Kwong, J.D. Andya, S.J. Shire, N. Bjork, K. Totpal, A.B. Chen, Identification of multiple sources of charge heterogeneity in a recombinant antibody, J. Chromatogr. B. Biomed. Sci. Appl. 752 (2001) 233–245.

[15] W. Zhang, M.J. Czupryn, Analysis of isoaspartate in a recombinant monoclonal antibody and its charge isoforms, J. Pharm. Biomed. Anal. 30 (2003) 1479–1490.

[16] A. Usami, A. Ohtsu, S. Takahama, T. Fujii, The effect of pH, hydrogen peroxide and temperature on the stability of human monoclonal antibody, J. Pharm. Biomed. Anal. 14 (1996) 1133–1140.

[17] Y. Mimura, E.A. Kabat, T. Tanaka, M. Fujimoto, K. Takeo, K. Nakamura, Microheterogeneity of mouse antidextran monoclonal antibodies, Electrophoresis 16 (1995) 116–123.

[18] D. Chelius, D.S. Rehder, P.V. Bondarenko, Identification and characterization of deamidation sites in the conserved regions of human immunoglobulin gamma antibodies, Anal. Chem. 77 (2005) 6004–6011.

[19] W. Wang, Instability, stabilization, and formulation of liquid protein pharmaceuticals, Int. J. Pharm. 185 (1999) 129–188.

[20] C. Aisenbrey, T. Borowik, R. Bystrom, M. Bokvist, F. Lindstrom, H. Misiak, et al., How is protein aggregation in amyloidogenic diseases modulated by biological membranes? Eur. Biophys. J. 37 (2008) 247–255.

[21] C.M. Dobson, Principles of protein folding, misfolding and aggregation, Semin. Cell Dev. Biol. 15 (2004) 3–16.

[22] J.F. Carpenter, T.W. Randolph, W. Jiskoot, D.J. Crommelin, C.R. Middaugh, G. Winter, et al., Overlooking subvisible particles in therapeutic protein products: Gaps that may compromise product quality, J. Pharm. Sci. 98 (2009) 1201–1205.

[23] E.Y. Chi, S. Krishnan, T.W. Randolph, J.F. Carpenter, Physical stability of proteins in aqueous solution: Mechanism and driving forces in nonnative protein aggregation, Pharm. Res. 20 (2003) 1325–1336.

[24] M.E. Cromwell, E. Hilario, F. Jacobson, Protein aggregation and bioprocessing, AAPS J. 8 (2006) E572–E579.

[25] S. Frokjaer, D.E. Otzen, Protein drug stability: A formulation challenge, Nat. Rev. Drug Discov. 4 (2005) 298–306.

[26] L. Morozova-Roche, M. Malisauskas, A false paradise – mixed blessings in the protein universe: The amyloid as a new challenge in drug development, Curr. Med. Chem. 14 (2007) 1221–1230.

[27] C.J. Roberts, Non-native protein aggregation kinetics, Biotechnol. Bioeng. 98 (2007) 927–938.

[28] T. Brody, Multistep denaturation and hierarchy of disulfide bond cleavage of a monoclonal antibody, Anal. Biochem. 247 (1997) 247–256.

[29] A.W. Vermeer, W. Norde, The thermal stability of immunoglobulin: Unfolding and aggregation of a multi-domain protein, Biophys. J. 78 (2000) 394–404.

[30] J.M. Finke, M. Roy, B.H. Zimm, P.A. Jennings, Aggregation events occur prior to stable intermediate formation during refolding of interleukin 1beta, Biochemistry 39 (2000) 575–583.

[31] W. Wang, Protein aggregation and its inhibition in biopharmaceutics, Int. J. Pharm. 289 (2005) 1–30.

[32] J.D. Andya, Y.F. Maa, H.R. Costantino, P.A. Nguyen, N. Dasovich, T.D. Sweeney, et al., The effect of formulation excipients on protein stability and aerosol performance of spray-dried powders of a recombinant humanized anti-IgE monoclonal antibody, Pharm. Res. 16 (1999) 350–358.

[33] A.J. Cordoba, B.J. Shyong, D. Breen, R.J. Harris, Non-enzymatic hinge region fragmentation of antibodies in solution, J. Chromatogr. B. Analyt. Technol. Biomed. Life Sci. 818 (2005) 115–121.
[34] A. Oliva, A. Santovena, J. Farina, M. Llabres, Effect of high shear rate on stability of proteins: Kinetic study, J. Pharm. Biomed. Anal. 33 (2003) 145–155.
[35] P.L. Privalov, Cold denaturation of proteins, Crit. Rev. Biochem. Mol. Biol. 25 (1990) 281–305.
[36] K. Baussay, C.L. Bon, T. Nicolai, D. Durand, J.P. Busnel, Influence of the ionic strength on the heat-induced aggregation of the globular protein beta-lactoglobulin at pH 7, Int. J. Biol. Macromol 34 (2004) 21–28.
[37] J.M. Sarciaux, S. Mansour, M.J. Hageman, S.L. Nail, Effects of buffer composition and processing conditions on aggregation of bovine IgG during freeze-drying, J. Pharm. Sci. 88 (1999) 1354–1361.
[38] C.B. Elias, J.B. Joshi, Role of hydrodynamic shear on activity and structure of proteins, Adv. Biochem. Eng. Biotechnol. 59 (1998) 47–71.
[39] P.T. Gomme, B.M. Hunt, O.C. Tatford, A. Johnston, J. Bertolini, Effect of lobe pumping on human albumin: Investigating the underlying mechanisms of aggregate formation, Biotechnol. Appl. Biochem. 43 (2006) 103–111.
[40] H.C. Mahler, R. Muller, W. Friess, A. Delille, S. Matheus, Induction and analysis of aggregates in a liquid IgG1-antibody formulation, Eur. J. Pharm. Biopharm. 59 (2005) 407–417.
[41] S.S. Yim, P.A. Shamlou, The engineering effects of fluids flow on freely suspended biological macro-materials and macromolecules, Adv. Biochem. Eng. Biotechnol. 67 (2000) 83–122.
[42] S. Kiese, A. Papppenberger, W. Friess, H.C. Mahler, Shaken, not stirred: Mechanical stress testing of an IgG1 antibody, J. Pharm. Sci. 97 (2008) 4347–4366.
[43] S. Beaufils, R. Hadaoui-Hammoutene, V. Vie, G. Miranda, J. Perez, E. Terriac, et al., Comparative behaviour of goat beta and alpha(s1)-caseins at the air–water interface and in solution, Food Hydrocolloid. 21 (2007) 1330–1343.
[44] C. Belhomme, E. vid-Briand, M.H. Ropers, C. Guerin-Dubiard, M. Anton, Interfacial characteristics of spread films of hen egg yolk phosvitin at the air–water interface: Interrelation with its charge and aggregation state, Food Hydrocolloid. 21 (2007) 896–905.
[45] H.H.J. de Jongh, P.A. Wierenga, Assessing the extent of protein intermolecular interactions at air–water interfaces using spectroscopic techniques, Biopolymers 82 (2006) 384–389.
[46] V. Sluzky, J.A. Tamada, A.M. Klibanov, R. Langer, Kinetics of insulin aggregation in aqueous solutions upon agitation in the presence of hydrophobic surfaces, Proc. Natl. Acad. Sci. U.S.A. 88 (1991) 9377–9381.
[47] V. Sluzky, A.M. Klibanov, R. Langer, Mechanism of insulin aggregation and stabilization in agitated aqueous solutions, Biotechnol. Bioeng. 40 (1992) 895–903.
[48] N.B. Bam, J.L. Cleland, J. Yang, M.C. Manning, J.F. Carpenter, R.F. Kelley, et al., Tween protects recombinant human growth hormone against agitation-induced damage via hydrophobic interactions, J. Pharm. Sci. 87 (1998) 1554–1559.
[49] L. Kreilgaard, L.S. Jones, T.W. Randolph, S. Frokjaer, J.M. Flink, M.C. Manning, et al., Effect of Tween 20 on freeze-thawing and agitation-induced aggregation of recombinant human factor XIII, J. Pharm. Sci. 87 (1998) 1597–1603.
[50] D.K. Chou, R. Krishnamurthy, T.W. Randolph, J.F. Carpenter, M.C. Manning, Effects of Tween 20 and Tween 80 on the stability of Albutropin during agitation, J. Pharm. Sci. 94 (2005) 1368–1381.

[51] D.J. Burgess, J.K. Yoon, N.O. Sahin, A novel method of determination of protein stability, J. Parenter. Sci. Technol. 46 (1992) 150–155.
[52] A.F. Henson, J.R. Mitchell, P.R. Musselwhite, The surface coagulation of proteins during shaking, J. Colloid Interf. Sci. 32 (1970) 162–165.
[53] H.L. Levine, T.C. Ransohoff, R.T. Kawahata, W.C. McGregor, The use of surface tension measurements in the design of antibody-based product formulations, J. Parenter. Sci. Technol. 45 (1991) 160–165.
[54] D. Vidanovic, A.J. Milic, M. Stankovic, V. Poprzen, Effects of nonionic surfactants on the physical stability of immunoglobulin G in aqueous solution during mechanical agitation, Pharmazie 58 (2003) 399–404.
[55] J.F. Carpenter, B.S. Kendrick, B.S. Chang, M.C. Manning, T.W. Randolph, Inhibition of stress-induced aggregation of protein therapeutics, Methods Enzymol. 309 (1999) 236–255.
[56] H.H. de Jongh, H.A. Kosters, E. Kudryashova, M.B. Meinders, D. Trofimova, P.A. Wierenga, Protein adsorption at air–water interfaces: A combination of details, Biopolymers 74 (2004) 131–135.
[57] Y.F. Maa, C.C. Hsu, Protein denaturation by combined effect of shear and air–liquid interface, Biotechnol. Bioeng. 54 (1997) 503–512.
[58] R.A. Ganzevles, K. Zinoviadou, T. van Vliet, M.A.C. Stuart, H.H.J. de Jongh, Modulating surface rheology by electrostatic protein/polysaccharide interactions, Langmuir 22 (2006) 10089–10096.
[59] M.R. Nichols, M.A. Moss, D.K. Reed, J.H. Hoh, T.L. Rosenberry, Rapid assembly of amyloid-beta peptide at a liquid/liquid interface produces unstable beta-sheet fibers, Biochemistry 44 (2005) 165–173.
[60] L.S. Jones, A. Kaufmann, C.R. Middaugh, Silicone oil induced aggregation of proteins, J. Pharm. Sci. 94 (2005) 918–927.
[61] J. Buijs, J.W.T. Lichtenbelt, W. Norde, J. Lyklema, Adsorption of monoclonal IgGs and their F(Ab′)(2) fragments onto polymeric surfaces, Colloid. Surface. B 5 (1995) 11–23.
[62] C.A. Haynes, W. Norde, Structures and stabilities of adsorbed proteins, J. Colloid. Interf. Sci. 169 (1995) 313–328.
[63] O. Joshi, J. McGuire, Adsorption behavior of lysozyme and Tween 80 at hydrophilic and hydrophobic silica–water interfaces, Appl. Biochem. Biotech. 152 (2009) 235–248.
[64] O. Santos, T. Nylander, M. Paulsson, C. Tragardh, Whey protein adsorption onto steel surfaces – effect of temperature, flow rate, residence time and aggregation, J. Food Eng. 74 (2006) 468–483.
[65] M. Rankl, T. Ruckstuhl, M. Rabe, G.R. Artus, A. Walser, S. Seeger, Conformational reorientation of immunoglobulin G during nonspecific interaction with surfaces, Chemphyschem. 7 (2006) 837–846.
[66] W. Wang, S. Singh, D.L. Zeng, K. King, S. Nema, Antibody structure, instability, and formulation, J. Pharm. Sci. 96 (2007) 1–26.
[67] V. Feingold, A.B. Jenkins, E.W. Kraegen, Effect of contact material on vibration-induced insulin aggregation, Diabetologia 27 (1984) 373–378.
[68] J.G. Biddlecombe, A.V. Craig, H. Zhang, S. Uddin, S. Mulot, B.C. Fish, et al., Determining antibody stability: Creation of solid–liquid interfacial effects within a high shear environment, Biotechnol. Prog. 23 (2007) 1218–1222.
[69] J.G. Biddlecombe, G. Smith, S. Uddin, S. Mulot, D. Spencer, C. Gee, et al., Factors influencing antibody stability at solid–liquid interfaces in a high shear environment, Biotechnol. Prog. 25 (2009) 1499–1507.

[70] N. Rathore, R.S. Rajan, Current perspectives on stability of protein drug products during formulation, fill and finish operations, Biotechnol. Prog. 24 (2008) 504–514.
[71] B. Sharma, Immunogenicity of therapeutic proteins. Part 2: Impact of container closures, Biotechnol. Adv. 25 (2007) 318–324.
[72] S.T. Tzannis, W.J.M. Hrushesky, P.A. Wood, T.M. Przybycien, Adsorption of a formulated protein on a drug delivery device surface, J. Colloid. Interf. Sci. 189 (1997) 216–228.
[73] C.B. Airaudo, A. Gayte-Sorbier, R. Momburg, P. Laurent, Leaching of antioxidants and vulcanization accelerators from rubber closures into drug preparations, J. Biomater. Sci. Polym. Ed. 1 (1990) 231–241.
[74] C.L. Bennett, S. Luminari, A.R. Nissenson, M.S. Tallman, S.A. Klinge, N. McWilliams, et al., Pure red-cell aplasia and epoetin therapy, N. Engl. J. Med. 351 (2004) 1403–1408.
[75] K. Boven, J. Knight, F. Bader, J. Rossert, K.U. Eckardt, N. Casadevall, Epoetin-associated pure red cell aplasia in patients with chronic kidney disease: Solving the mystery, Nephrol. Dial. Transplant. 20 (Suppl. 3) (2005) iii33–iii40.
[76] F. Locatelli, L. Del Vecchio, P. Pozzoni, Pure red-cell aplasia "epidemic" – mystery completely revealed? Perit. Dial. Int. 27 (Suppl. 2) (2007) S303–S307.
[76a] S.K. Singh, N. Afonina, M. Awwad, K. Bechtold-Peters, J.T. Blue, D. Chou, et al., An industry perspective on the monitoring of subvisible particles as a quality attribute for protein therapeutics, J. Pharm. Sci. 99 (2010) 3302–3321.
[76b] L.O. Narhi, Y. Jiang, S. Cao, B. Kalman, D. Shnek, A critical review of analytical methods for subvisible and visible particles, Curr. Pharm. Biotech. 10 (2009) 373–381.
[77] A.K. Tyagi, T.W. Randolph, A. Dong, K.M. Maloney, C. Hitscherich Jr., J.F. Carpenter, IgG particle formation during filling pump operation: A case study of heterogeneous nucleation on stainless steel nanoparticles, J. Pharm. Sci. 98 (2009) 94–104.
[78] M.J. Akers, S.L. Nail, W. Saffell-Clemmer, Top ten hot topics in parenteral science and technology, PDA J. Pharm. Sci. Technol. 61 (2007) 337–361.
[79] G.A. Sacha, W. Saffell-Clemmer, K. Abram, M.J. Akers, Practical fundamentals of glass, rubber, and plastic sterile packaging systems, Pharm. Dev. Technol. 15 (2010) 6–34.
[80] U.S. Department of Health and Human Services Food and Drug Administration, Center for Drug Evaluation and Research (CDER), Center for Biologics Evaluation and Research (CBER), Guidance for industry: Container closure systems for packaging human drugs and biologics, Chem. Manufact. Control. Document. (1999) 1–56.
[81] R.T. Wood, Validation of elastomeric closures for parenteral use: An overview, J. Parenter. Drug Assoc. 34 (1980) 286–294.
[82] M.J. Treuheit, A.A. Kosky, D.N. Brems, Inverse relationship of protein concentration and aggregation, Pharm. Res. 19 (2002) 511–516.
[83] B.S. Chang, S. Hershenson, Practical approaches to protein formulation development, Pharm. Biotechnol. 13 (2002) 1–25.
[84] M.C. Manning, D.K. Chou, B.M. Murphy, R.W. Payne, D.S. Katayama, Stability of protein pharmaceuticals: An update, Pharm. Res. 27 (2010) 544–575.
[85] K.J. Kim, V. Chen, A.G. Fane, Some factors determining protein aggregation during ultrafiltration, Biotechnol. Bioeng. 42 (1993) 260–265.
[86] T.J. Narendranathan, P. Dunnill, The effect of shear on globular-proteins during ultrafiltration – studies of alcohol-dehydrogenase, Biotechnol. Bioeng. 24 (1982) 2103–2107.
[87] D. Mockel, E. Staude, M.D. Guiver, Static protein adsorption, ultrafiltration behavior and cleanability of hydrophilized polysulfone membranes, J. Membrane Sci. 158 (1999) 63–75.

[88] K. Ahrer, A. Buchacher, G. Iberer, A. Jungbauer, Thermodynamic stability and formation of aggregates of human immunoglobulin G characterised by differential scanning calorimetry and dynamic light scattering, J. Biochem. Biophys. Methods 66 (2006) 73–86.

[89] H.C. Ahler, M. Printz, R. Kopf, R. Schuller, R. Muller, Behaviour of polysorbate 20 during dialysis, concentration and filtration using membrane separation techniques, J. Pharm. Sci. 97 (2008) 764–774.

[90] W. Norde, C.E. Giacomelli, BSA structural changes during homomolecular exchange between the adsorbed and the dissolved states, J. Biotechnol. 79 (2000) 259–268.

[91] K.A. Pikal-Cleland, N. Rodriguez-Hornedo, G.L. Amidon, J.F. Carpenter, Protein denaturation during freezing and thawing in phosphate buffer systems: Monomeric and tetrameric beta-galactosidase, Arch. Biochem. Biophys. 384 (2000) 398–406.

[92] G.B. Strambini, E. Gabellieri, Proteins in frozen solutions: Evidence of ice-induced partial unfolding, Biophys. J. 70 (1996) 971–976.

[93] B.A. Kerwin, M.C. Heller, S.H. Levin, T.W. Randolph, Effects of Tween 80 and sucrose on acute short-term stability and long-term storage at −20 degrees C of a recombinant hemoglobin, J. Pharm. Sci. 87 (1998) 1062–1068.

[94] S. Hermeling, D.J. Crommelin, H. Schellekens, W. Jiskoot, Structure-immunogenicity relationships of therapeutic proteins, Pharm. Res. 21 (2004) 897–903.

[95] R.J. Harris, S.J. Shire, C. Winter, Commercial manufacturing scale formulation and analytical characterization of therapeutic recombinant antibodies, Drug Develop. Res. 61 (2004) 137–154.

[96] J.Y. Zheng, L.J. Janis, Influence of pH, buffer species, and storage temperature on physicochemical stability of a humanized monoclonal antibody LA298, Int. J. Pharm. 308 (2006) 46–51.

[97] T. Garcia, G. Cook, R. Nosal, PQLI key topics – criticality, design space, and control strategy, J. Pharm. Innov. 3 (2008) 60–68.

[98] S.E. Bondos, A. Bicknell, Detection and prevention of protein aggregation before, during, and after purification, Anal. Biochem. 316 (2003) 223–231.

[99] E.V. Kudryashova, A.J.W.G. Visser, H.H. de Jongh, Reversible self-association of ovalbumin at air–water interfaces and the consequences for the exerted surface pressure, Protein Sci. 14 (2005) 483–493.

[100] E.V. Kudryashova, A.J.W.G. Visser, A. van Hoek, H.H.J. de Jongh, Molecular details of ovalbumin-pectin complexes at the air/water interface: A spectroscopic study, Langmuir 23 (2007) 7942–7950.

[101] E.D. Breen, J.G. Curley, D.E. Overcashier, C.C. Hsu, S.J. Shire, Effect of moisture on the stability of a lyophilized humanized monoclonal antibody formulation, Pharm. Res. 18 (2001) 1345–1353.

[102] X. Tang, M.J. Pikal, Design of freeze-drying processes for pharmaceuticals: Practical advice, Pharm. Res. 21 (2004) 191–200.

[103] J.D. Andya, C.C. Hsu, S.J. Shire, Mechanisms of aggregate formation and carbohydrate excipient stabilization of lyophilized humanized monoclonal antibody formulations, AAPS Pharm. Sci. 5 (2003) E10.

[104] M. Maury, K. Murphy, S. Kumar, A. Mauerer, G. Lee, Spray-drying of proteins: Effects of sorbitol and trehalose on aggregation and FT-IR amide I spectrum of an immunoglobulin G, Eur. J. Pharm. Bio. Pharm. 59 (2005) 251–261.

[105] J.L. Cleland, X. Lam, B. Kendrick, J. Yang, T.H. Yang, D. Overcashier, et al., A specific molar ratio of stabilizer to protein is required for storage stability of a lyophilized monoclonal antibody, J. Pharm. Sci. 90 (2001) 310–321.

[106] Z. Yu, K.P. Johnston, R.O. Williams III, Spray freezing into liquid versus spray-freeze drying: Influence of atomization on protein aggregation and biological activity, Eur. J. Pharm. Sci. 27 (2006) 9–18.
[107] Y.F. Maa, P.A. Nguyen, S.W. Hsu, Spray-drying of air–liquid interface sensitive recombinant human growth hormone, J. Pharm. Sci. 87 (1998) 152–159.
[108] S.D. Webb, S.L. Golledge, J.L. Cleland, J.F. Carpenter, T.W. Randolph, Surface adsorption of recombinant human interferon-gamma in lyophilized and spray-lyophilized formulations, J. Pharm. Sci. 91 (2002) 1474–1487.
[109] S. Schule, T. Schulz-Fademrecht, P. Garidel, K. Bechtold-Peters, W. Frieb, Stabilization of IgG1 in spray-dried powders for inhalation, Eur. J. Pharm. Bio. Pharm. 69 (2008) 793–807.

Glossary

Absorption (spectroscopy): here, referring to spectroscopic techniques which measure the absorption of electromagnetic radiation by a material or biomolecule. Generally used to characterize a sample or quantify the amount of sample present. See also UV-vis spectroscopy.

Actin: exists in eukaryotic cells in either globular or fibrillar forms, the latter providing a major component of the cell cytoskeleton, and is anchored to many other proteins including those associated with the plasma membrane.

Active (drug) targeting: the exploitation of a native receptor to localize an administered drug. Typically involves the conjugation of the drug with a ligand or monoclonal antibody. Alternatively, it can be formulated in a delivery vehicle, which specifically binds a chosen cell surface receptor.

Adherens junction: a continual circumventing belt around each cell, organized in a monolayer (e.g. the epithelium) formed by proteins known as cadherins which are anchored to actin fibrils; maintains cell morphology and contact. See also Tight junction.

Adsorption: here, the localization (possibly reversible) of a biomolecule at a surface which represents the solid—liquid, liquid—liquid or gas—liquid interface.

Amino acid: the building blocks of peptides and proteins through their concatenation into chains. They have primary amine groups (except the imino acid Proline) and carboxyl groups and a unique side chain at the chiral (α-) carbon. There are 20 native amino acids encoded by genes.

Amphiphile: generally a multi-block co-polymer or lipidic surfactant, which possesses both hydrophilic and hydrophobic regions and is therefore able to self-assemble in aqueous environments.

Analytical ultracentrifugation (AUC): in this context, the centrifugation of proteins at very high gravitational acceleration (105—106 *g*) while monitoring the absorbance bands in the UV-vis region relating to the protein species across the axis of rotation in real time.

Anionic polymers: polymers which have a net negative charge at neutral pH.

Atomic force microscopy (AFM): a relatively new form of microscopy which yields both information about a material's topology at the nanoscale and information about the material's elasticity and/or hardness at the surface. Different modes of data acquisition are required for these measurements.

Bioavailability: describes the amount of free drug that is available in the systemic circulation; distinct from the pharmacological availability.

Biocompatibility: with reference to polymeric vehicles, indicates the extent to which a material can integrate *in vivo* without causing toxicity at the level of the cell/tissue/organism.

Biodegradable: a property which involves the erosion and fragmentation of a material *in vivo*, generally involving hydrolysis or enzyme-catalyzed cleavage, followed by elimination from the body. May also refer to cleavage of a covalent bond between a drug and conjugated moiety, sometimes performed by interaction of the conjugate at a receptor/enzyme target.

Biopharmaceutical: a therapeutic drug based on a biomolecule (e.g. protein, DNA, RNA, viral particles) produced using biotechnology.

Bioprocessing: a collective term involving several areas, from genetic engineering of cells to materials engineering, relating to the downstream processes required to manufacture, package and supply biomolecular therapeutics at an industrial scale.

Block copolymers: polymers which have two or more repeating units (monomers) organized into distinct groupings (blocks). See also Pluronic.

Blood–brain barrier (BBB): an endothelial cell layer with specialized tight junctions generating a tightly-regulated barrier between the blood and cerebrospinal fluid (CSF) in the central nervous system (CNS).

Blood–cerebrospinal fluid barrier: similar in function to the BBB but distinct in location to the tight junctions of the epithelial cells on the surface of the choroid plexus.

Bone engineering: the design and cellular/materials engineering of a synthetic tissue or scaffold which resembles, or comes to resemble, bone, in terms of its function and property.

Caco-2 cells: an immortalized human epithelial colorectal adenocarcinoma cell line used as a model of the intestinal epithelium for study of drug efflux or drug permeability across the intact cell monolayer.

Cationic polymers: polymers which have a net positive charge at neutral pH.

Cell culture: here, the nurtured proliferation of mammalian cells under highly controlled conditions, which may involve cell differentiation and the generation of a particular tissue type.

Cell differentiation: the process of cell specialization involving changes in cell morphology, proliferation, behavior and function.

Cell penetrating peptide: generally short synthetic peptides, whose sequence may have been inspired from nature, often containing poly-arginine. They facilitate the intracellular transport of a cargo, e.g. a therapeutic protein. Their mechanism of cell entry and internalization is a matter of debate.

Chelating agents: chemicals forming ionic complexes with metal ions, particularly divalent metal ions; used *in vitro* to disrupt the cell–cell adherens junctions.

Circular dichroism: the differential absorption of the left and right circularly polarized components of plane-polarized radiation by a chiral compound.

Coacervation: relevant to drug formulation wherein a drug is encapsulated in a droplet of polymer-concentrate (the coacervate) which has phase separated, and is therefore distinct, from the bulk solution. Droplets may be 1 to 100 μm in diameter.

Collapse temperature (Tc): relates particularly to freeze-dried cakes – the temperature at which the material softens to the point of not being able to support its own structure.

Colloid: sub-visible droplets or particles suspended in a medium. They range between 10 and several hundred nanometers in diameter; they may nowadays be called nanodroplets or nanoparticles.

Complementary DNA (cDNA): a DNA strand generated *in vitro* from a messenger RNA (mRNA) template by an enzyme called reverse transcriptase.

Copolymer: a polymer synthesized from one or more monomers. See also Block co-polymer.

Critical micelle concentration (CMC): the concentration above which amphiphilic molecules in bulk solution begin to aggregate through self-assembly into micelles, stabilized through intermolecular forces and entropic effects.

Cryoprotectant: an excipient added to a liquid protein formulation to attenuate protein unfolding and denaturation during freezing.

Cytokine: a broad ranging term for peptides, polypeptides and proteins secreted by the immune system and nervous tissue which have a correspondingly large number of cell regulatory functions *in vivo*. The term overlaps with "hormones", and often refers to the immunoregulatory interleukins and interferons.

Denaturation: the loss of the identifiable structural elements that are characteristic of a particular protein, therefore involving unfolding of the protein's higher order structure and leading to aggregation.

Differential scanning calorimetry (DSC): a classical thermal analytical technique measuring the difference in heat flow to a control pan and a test pan containing the sample under examination. Typically used to identify endo- and exothermic events such as melting, crystallization, glass transition and aging.

Double emulsion: an emulsion of two immiscible liquid phases of the type water-in-oil-in-water (w/o/w) or oil-in-water-in-oil (o/w/o). May also include two immiscible oils of the type water-in-oil-in-oil (w/o/o).

Drug targeting: originally relating to the "magic bullet" concept proposed by Paul Ehrlich, but now generally describes the targeting and localization of a drug to a particular site within the organism following administration, with the intention of minimizing side-effects while increasing pharmacological activity where needed (generally the diseased tissue). See also Active targeting.

Dry powder inhaler (DPI): a hand held medical device that is used to deliver a precise dose of drug in the form of a dry powder to the lung during inspiration.

Dynamic light scattering: an analytical technique for measuring the hydrodynamic diameter of micron/sub-micron sized particles/macromolecules suspended in a liquid medium; exploits the fluctuations in the intensity of a scattered, incident light beam as a function of the velocity of the particles' Brownian motion.

Efflux pump: transmembrane proteins that transport drugs out of the cell membrane into the extracellular environment; particularly relevant to some anti-cancer therapies. See also P-glycoprotein.

Elastic modulus (G′): relates to the non-permanent deformation of a material, yielding a measure of its "hardness" by calculating the stress:strain ratio (applied force per unit area: deformation).

Emulsion: a dispersion of one liquid within another, immiscible liquid.

Endocytosis: the process by which cells sample their environment involving internal invagination and pinching-off of the plasma membrane, capturing the fluid/solutes originally at the cell surface. The process may occur through pinocytosis, phagocytosis or, as relevant to drug delivery, in response to receptor occupancy or non-specific adsorption of particles to the cell membrane.

Endosome: an intracellular vesicle (a bilayer membrane bound compartment inside eukaryotic cells), produced as a result of endocytosis which facilitates transport of the internalized molecules to the lysosome for degradation, or recycling back to the plasma membrane.

Endotoxin: a heat stable toxin (lipopolysaccharide or lipoprotein) derived from the bacterial cell wall during disruption of a bacterial cell, causing an immune reaction that generally results in fever.

Enterotoxin: a subset of exotoxins that are produced by bacteria within and act on the intestinal wall.

European Medicines Agency (EMA): the agency responsible for the scientific evaluation of medicines developed by pharmaceutical companies for use in the European Union.

Exotoxin: a protein secreted by a microorganism that disrupts cell integrity or function, some are fatal at very low doses, most are heat labile.

Extractables: chemical substances that arise as a consequence of exposure of a pharmaceutical package to various solvents, including aqueous buffers, under accelerated storage conditions. They may be toxic and are therefore regulated.

Fab fragment: the fragment antigen-binding region of an antibody which consists of the variable domains and neighboring constant domains of the heavy and the light chains; generated by papain digest of an antibody above the "hinge".

$F(ab')_2$ fragment: disulphide bridged Fab fragments generated by pepsin digestion of an antibody just below the hinge region.

Fc region: the fragment crystallizable region of an antibody which consists of the constant domains of the heavy chains below the hinge; binds the cell surface Fc receptors.

Food and Drug Administration (FDA): the US agency responsible for the scientific evaluation of medicines developed by pharmaceutical companies for use in the USA.

First order drug release: a drug release profile described by the release of a constant fraction of the drug into the body per unit time, wherein the release is proportional to the amount of drug residing in the drug reservoir.

Fluorescence: when a chemical absorbs light at a specific wavelength exciting an electron to a higher quantum state, then emitting a photon of light at a slightly longer wavelength on relaxation of the electron back to its ground state.

Formulation: the process by which a drug is made into a medicine suitable for administration to humans (in this context).

Fourier transform infrared spectroscopy (FTIR): a modern infrared spectroscopy technique that uses a broadband light source with a Fourier transform to convert the raw data into a spectrum. See also Infrared spectroscopy.

Freeze-dried cake: the porous, dry powder remaining after lyophilization of (typically) a protein solution containing excipient(s).

Freeze drying: see Lyophilization.

Gastro-intestinal tract: a collective term here referring to the stomach, small intestine and large intestine.

Gel permeation chromatography (GPC): see Size exclusion chromatography.

Glass transition temperature (Tg): for amorphous materials (particularly polymers), the temperature above which the molecules exist in a state of increased mobility, called the "rubbery" state; below the Tg, the molecules exist in a "glassy", less mobile state. The glassy state is associated with an increased heat capacity, higher density and a more brittle nature.

High-performance liquid chromatography (HPLC): a chromatographic technique performed at high pressures; there are various column types, but the method commonly involves resolution of a mixture of compounds by phase partition between a hydrophobic stationary phase and relatively polar mobile phase. See also Hydrophobic forces, Size exclusion chromatography and Ion exchange chromatography.

Hydrodynamic diameter: the apparent diameter of particle or protein which is hydrated and suspended in aqueous media.

Hydrogels: a hydrophilic polymer which expands in aqueous media to an extent that is determined by intermolecular cross-linking, in this context, relating to the release of encapsulated drug. The polymers may be pH and/or temperature sensitive, leading to a triggered drug release in response to a relevant change in the surrounding media.

Hydrophilic: a molecule possessing an affinity for water via formation of hydrogen bonds, typically leading to dissolution.

Hydrophobic: non-polar or neutral molecules which are "repelled" from water owing to their inability to form hydrogen bonds (and are therefore lipophilic). See also Hydrophobic effect and Hydrophobic forces.

Hydrophobic effect: often used in relation to the Hydrophobic force, but specifically relates to the disruption of the hydrogen bonding network in water by neutral, non-polar molecules, creating a so-called solvation "shell" or "cage" around the non-polar molecule, and is therefore an entropic effect driven by the tendency to increase the mobility and disorder of the water molecules. Used to explain the cloud point of surfactants. Important in protein folding since it explains the drive to bury hydrophobic amino acids within the protein core, minimizing exposure to the aqueous environment; also underpins the "cold denaturation" of proteins.

Hydrophobic forces (Hydrophobic interactions): these are dispersive interactions resulting from London's dispersion forces: the induced dipole–dipole interactions arising from shifts in electron density between neutral molecules. See also van der Waal's forces and Hydrophobic effect.

Hydrogen bond (H-bond): interaction between a hydrogen atom and a strongly electronegative atom, particularly oxygen or nitrogen in the case of proteins. Intramolecular H-bonds are key to the stabilization of protein secondary structure motifs and higher order structure.

Infrared spectroscopy: an analytical technique which measures the absorption of light in the infrared region by a sample, in which the frequency of the absorbed radiation matches the frequency of a particular vibration of a bond. In this context, used to characterize changes in secondary structure of a protein in the solid state.

Ion exchange chromatography: in this context, the resolution of a mixture of proteins according to their isoelectric point (or localized net charge on a particular domain) via their propensity for exchange with counterions of a charged resin column.

Iontophoresis: a non-invasive, electrically assisted technique in which a physiologically acceptable amount of electric current is used to facilitate transdermal delivery of charged and neutral molecules.

Interferons: a subset of cytokines, they are glycoproteins produced by the immune system in response to infection by microorganisms. Have been exploited to treat specific cancers and hepatitis.

Interleukins: a subset of cytokines, they are peptides and proteins produced by numerous cells of the immune system which have various regulatory roles.

Internalization: here relating to various processes by which a drug may gain entry into the cytoplasm of a mammalian cell, for example by fusion with a cell penetrating peptide or endocytosis.

Immunoliposomes: liposomes that contain lipids covalently conjugated to specific antibodies for immunotargeting.

Interface: the boundary between two phases (liquid–solid or liquid–air) or two immiscible liquids (liquid–liquid).

Intrinsic viscosity: in this context, relating to the measure of a macromolecule's contribution to the increase in the viscosity of a solvent in which it is dissolved.

Isoelectric point (pI): for a protein molecule with multiple charged groups, this is the pH at which the protein carries no net electrical charge in that particular buffer.

Leachable: a subset of extractables; chemicals which migrate into a drug product from packaging under standard conditions of temperature and pressure over time.

Liposome: vesicles composed of phospholipids (as distinct from niosomes) mixed with other lipids or cholesterol. May have one or more lipid bilayers with an aqueous core encapsulating a drug.

Liquid formulation: the process by which a drug is made into a medicine for administration in liquid form (e.g. syrups, injectables).

Lyophilization: a drying process involving the sublimation of water from a frozen solution (here containing protein and excipients) at low temperatures under a vacuum.

Lyoprotectant: an excipient added to a liquid protein formulation to attenuate protein unfolding and denaturation during freeze-drying, particularly during the removal of bound water.

Mass spectrometry: an analytical technique which measures the mass-to-charge ratio of charged molecules.

Medical device: a product which is used for clinical benefit in a patient, either for delivery of a drug, or physical intervention *per se*, or a combination of both (see also Tissue regeneration), or diagnostics.

Mesoscale: poorly defined length scale but generally above the upper limit of the nanoscale (100 nanometers) and below 1 micrometer. In pharmaceutics, particles in this size range may also be referred to simply as sub-micron.

Metered dose inhaler: a hand-held medical device that is used to deliver a precise dose of drug in the form of an aerosol of liquid droplets to the lung during inspiration.

Micelle: colloidal systems composed of a single lipid layer self-assembled into a sphere, the orientation of the surfactant lipids is dependent on the surrounding media: in aqueous media a hydrophobic core is formed, but in non-polar solvents a reverse micelle is formed with a hydrophobic shell.

Microneedles: here relating to transdermal drug delivery, an array of needles penetrating only as far as the epidermis, avoiding the pain receptors in the dermis.

Microporation: the creation of micron-sized channels through the skin to assist transdermal delivery of molecules.

Microsphere: a small sphere typically between 1 and 500 μm in diameter with a porous inner matrix and variable surface topography. Drug is generally dispersed throughout the inner matrix. Alternative terms such as microcapsules and microparticles are also encountered in the literature.

Mucoadhesion: similar to bioadhesion but relating to the specific adhesion of a macromolecule to the layer of mucus at mucosal surfaces, particularly of the gastro-intestinal tract or nose.

Nanoscale: length scales which are most generally noted as being between 10 to 100 nanometers (10^{-8} to 10^{-7} m).

Nanoparticles: particles which have sizes within the nanoscale range, though in pharmaceutics may refer to particles several hundreds of nanometers in diameter (see Mesoscale). May include hollow nanocapsules or solid/porous nanospheres.

Niosome: a non-ionic surfactant vesicle often with cholesterol or other lipids incorporated into the lipid bilayer, generally micron sized and similar to liposomes.

Nuclear localization motif (NLS): a short peptide motif that trafficks a cargo through the cytoplasm towards and into the nucleus of a eukaryotic cell.

Oligonucleotide: typically, short polymers of around 20–40 nucleic acids, though can be synthesized to be much longer. Have therapeutic potential but are now largely superseded by interfering RNA technology.

Paracellular transport: a mode of transport wherein a drug passes between the neighboring cells in a monolayer (e.g. epithelium).

Passive targeting: as distinct from Active targeting, referring to the localization of a drug or drug carrier system to a specific tissue(s) through an inherent physical effect, or exclusion from other sites.

Patient compliance: the extent to which patients follow verbal and written instructions for the correct usage of their medication.

PEGylation: the covalent conjugation of a poly(ethylene glycol) (PEG) chain of varying size to a biopharmaceutical or vesicle. In the case of the latter, PEGylation of liposomes generates so-called "stealth liposomes" which better evade the immune system.

Peptide: a short polymer composed of amino acids linked together by amide bonds. The definition of size is broad: typically up to 50 amino acids, but may be longer.

P-glycoprotein: a member of the ATP-binding cassette (ABC) cell membrane transporter family, functioning as an efflux pump in the intestinal epithelium and involved in multi-drug resistance.

Photon correlation spectroscopy: see Dynamic light scattering.

Plasma membrane: the bilayer membrane composed of phospholipids, cholesterol and other lipids, and harboring proteins, which describes the cell perimeter. Artificial membranes can be formed by the self assembly of amphiphiles. See also Liposome.

Pluronics: amphiphilic triblock copolymers with the architecture: poly(ethylene oxide)-poly (propylene oxide)-poly(ethylene oxide) (PEO-PPO-PEO). Have various properties dependent on the molecular weight and size ratio of the PEO and PPO blocks. Trade name of the poloxamers of the BASF Group.

Poloxamer: see Pluronics.

Polyelectrolyte: polymers composed of repeating monomers which have charged side chains, thereby generating a strong net charge at neutral pH. May include permanently charged groups such as quaternary amine groups.

Polymer: a large molecule (macromolecule) consisting of a chain of covalently linked monomers which need not be identical and may harbor pendent groups.

Polypeptide: as for peptide but above the upper limit in terms in amino acid number – generally around 100 amino acids. Their size may be usefully related to the upper limit of solid phase peptide synthesis.

Primary structure: the order of amino acids in a peptide, polypeptide or protein; convention begins the order with the amino acid at the N- (amine) terminus, and finishes at the C-(carboxy) terminus.

Protein: a molecule consisting of more than ~100 amino acids in length; there may be some overlap with the term polypeptide. Typically folded in native conditions into a specific 3-dimensional conformation, possibly in distinct domains (as for antibodies); may be post-translationally modified in eukaryotic cells.

Protein aggregation: the precipitation of protein molecules out of aqueous solution following unfolding as a consequence of chemical or mechanical denaturation. Generally leads to visible particles, but sub-visible particles can be observed by UV-vis spectroscopy and other techniques such as light scattering.

Protein data bank (PDB): the worldwide repository of 3-dimensional structures of large biological molecules, including proteins and nucleic acids (www.rcsb.org).

Protein structure: a general term encompassing primary, secondary, tertiary and quaternary structure.

Protein transduction domains: see Cell penetrating peptide.

Pulmonary delivery: the delivery of a drug to the lungs via inhalation.

Quaternary structure: relating to proteins only, the arrangement through non-covalent forces and/or disulphide bridges of folded domains or coiled coils into a multi-domain (or multi-subunit) complex.

Raman spectroscopy: an analytical technique which measures the shift in energy as an incident light beam interacts with the molecular vibrations of a molecule through inelastic scattering (not transmittance as in Infrared spectroscopy). The intensity of the shifts are plotted against wave number as a spectrum, which is characteristic for a particular compound.

Receptor mediated endocytosis: see Endocytosis.

Reticuloendothelial system (RES): comprises the phagocytic cells located in the spleen, lymph nodes and liver (where they are termed Küpffer cells), and therefore function as a part of the immune system removing/sampling circulating antigens.

Scanning Electron Microscopy (SEM): an imaging technique involving the rastering of high energy electron beams across the surface of a sample, made conductive with a gold coating, revealing the sample's topography at the nanoscale.

Secondary structure: the most basic level of 3-dimensional structure of a protein and chiefly classified into α-helices, β-strands and β-turns. Does not describe the global fold of the protein – see Tertiary structure.

Self assembly: a process involving discrete molecules (e.g. amphiphilic polymers) aggregating in a controlled fashion to generate ordered, large-scale complexes (or supramolecular structures).

Shear stress (fluids): for a fluid traveling along a solid boundary, a shear stress arises as a consequence of the zero speed of the fluid immediately at this boundary but flow at some distance away (the layer in between being the boundary layer).

Side-effect: an unwanted, detrimental effect of a drug which may in some cases prevent or be limiting to patient compliance.

Size exclusion chromatography (SEC): a chromatographic technique which involves the resolution of a mixture of molecules based on the differences in their distribution through a porous matrix – related to the molecules' hydrodynamic diameters or molecular weights following appropriate calibration.

Small interfering RNA (siRNA): short double stranded RNA molecules which downregulate specific gene targets; may have therapeutic potential.

Solid state formulation: the process by which a drug is made into a medicine for administration as a solid dosage form (e.g. tablets).

Sonophoresis: the use of ultrasound to drive molecules into and across skin.

Spray-drying: a method used to generate a dry powder via the vaporization of a solvent from a suspension or solution sprayed into a stream of hot gas (drying chamber).

Sub-micron: see Mesoscale

Supercritical carbon dioxide: carbon dioxide that exists in a liquid form at or above its critical temperature and pressure.

TAT peptide: amino acid residues 48–60 from the full length TAT (transactivator of transcription) protein, showing an elevated internalization efficacy over the parental protein. See also Cell penetrating peptide.

Tight junction (zonula occludens): a continual circumventing belt around each cell organized in a monolayer (e.g. the epithelium) formed by the transmembrane occludins and claudins, anchored to the cell cytoskeleton. Brings neighboring cells into intimate contact and regulates the paracellular passage of molecules and ions. See also Adherens junction.

Tissue engineering: a broad field of research and development including cell biology, materials and engineering; focused on mimicking the biological and mechanical environments experienced by cells *in vivo*, in order to generate synthetic, functional tissue to repair/replace diseased or damaged tissue.

Transcellular transport: a mode of transport wherein a drug passes through the cells in a monolayer (e.g. epithelium).

Transepithelial electric resistance (TEER): measurement of the resistance to the flow of ions across a confluent cellular monolayer, for example, used to assess the integrity of the cell–cell tight junctions.

Transmission Electron Microscopy (TEM): an imaging technique involving the interaction and scattering of high energy electron beams through an ultrathin sample (stained with heavy metals), then focused onto a detector.

Ultraviolet-visible (UV-vis) spectroscopy: an analytical technique measuring the absorption of light across the UV-vis region following transmission through a sample.

van der Waal's interactions: the sum attractive force between neutral, non-polar molecules (may be intramolecular in the case of proteins or polymers) arising from permanent or transient dipole–dipole interaction. See also Hydrophobic force.

Vesicle: a discrete structure of self assembled amphiphilic lipids composed of a spherical bilayer encapsulating an aqueous core. See also Liposome and Niosome.

Wavenumber: as regards spectroscopy, the reciprocal of wavelength ($1/\lambda$), typically in units of cm^{-1}.

Zero order drug release: a constant release of drug over time.

Zonula adherens: see Adherens junction.

Zonula occludens: see Tight junction.

Appendix 1

Calculated Properties of Some Clinically Relevant Peptides and Proteins

Peptide/Protein (mature sequence)	Sequence MW	No. amino acids	No. polypeptide chains	Isoelectric point (pI)	Molar extinction coefficient (M^{-1} cm^{-1}) at 280 nm
Calcitonin, human	3421	32	1	7.09	1400
Calcitonin 1, salmon	3433	32	1	8.59	1400
Cyclosporine-A	1203	11	1 (cyclic)	–	–[a]
Erythropoietin, human	18396 (~34000[b])	166	1	8.48	22430
Insulin A chain, human	2384	21	1	3.53	2800
Insulin B chain, human	3430	30	1	7.28	2680
Insulin, human	5808	51	2	5.59	5480
Fab from human IgG (generalized)	~48000	~440	2	–	–
$F(ab')_2$ from human IgG (generalized)	~100000	~900	4	–	–
Fc from human IgG (generalized)	~48000	~440	2	–	–

Immunoglobulin A, human (generalized)	~300000[c]	~2500[c]	9[d]	4.5–6.5	–
Immunoglobulin E, human (generalized)	~150000	~1250	4	~5–6	–
Immunoglobulin G, human (generalized)	~145000	~1250	4	5–9[e]	–
Interferon alpha-2B, human	19300	166	1	6.34	18020
Interferon beta-1A, human	20000 (22500[b])	166	1	8.78	29990
Interleukin-2, human	15414	133	1	7.33	9650
Serum albumin, human	66472	585	1	5.96	30770
Serum albumin, bovine	66433	583	1	5.86	39020
Somatotropin, human growth hormone 1 (hGH)	22125	191	1	5.41	16170

[a] *partition coefficient (LogP) = 4.12*
[b] *glycosylated form*
[c] *dimeric form*
[d] *approx. one J-chain links two IgA monomers*
[e] *variable and dependent on class: IgG1 ~ 8–9, IgG2 ~ 6–7, IgG3 ~ 7–9, IgG4 ≤6–8*

Commonly Encountered Units and Abbreviations

AU	absorbance units
BSA	bovine serum albumin
CD	circular dichroism
CLSM	confocal laser scanning microscopy
D	diffusion coefficient
d_a	aerodynamic diameter
Da	Dalton ($\equiv$ one unified atomic mass unit)
DCM	dichloromethane
°C	degree Celsius
ΔC_p	heat capacity change at constant pressure
d_g	geometric mean weight diameter
dL/g or dl/g	deciliters per gram
DLS	dynamic light scattering
DNA	deoxyribonucleic acid
DSC	differential scanning calorimetry
EMA	European Medicines Agency
Fab	Fragment antigen binding
Fc	Fragment crystallization
FDA	(US) Food and Drug Administration
FITC	fluorescein isothiocyanate
FTIR	Fourier transform infrared spectroscopy
g	gram
g	gravitational acceleration
ΔG	Gibbs free energy
$\Delta G_{f \rightarrow u}$	Gibbs free energy change between the folded and unfolded states of a protein
g/cm^3	gram per cubic centimeter
g/mL	gram per milliliter
GdmCl	guanidine hydrochloride

h	hour
ΔH	enthalpy change
ΔH_v	Van't Hoff enthalpy change
HCl	hydrochloric acid
HPLC	high performance (or high pressure) liquid chromatography
HSA	human serum albumin
i.v.	intravenous
ID	inner diameter
IgG	immunoglobulin G
K	Kelvin
kDa	kiloDalton
L or l	liter
mAb	monoclonal antibody
μm	micrometer
min	minute
mL or ml	milliliter
mM	millimolar
mol	mole
M_r	relative molecular mass
MSLI	multi-stage liquid impinger
mTorr	milliTorr
MW	molecular weight
Na_2HPO_4	disodium hydrogen phosphate
NaCl	sodium chloride
nm	nanometer
OD	optical density
PAGE	polyacrylamide gel electrophoresis
P_{app}	apparent permeability coefficient
pI	isoelectric point
pK_a	$-\log_{10}$ acid dissociation constant
p.o.	"by mouth"
PBS	phosphate buffer saline
PEG	polyethylene glycol
PLA	poly(D,L-lactic acid)
PLGA	poly(D,L-lactic-co-glycolic acid)
PVA	polyvinyl alcohol
RNA	ribonucleic acid
rpm	rotations per minute
s	second
ΔS	entropy change
S.D.	standard deviation
SDC	deoxycholic acid, sodium salt
SDS	sodium dodecyl sulphate

SEM	scanning electron microscopy
T	absolute temperature
TEER	transepithelial electrical resistance
TEM	transmission electron microscopy
T_c	collapse temperature
T_g	glass transition temperature
T_g'	glass transition temperature prime
v/v	volume by volume (%)
w/o	water-in-oil emulsion
w/o/w	water-in-oil-in-water emulsion
w/v	weight by volume (%)

Index

FSC
www.fsc.org
MIX
Papier | Fördert
gute Waldnutzung
FSC® C083411

Kolibri 360 GmbH
Ferdinand-Jühlke-Straße 7
99095 Erfurt, Deutschland
produktsicherheit@kolibri360.de